# SOURCE AND CHANNEL CODING
# An Algorithmic Approach

## THE KLUWER INTERNATIONAL SERIES IN ENGINEERING AND COMPUTER SCIENCE

COMMUNICATIONS AND INFORMATION THEORY

*Consulting Editor:*
**Robert Gallager**

Other books in the series:

*Digital Communication.* Edward A. Lee, David G. Messerschmitt
ISBN: 0-89838-274-2

*An Introduction to Cryptology.* Henk C.A. van Tilborg
ISBN: 0-89838-271-8

*Finite Fields for Computer Scientists and Engineers.* Robert J. McEliece
ISBN: 0-89838-191-6

*An Introduction to Error Correcting Codes With Applications.*
Scott A. Vanstone and Paul C. van Oorschot
ISBN: 0-7923-9017-2

*Source Coding Theory.* Robert M. Gray
ISBN: 0-7923-9048-2

*Switching and Traffic Theory for Integrated Broadband Networks.*
Joseph Y. Hui
ISBN: 0-7923-9061-X

*Advances in Speech Coding,* Bishnu Atal, Vladimir Cuperman and
Allen Gersho
ISBN: 0-7923-9091-1

# SOURCE AND CHANNEL CODING
# An Algorithmic Approach

*by*


**John B. Anderson**
*Rensselaer Polytechnic Institute*

**Seshadri Mohan**
*Bell Communications Research*



**KLUWER ACADEMIC PUBLISHERS**
**Boston/Dordrecht/London**

**Distributors for North America:**
Kluwer Academic Publishers
101 Philip Drive
Assinippi Park
Norwell, Massachusetts 02061 USA

**Distributors for all other countries:**
Kluwer Academic Publishers Group
Distribution Centre
Post Office Box 322
3300 AH Dordrecht, THE NETHERLANDS

**Library of Congress Cataloging-in-Publication Data**

Anderson, John B., 1945-
Source and channel coding : an algorithmic approach / by John B. Anderson, Seshadri Mohan.
p. cm. -- (The Kluwer international series in engineering and computer science ; SECS 150. Communications and information theory)
Includes bibliographical references and index.
ISBN 0-7923-9210-8 (acid-free paper)
1. Coding theory. I. Mohan, Seshadri. II. Title. III. Series: Kluwer international series in engineering and computer science ; SECS 150. IV. Series: Kluwer international series in engineering and computer science. Communications and information theory.
QA268.A5 1991
003'.54--dc20 91-4966
CIP



*Printed on acid-free paper.*

Printed in the United States of America

# Contents

## 3 Error Control Coding 77



## 4 Trellis Codes 199



## 5 Reduced Trellises 275

# Preface

How should coded communication be approached? Is it about probability theorems and bounds, or about algorithms and structures? The traditional course in information theory and coding teaches these together in one course in which the Shannon theory, a probabilistic theory of information, dominates. The theory's predictions and bounds to performance are valuable to the coding engineer, but coding today is mostly about structures and algorithms and their size, speed and error performance. While coding has a theoretical basis, it has a practical side as well, an engineering side in which costs and benefits matter. It is safe to say that most of the recent advances in information theory and coding are in the engineering of coding. These thoughts motivate the present text book: A coded communication book based on methods and algorithms, with information theory in a necessary but supporting role.

There has been much recent progress in coding, both in the theory and the practice, and these pages report many new advances. Chapter 2 covers traditional source coding, but also the coding of real one-dimensional sources like speech and new techniques like vector quantization. Chapter 4 is a unified treatment of trellis codes, beginning with binary convolutional codes and passing to the new trellis modulation codes. There is much confusion today about what terms like modulation code and trellis mean and about what is old and what is new, and we especially hope to shed some light on such matters. Chapters 5 and 6 discuss the important practical topic of how much of a code book a decoder needs to search in order to find a good decoding; Chapter 7 is a brief look at VLSI structures for this job. Most of this material has not previously appeared in a text book.

Our aim has been to produce a first-year graduate coding text book of moderate size, and this has meant explaining a limited number of topics fully rather than many topics briefly. Many important but smaller topics are left out of this book, and we mean no slight to the talented researchers who work on them. Particularly, we emphasize channel coding over source coding, and trellis and tree coding over block coding. Our rationale is that this bias reflects advances in recent years and advances to come. About the references cited in the text, we follow a priority that has meant injustice to some. Our first aim has been to refer to a source of further reading, and so readability and information often take precedence in our citations. Establishing the discoverer of an idea has taken second priority and listing recent contributions a third.

A one-term introduction to coding and information theory for senior undergraduate or first year graduate students can be based on Chapters 1–4, with topics from later chapters added to suit the instructor. This yields a course that stresses channel coding over source coding in a 2:1 ratio. While the text does not cover information theory in detail, all the basics are present, including information measures, mutual information and capacity, and the basic source and channel coding theorems. Particular attention has been paid to the problems, exercises and examples in Chapters 2–4. The prerequisites for these chapters are good courses in probability theory, communications systems, and linear systems. The book depends to some degree on signal space theory and on modern algebra, but we review the needed material in Appendices A and B. Although no further background is required, a strong mathematical aptitude is always a plus for those who would learn coding.

The ideas behind this book, both pedagogical and technical, evolved over many years and we owe a great debt to our students and colleagues over that time. It is a particular pleasure to acknowledge the encouragement and support of Prof. Robert Gallager, the consulting editor of this series, of our publishers at Kluwer, Robert Holland and Carl Harris, and of our editorial assistant, Rose Luongo. Our institutions, Rensselaer Polytechnic Institute and Bellcore, Inc., provided time and technical support. We owe a particular debt to Virginia Palmer, who typeset most of the initial pages. Her cheerful persistence with the job was appreciated by both of us. Several colleagues read and made suggestions on parts of the text; these include Sandeep Chennakeshu, William Pearlman, Stanley Simmons, Ivonete Markman, Amir Said, Kumar Balachandran and Frank Gozzo. Finally, we recognize the enduring support of our families, who felt as we do, that the writing of a book like this one is a worthwhile effort.

*John B. Anderson*
*Troy, New York*

*Seshadri Mohan*
*Red Bank, New Jersey*

# SOURCE AND CHANNEL CODING
## An Algorithmic Approach

# CHAPTER 1

# Introduction

Before beginning an organized study of coding algorithms we should prepare the way by reviewing the mathematical ideas and the historical developments that led to the present day subject. In the field of communication theory we need to know about information theory and coding in particular and about signals, their energy and their bandwidth. The engineering events of importance to us are the development of algorithms and the advent of cheap, large-scale integrated (or LSI) digital circuits.

## 1.1 THE HISTORICAL CONTEXT

The field of coded communication algorithms stands at the confluence of three evolving disciplines. The first of these is communications theory, particularly coding theory and signal space theory, which have given us more sophisticated and powerful ideas about how to communicate efficiently. The others are the study of algorithms and the design of signal processors that are simultaneously cheap, small, fast and reliable.

Taken together, these disciplines have a most powerful synergism. Theory inspires us as to what we might do, algorithms are the language of expression, and processors provide the implementation.

An algorithm is a step by step procedure to attain an end. It is an ancient idea and we all learn at least a few algorithms like simple addition in our earliest schooling. Still, a quantum jump in our concept of an algorithm occurred in the mid-20th century with the introduction of computers. Most of us now think of an algorithm as a *sequential program*, to use computer terminology, that is, as a strict list of instructions in a fixed language that contains a branch-on-condition command. We think of algorithms as residing in computers, which are machines that fetch instructions from a store and execute them one by one. In principle there seems to be no limit to the complexity of the task that this kind of machine can carry out, although the requirement that the program be sequential may also make it inefficient. Nonetheless, stored sequential program algorithms have been more than enough to make complicated coded communication practical. New ideas about what is an algorithm are appearing; for instance an algorithm may have parts that run concurrently, and we might design a computer that works from nonsequential programs.

Large-scale integrated circuits with microscopic feature size provide a cheap means to implement sequential algorithms. They also implement the many kinds of processors that do not follow programs. These include pipeline processors, in which signals flow through at will, and parallel processors, which have several flows, and systolic arrays, to name a few. Processors like these are usually cheaper and faster for filters, adders, multipliers, and many of the coding functions that appear in later chapters.

Making a choice between hardware and computer software is a common decision in the design of coding algorithms. Deciding which is cheaper can be tricky. Dedicated hardware may be cheaper in the sense that it adds more numbers or filters more data per second for the same LSI circuit size, but the design of special LSI hardware is much more expensive than writing a program for an existing LSI computer chip. This software-hardware dilemma is often present in coding processors. As a generalization, we can say that software is easier, but experience shows that the same function in hardware is 10-100 times faster if we constrain ourselves to the same LSI device count and technology and the same cleverness of design. The speed required often decides the issue.

While everyone has heard of the LSI chip, it is perhaps not widely appreciated how revolutionary the chip really is. The cost of a single vacuum tube device in the 1950s was about five late-1980s dollars.

Transistor circuits with separate components soon appeared. These were eventually cheaper, far more reliable, and consumed and then radiated as heat far less power. Cost and power dropped further orders of magnitude in the 1960s when photolithography was used to make integrated circuits. Yet more orders have occurred since as we apply lithography to ever smaller feature sizes. The total of all these advances is perhaps a million-fold reduction in the cost of a single device.

In human history it is not often that an everyday object drops so much in price. In the economic realm costs are parts of an economic equation, in which a strong change in one part leads to similar charges in other parts. It is likely that we have passed only part way through this particular upheaval. In the near future we can look forward to even cheaper LSI circuits. Computer-aided design tools and new chip architectures will make custom designs much easier. These factors all motivate this book. We need to lay out clearly the component algorithms and signal processings that are needed in coded communication.

Of equal importance to our story have been the advances in communications theory that have occurred since the 1940s. These give us a deeper understanding of the communication process and show us how to make it cheaper and more reliable. The most famous of these advances was a series of papers by Claude Shannon, beginning with his landmark 1948 paper "A Mathematical Theory of Communication" [1-1]. In Shannon's concept of communication there were three parts. *Source coding* studies information measurement and its conversion into different forms; Shannon extended a measure of information content, which had been proposed earlier, called entropy, and showed that it was always possible in principle to express information in a number of symbols equal to its entropy. *Channel coding* studies how much information a channel can transfer without error. Shannon discovered a computable quantity called channel capacity and proved that for a given channel error-free communication was possible at information rates below capacity but not above. The third part of Shannon's theory was *rate-distortion theory* [1-2]. Here not enough information is available to ensure perfect reproduction of an information source, but one wishes to do the best job possible with the symbols available. Shannon gave a precise relationship between the number of these symbols and the least attainable message distortion.

In 1948 these ideas were for the most part new and they revolutionized communications theory. It had been thought, for instance, that information could not be transmitted through a noisy channel with perfect reliability, no matter how small the noise. One can gain some sympathy for this view by simply listening to an AM or FM transmission: The weaker the noise the better the transmission but there is always distortion and it is directly coupled to the noise in the channel. Shannon showed on the other hand that if the information is gathered into long blocks with a certain coded structure, then a receiver can decide the information perfectly with high probability even if the structure is blurred by noise. The probability can be driven toward certainty by increasing the length of the information blocks, so long as the flow of symbols per unit time is less than the channel's capacity. Shannon's work was full of new insights like this. We will review his work further in the next section.

A second great advance in communications theory crystallized around the 1947 Ph.D. thesis of Kotelnikov [1-8]. A related contribution was the invention of the matched filter by Turin [1-9]. Today we call the field of these works *signal space theory.* It gives us an analytical framework for replacing signals that are continuous functions of time with vectors and components of vectors. In this way we can handle continuous signals in a way at least roughly similar to the way we handle discrete symbols in discrete-time channels. Continuous-time channel noise, too, may be reduced to the same vectors, and the theory shows that we can ignore all the effects of noise except these vectors. The final achievement of the theory is to show how to build a maximum likelihood receiver.

Some of the details of signal space theory are reviewed in Appendix A. For those who want more, we can recommend the classic text by Wozencraft and Jacobs [1-10].

Taken together, the Shannon theory and signal space theory set an array of limits to communication that serve as benchmarks to measure machines that we may build. With them, we can judge the effectiveness of a speech digitizer or a practical error correction scheme. Second, they suggest a variety of structures for communications processors and algorithms. Some of these, like the signal space maximum likelihood receiver, may be too large in some applications; for these we might develop simplifications that have, we hope, nearly the same performance. Some ideas for processors come from quite different fields that we have not just mentioned. Examples are error correction schemes based on algebraic group theory and code book search algorithms borrowed from computer science.

## 1.2 INFORMATION AND CODING

In this section and the next we will review some basics about information and coding and about signal energy and bandwidth. These form the basic commodities in electrical communication. It is helpful here to use some economic analogies. We can think of ourselves as "paid" for each unit of information that we reliably transmit. Our "costs" for the transmission are the energy in the signal and the width of scarce spectrum that it occupies. We will always think of energy and bandwidth in terms of per unit of information carried. In economic terms it could be said that we are counting costs per unit of revenue. There are other costs. The major one is really the subject of this book, the complexity of equipment.

### Information and Entropy

In the view of Shannon information theory an *information source* is defined in terms of a probability distribution. A simple example is the rolling of a six-sided fair die. Each face has probability 1/6 of coming up and we view the information in the outcome as related to this probability distribution. The favored view in information theory is that the information should measure the uncertainty in the outcome, a view with which dice players will sympathize. But only the distribution is important, not the physical fact of the die; any object with six equiprobable outcomes will do.

The model we will use for transmission of data through channels in Chapters 3–5 provides another example of the die-type information source. The data there will always be thought of as having independent and equiprobable outcomes, two for a binary source, four for a quaternary source, and so on.

Information theorists say that an information measure should have three qualities. These are all in keeping with our daily experiences:

1. The information in separate, independent outcomes should add;
2. Information should be proportional to the uncertainty of the source outcome;
3. The quantity of information should relate to the number of symbols needed to define the outcome.

A measure that satisfies these three is the expression $\log[1/P(A)]$, with $A$ the outcome of the source and $P(A)$ its probability. We can easily verify the three axioms for this measure. Consider the third axiom and the outcome $A$ of an equiprobable quaternary information source. Ordinarily we will use base-2 logarithms to measure information; with this base the information in the quaternary outcome is $\log 4$, or 2. Since the binary outcome has information $\log 2 = 1$, this suggests that a quaternary source produces the information of two binary symbols, which matches our experience.

Thus far we have discussed the information in a particular source outcome, but information theory normally works with a more meaningful measure, the *expected* information in the outcome of a source. This is called the *entropy* of the source. It is defined by

$$H(Z) = \sum_j P(z_j) \log \left( \frac{1}{P(z_j)} \right) . \qquad (1.1\text{-}1)$$

Here and throughout the book capital $Z$ denotes a random variable that takes on values $z_j$ with probability $P(z_j)$. The dimensions of entropy (or of information) are bits/source symbol. Note that bits in this context is a real-number measure of something, not a two-valued symbol.

The entropy of our quaternary data source is 2 bits/symbol, the same as was the information in an outcome. This is because all the outcomes have the same information as the average. The six-sided die has entropy 2.58.

In our Chapter 2 discussion on source coding, we will consider sources that are quite a bit more general. For instance, the outcomes of a uniform-bin PCM quantizer are not generally uniform in their probabilities (some profit can be made from this) and natural sources like speech are markedly nonuniform. The next level of generalization from the equiprobable source is the *binary symmetric source*, or BSS. This produces an independent sequence of symbols that take on two outcomes, $z_o$ with probability $p$ and $z_1$ with probability $1 - p$. The entropy of this source has a special name, the *binary entropy function*, and is defined by

$$h_B(p) = -p \log p - (1-p) \log(1-p). \qquad (1.1\text{-}2)$$

This important function will appear in many places in this book, some of them surprising. It is plotted in Figure 1.2-1. We can see that this source has almost no average information when $p$ is small or near 1; this is because the outcome is almost certain to be $z_o$ or $z_1$. Maximum entropy, and uncertainty, are reached at $p$ equal 0.5. For a general discrete random variable $Z$, $H(Z)$ satisfies

$$0 \leq H(Z) \leq \log J , \qquad (1.1\text{-}3)$$

where $J$ is the number of outcomes of $Z$. $H(Z) = \log J$ when all the outcomes are equiprobable. The proofs of these facts can be found in the texts at the end of the chapter.

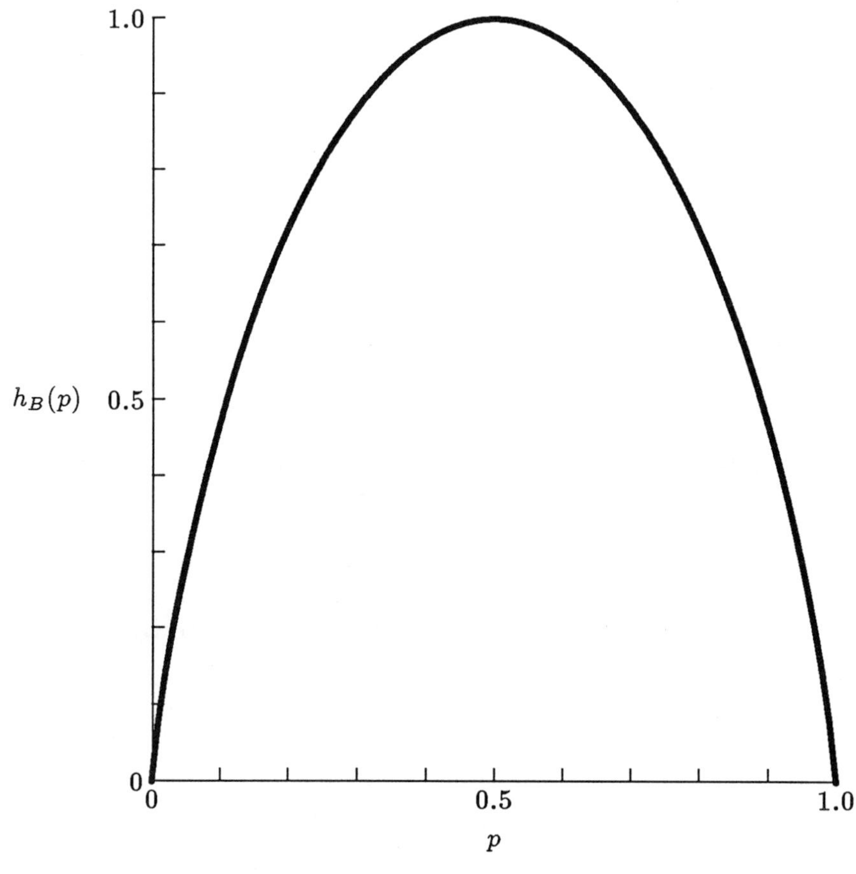


**Figure 1.2-1.** The binary entropy function.

## Conditional Information

Another type of information source that will appear in Chapter 2 is the *correlated* source, in which successive source outputs are dependent. Most natural sources like text or samples of speech or pictures are like this. We expect that each new sample here will contribute less new information than an independent sample would, since we already know something about it from the old samples.

Let a source output $Z$ depend on another random variable $W$, which could be the previous source output. The conditional distribution of $Z$

is $P(z|w)$, with $w$ the outcome of the second variable. If we substitute $P(z|w)$ into (1.1-1) and take the average over $Z$, we get

$$H(Z|w) = \sum_j P(z_j|w) \log \left( \frac{1}{P(z_j|w)} \right) . \qquad (1.1\text{-}4)$$

This is called the *conditional entropy* of $Z$; it is conditioned on the outcome $w$.

As an example, we can consider the transmission of pixels from an image that contains only black or white regions, with the pixels modeled as follows. A white pixel $Z$ follows a white one with probability 0.95, and similarly for black following black; an opposite pixel follows with probability 0.05. (This is called a first-order Markov chain model.) Whether the previous pixel $W$ is ($w$ = black) or ($w$ = white), eq. (1.1-4) gives 0.29 bits/pixel for the conditional entropy of $Z$. Without the conditioning we only know the marginal distribution of $Z$, which we can assume is the equiprobable one; from (1.1-1) the entropy of this is 1. The example shows that recognizing the correlation present in a source can greatly reduce the new information with which we will have to deal. A sophistication of the model in this example provides the standard model for speech and gray-scale images in Chapter 2.

Another kind of entropy called the *equivocation* is defined to be the average of the conditional entropy over the condition $w$. It is given by

$$H(Z|W) = \sum_i H(Z|w_i)\ P(w_i). \qquad (1.1\text{-}5)$$

This entropy accounts for all the values that $W$ can take. In the example the equivocation is still 0.29, because of the symmetry in the black-white probabilities.

## Coding

Coding is the conversion of information to another form for some purpose. In a kind of source coding called noiseless source coding, the purpose is lowering the redundancy in the information. Noiseless source coders try to remove anything repetitive or predictable in the probabilistic sense from a symbol stream, ending with a shorter stream of symbols that still perfectly represents the information. The extent to which the stream can be shortened without damage to the information is given by Shannon's First Theorem, which will be discussed in Section 2.1. The place of source coding, and of some other kinds of coding to come in a

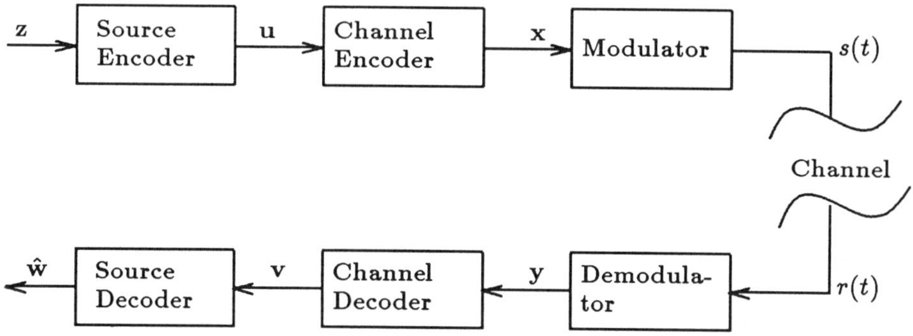


**Figure 1.2-2.** Block diagram of a general communications system

general communications system are shown in Fig. 1.2-2, which also shows the sequence notation we will use.

Unless the coding is very simple, it will involve forming the information stream into blocks. This grouping of data is crucial to the efficiency of all types of coders. It also means that coding machines impose a delay in the stream, since the machine must view all the information in a group before it carries out the coding. Theorems about the ultimate limits to coding assume that a sequence of unbounded length is considered during the coding. Often the idea that a sequence of data have an affinity is a natural one, even when not required by a coding process. Examples are samples of a correlated data source like speech or samples of a channel waveform that have passed through a filter.

Another kind of source coding is rate-distortion coding. Here the object is to remove information and therefore suffer distortion, because the loss is imperceptible or the information is not worth the cost of handling it. In designing this kind of coder we try to keep the distortion as small as possible for a given quantity of information at the output. Most practical source coding problems are of the rate-distortion type, perhaps because the ideas of cost and relative value are important in the practical world. Whether or not we wish to admit it, all analog sources like speech or pictures contain distortion because we cannot represent analog samples in finitely many bits. Nor can we perceive distortion below a threshold. For each information source and measure of distortion Shannon's Rate-Distortion Theorem gives the least distortion possible for coders with a given output rate of information. Rate-distortion theory underlies Chapter 2 beginning with Section 2.3.

It is useful to place in context a simple coder like a PCM analog-to-digital converter chip. PCM converts analog samples to binary form one at a time, each without reference to the others. The binary code represents which of a set of analog levels lies closest to the sample. By comparison, a rate-distortion coder takes a group of samples and compares them to extended words in a code book. These words are patterns of outputs that are typical of the source; even if they are not exact source sequences they are close to the source sequence. The words take into account redundancy and correlation among the source outputs.

The final type of coder in this book is the channel coder. Here the object is to correct channel errors, or more generally, to defeat channel noise. Here are some older definitions of channel coding and modulation, adapted from [1-11]. These will help us place in perspective some of the newer types of channel coding.

**Channel encoding:** The application of redundant symbols to correct data errors.

**Modulation:** Conversion of symbols to a waveform for transmission.

**Demodulation:** Conversion of the waveform back to symbols, usually one at a time.

**Decoding:** Using the redundant symbols to correct errors.

These functions appear in Fig. 1.2-2.

Today it is not so easy to distinguish coding and modulation. Many codes do not add parity check symbols, nor are their words best thought of as made up of symbols. Demodulators that work with one symbol at a time are getting rarer. We will save these older definitions for the discrete memoryless channels of Chapter 3. The codes there are accurately described as correcting errors through the addition of parity checks to the data stream and the modulator has disappeared into the channel model.

For Chapters 4 and 5 we need a more modern definition of coding and modulation as one process. We will define it to be the *imposition of one of a set of signal patterns* onto the transmitted signal. These patterns could be chosen to defeat noise, to curtail signal bandwidth (as discussed in the next section), or for quite a different purpose like encryption or multiple access. The decoder has available the same set, and it chooses the pattern closest to the noisy received signal. In so doing it removes the noise. We have fused decoding and demodulation into one process.

We cannot do justice to information theory in a book about coding algorithms. In the succeeding chapters we will take insight from the theory as needed and prove some of the fundamental theorems in a simple way where appropriate. For those wanting more, we recommend among the older texts the ones by Gallager [1-3] or Viterbi and Omura [1-4] or the very readable reprinting of Shannon's papers [1-5]. Books that are older or out of print but worth the effort to find are Abramson's introduction to information theory [1-6] or Berger's *Rate-Distortion Theory* [1-7].

## 1.3 BANDWIDTH AND ENERGY

### Signal Bandwidth

The Fourier transform provides the standard tool for measuring the bandwidth of continuous signals. Occasionally we can make use of the outright transform of a particular signal, but more often an averaged spectrum, called the average power spectral density, or PSD, is used to measure bandwidth. This is the average of the spectra of all the individual signals that can occur. Calculation of the power spectral density† is often not easy; a powerful, modern approach appears in [1-11].

It must be said that the power spectral density of an information bearing signal can be a misleading measure and its meaning in stochastic process theory is dubious. Nonetheless, it is informative and it matches intuition most of the time. The PSD shows the average location of a modulated signal. Figure 1.3-1 shows the PSD of uncoded quaternary phase shift keying (QPSK). The pattern there is typical of all modulated signals, whether coded or not. A strong central part of the spectrum called the main lobe carries most of the average signal energy, while a minor part of the energy appears in a series of rapidly descending side lobes. As a rule a signal is not much affected by channel filtering if its main lobe is left untouched. The significance of side lobes is that they interfere with adjacent channels, especially for receivers that are geographically near the transmitter. Thus the design of coded signals from a spectral point of view can focus on the main lobe, the side lobes, or both, depending on the situation.

†We will suppress the word "average"; in fact the average PSD is simply called the spectrum in the literature.

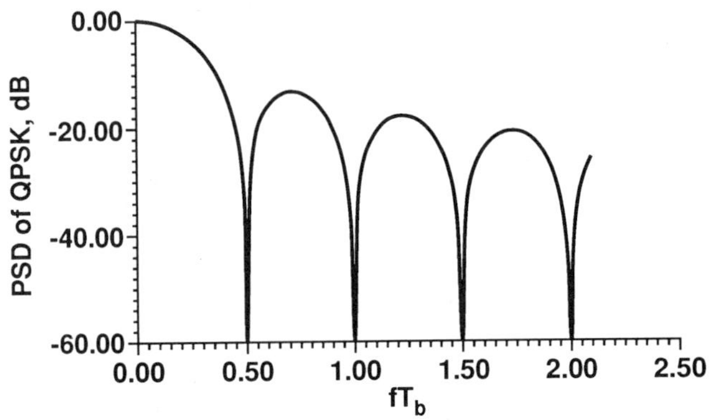


**Figure 1.3-1.** The PSD of uncoded QPSK, relative to the carrier. $T_b$ is the time devoted to a data bit.

During any one time period, however, the short term Fourier signal transform may vary widely from the PSD average. Short bursts of energy can lie far outside the main lobe location. One application for short term analysis is to find the receiver error probability for a filtered signal. The method consists of finding the short data sequence whose signal is most impaired by the filter.

Another short term effect of band limitation is intersymbol interference, or ISI. This is the name given to the smearing effect of a filter whereby signal activity that used to be confined to one time interval is dispersed into others. The smearing is caused by the non-zero pulse response of a filter. At least one class of filters, the Nyquist class, exists that has a long pulse response but does not cause ISI in a sense that matters at the receiver (see refs. [1-11] to [1-13]), but these filters can exert only a limited bandwidth reduction, and for most filters and for all really narrow filtering, ISI is inescapable.

In Chapter 4 we will study receivers that attempt to remove ISI from signals and thereby improve their error performance. The approach is to model ISI as a trellis channel code and then "decode" a likely transmitted signal.

This modeling of ISI as a code is part of a trend in modern coding theory to include bandwidth in the design of codes. Originally, codes manipulated symbols that of themselves had neither time duration nor bandwidth. A connection to continuous-time signals was made by constructing a signal space from the signals and assigning an independent dimension of the space to each symbol. A simple example would be a Hamming code combined with a binary PSK modulator; the code, which is made up of symbols, corrects single errors in the data symbols, while the PSK modulator carries each code symbol as an independent function of time. In Appendix A we give the standard relating of signal space dimensions to bandwidth. The result is simple enough to give now: A collection of signals of length $\tau$ seconds and bandwidth $(-W, W)$ Hz can represent $2W\tau$ independent dimensions as $W$ and $\tau$ grow large.

To the extent that symbolic codes add parity check symbols, they also expand bandwidth. If a code has rate $R$ data symbols per code symbol, it will need a ratio $1/R$ more code symbols, which will consume $1/R$ more signal space dimensions, which will consume $1/R$ times the bandwidth needed by the uncoded data. In recent years, coding with symbols has increasingly been extended to coding with pieces of continuous-time signals. Where code words used to consist of patterns of data and parity check symbols, they nowadays are often patterns, for instance, of signal phases and amplitudes, without added symbols. This kind of code can achieve error protection without bandwidth expansion and can even be used for bandwidth *compaction.* Its bandwidth is directly the bandwidth of its words, with no recourse to counting dimensions. In this book, the shift from parity symbols to continuous-time patterns occurs after Section 4.2.

## Signal Energy and Receiver Errors

The other fundamental measure of a signal is its energy. It has been known since the time of Marconi that raising the energy of a signal increases its reliability. In this book our channel signalling is digital, and for us the probability of receiver error is what improves. We need to be precise about this. In coded communication errors occur during error *events*, that is, data errors occur in groups during times when the receiver is not able to overcome channel noise. Added signal energy reduces the probability of these events. We can understand the effect of signal energy by studying the form of this probability.

At some signal energy levels, a coded system can have a worse data error rate than an uncoded one, and in fact this normally happens if the energy is small enough. The reason is that coded systems have long error events containing multiple data errors, while simple modulations have typically only one data error per event. More complicated codes have longer events. In all cases, however, the *event* probability declines as the energy grows and ordinarily the coded system has fewer data errors at high enough signal energy.

The probability of a particular event in Gaussian noise is almost always of the form

$$P_{\mathrm{ev}} = Q\left(\sqrt{\frac{d^2 E_b}{N_0}}\right), \tag{1.3-1}$$

where $Q(\ )$ represents the Gaussian integral

$$Q(x) = \frac{1}{\sqrt{2\pi}} \int_x^\infty \exp\left(\frac{-u^2}{2}\right) du, \qquad x > 0. \tag{1.3-2}$$

This is derived in Appendix A. $d^2$ is the signal space *square distance* between the transmitted data signal and the signal that constitutes the error event. $d^2$ must be normalized properly, as is done in Chapter 4. For codes made up of continuous-time signals, $d^2$ is a direct signal space distance. For codes made up of symbols like bits, we can get $d^2$ by considering the modulator signals that the bits produce; with the simple cases of BPSK and QPSK modulation $d^2$ is proportional to the Hamming distance, i.e., to the number of bits different, between the transmitted code word and the hypothetical receiver output. We will develop these ideas much further in Chapters 3 and 4.

The remaining quantities in (1.3-1) are the measure of signal energy $E_b$ and the Gaussian power spectral noise density, denoted by $N_0$. It is clear that the size of $E_b$ matters less than the ratio of $E_b$ and $N_0$. Thus the channel noise should be reduced as much as possible before any kind of coding is considered. The quantity $N_0$ has several interpretations. $N_0/2$ is the white noise power per Hertz of bandwidth (so that $N_0$ is the power counting both positive and negative frequencies); $N_0/2$ is also the variance of a sample of the noise. For its part, the energy parameter $E_b$ is normalized by the number of data *bits* carried in a signal interval. This means that the total interval energy $E_s$ is divided by the number of data bits per signal interval; if the interval energy varies, $E_s$ is taken as the average.

In a quaternary modulation, for example, $E_b$ is $E_s/2$ because two data bits are carried in one quaternary symbol. The normalizing is in keeping with the idea in Section 1.2 that energy is a cost per unit of information carried. Normalizing also allows us to compare schemes that carry different numbers of data bits in an interval and coded schemes that do not carry an integer number of bits.

## Bandwidth, Energy and Complexity Tradeoffs

These are the three basic costs of communication. A useful insight to have about them is that bandwidth is usually the most expensive; the information carrying capacity of a signal is most sensitive to bandwidth, the cost of processing is declining daily, and the political authorities, for their part, are loath to part with spectrum space.

Some feeling for the expensiveness of bandwidth can be gained from the Shannon capacity for a white Gaussian noise channel with bandwidth $(-W, W)$ Hz [1-1], [1-4], [1-10]

$$C = W \log_2 \left(1 + \frac{2E_d}{N_0}\right) \qquad \text{bits/s.} \tag{1.3-3}$$

Here $E_d$ is the signal energy per signal space dimension and dimensions appear at the rate of $2W$ per second. We see that a reduction of bandwidth is a linear reduction of channel capacity, but capacity falls only as the log of energy. Even though (1.3-3) is only a theoretical number, all sorts of practical systems seem to follow it in their sensitivity to energy and bandwidth, in the sense that for a set system a reduced bandwidth will worsen the error performance far more than a reduced energy.

Coding provides the means of patterning signals so as to reduce their energy or bandwidth consumption for a given error performance. But coding means complex signals, transmitters, and especially receivers. It follows that the three costs energy, bandwidth, and processing complexity all trade off against each other: For a fixed error performance, a reduction on one can be made up by an increase in the other two. There is an economic lesson in this. If the cost of complexity constantly decreases while the other commodities do not, we should work toward more subtle and sophisticated signalling schemes.

The subjects of signal spectra, ISI, and digital modulation come under the heading of digital communication theory, and are not part of this book. Readers needing further background reading are directed to the list below. Of the books listed there, Stark et al. [1-12], Peebles [1-13], and Carlson [1-14] are modern introductory texts, while [1-11] is an advanced monograph.

# *REFERENCES*

What follows is for the most part a list of selected reading in the prerequisites for this book. The fields represented are information theory, including rate-distortion theory, digital modulation theory, and stochastic processes.

[1-1] C. E. Shannon, "A Mathematical theory of communication," *Bell System Technical J.*, vol. 27, pp. 379–423 and 623–656, July 1948.

[1-2] C. E. Shannon, "Coding theorems for a discrete source with a fidelity criterion," *IRE Natl. Conv. Record*, Part 4, pp. 142–163, 1959.

[1-3] R. G. Gallager, *Information Theory and Reliable Communication*, Wiley, New York, 1968.

[1-4] A. J. Viterbi and J. K. Omura, *Principles of Digital Communication and Coding*, McGraw-Hill, New York, 1979.

[1-5] C. E. Shannon and W. Weaver, *Mathematical Theory of Communication*, Univ. of Illinois Press, Urbana, Ill., 1963.

[1-6] N. Abramson, *Information Theory and Coding*, McGraw-Hill, New York, 1963.

[1-7] T. Berger, *Rate Distortion Theory*, Prentice-Hall, Englewood Cliffs, N.J., 1971.

[1-8] V. A. Kotelnikov, *The Theory of Optimum Noise Immunity*, Ph.D. Dissertation, Molotov Energy Institute, Moscow, USSR, Jan. 1947; reprinted under the same title in English by Dover, New York, 1968.

[1-9] G. L. Turin, "An introduction to matched filters," *IRE Trans. Information Theory*, vol. IT-6, pp. 311–329, June 1960.

[1-10] J. M. Wozencraft and I. M. Jacobs, *Principles of Communications Engineering*, Wiley, New York, 1965.

[1-11] J. B. Anderson, T. Aulin, and C.-E. Sundberg, *Digital Phase Modulation*, Plenum, New York, 1986.

[1-12] H.Stark, F. B. Tuteur, and J. B. Anderson, *Modern Electrical Communications*, 2nd ed., Prentice-Hall, Englewood Cliffs, N.J., 1988.

[1-13] P. Z. Peebles, Jr., *Digital Communication Systems*, Prentice-Hall, Englewood Cliffs, N.J., 1987.

[1-14] A. B. Carlson, *Communication Systems*, 3rd ed., McGraw-Hill, New York, 1986.

# CHAPTER 2

# Source Coding Algorithms

We look now at algorithms for source coding. Source coding theory sets the ultimate limits for this coding and it suggests many of the algorithms that are in practical use. But some schemes are motivated by and perform well because of peculiarities of the data source that our theories do not describe. A classic example of this is the speech source.

For forty years speech digitizers have been at the center of source coding. Many of our examples will be taken from this field and a number of our algorithms were first devised as speech coders. Speech coding illustrates the fact that source coding must be viewed both as an art and a science. Historically, speech coders were based on mathematical principles, first those of least mean-square prediction and later those of coding theory. Today these are still important, but really effective coders play artful tricks with human perception. Lessons learned from speech coding are applied to sources like music, radar, sonar and medical waveforms.

The coding of two-dimensional sources like images has lagged somewhat in sophistication, compared to these one-dimensional sources. Image coding tends to work at a much higher speed that makes complex coding schemes less attractive. Until recently, image coders have been line-by-line coders, an approach that views the image as a kind of one-dimensional source. For these reasons we will emphasize image coding less than the other kinds.

Our story begins with coding theory and continues with applications. Sections 2.1 and 2.2 discuss noiseless source coding and a few applications, but most of the chapter is based on rate-distortion theory, which is introduced in Section 2.3. The algorithms in this chapter are primarily simple ones. We will return to more complex schemes in Chapter 6, when we take up sequential coding. As a rule relatively simple coding schemes are more effective and more common in source coding than they are in channel coding.

## 2.1 SOURCE CODING PRELIMINARIES

A source code is a mapping from sequences of source outcomes $\mathbf{Z}$ to another set of sequences $\mathbf{W}$. Presumably sequence $\mathbf{W}$ is shorter and has less redundancy than $\mathbf{Z}$. In the form of a set mapping we could denote this as

$$\mathcal{C} : \mathcal{Z} \to \mathcal{W}; \tag{2.1-1}$$

here $\mathcal{C}$ is the code mapping, $\mathcal{Z}$ is the set of possible source outcomes $\mathbf{Z}$ and $\mathcal{W}$ comprises all the code words $\mathbf{W}$. We use the bold face vector notation to emphasize that $\mathbf{W}$ and $\mathbf{Z}$ are sequences of symbols. In line with our discussion about information in Section 1.2, the sequence $\mathbf{Z}$ has a probability distribution and as a consequence of the code mapping, so does $\mathbf{W}$. Thus we can define average lengths $\Lambda_{\mathbf{z}}$ and $\Lambda_{\mathbf{w}}$ for both of these. The *efficiency* of the code is then $\Lambda_{\mathbf{w}}/\Lambda_{\mathbf{z}}$ output symbols/source symbol. When comparing the efficiencies of different codes, it is much easier if the output symbols are converted to an equivalent number of bits. If B is the size of the output symbol alphabet, this conversion yields the normalized efficiency

$$\eta = \Lambda_{\mathbf{w}} \frac{\log_2 B}{\Lambda_{\mathbf{z}}} \qquad \text{bits/source symbol.} \tag{2.1-2}$$

When the source sequences are all of the same length $K$ and the code words are all of length $N$, the code is called a *block code*. Otherwise, it is a *variable length* code; these codes can be block-to-variable, variable-to-block, or variable-to-variable, depending on the lengths of the source and code sequences. Most of the source codes in this book (and all

**Table 2.1-1.** A RATE 0.6 BLOCK CODE FOR THE SKEWED BINARY SOURCE IN EXAMPLE 2.1-1. IF $P(0) = 0.05$, THE DECODER MAKES AN ERROR WITH PROBABILITY 0.018.

| Source Sequence | Probability | Code Word |
|---|---|---|
| 11111 | 0.774 | 000 |
| 01111 | 0.041 | 001 |
| 10111 | 0.041 | 010 |
| 11011 | 0.041 | 011 |
| 11101 | 0.041 | 100 |
| 11110 | 0.041 | 101 |
| 11001 | 0.002 | 110 |
| 10011 | 0.002 | 111 |
| All the rest | 0.018 | 111 |

the channel codes) are block codes. The efficiency expression (2.1-2) for block codes has the simple form

$$\eta = N \log \frac{B}{K} \qquad \text{bits/source symbol.} \tag{2.1-3}$$

This ratio is a property of the block code only, and not the data source. It is called the *rate* of the code. Technically the block code rate is dimensionless, but the dimensions in (2.1-3) help us keep the coding process straight. If both the source and code alphabets are binary the rate is just $N/K$ output symbols per input, and the symbol compression function of the coder is clear. Variable-length codes have an efficiency (2.1-2) that depends on the source distribution as well as the code, so we cannot define a rate for them.

A simple block code example will help fix ideas.

**Example 2.1.1.**

A strongly skewed independent binary source has $P(1) = 0.95$ and $P(0) = 0.05$. This could model the pixels along a scan line in a black and white line drawing, with the 0 symbol representing a black pixel (another model was given in Section 1.2). The 8-word source code in Table 2.1-1 has binary-symbol code words of length 3 that represent sequences of 5 outcomes. If any other sequence appears besides the eight shown, it is encoded as 111, which, of course, is an encoding error. The probability of this is 0.018. The rate of the code from (2.1-3) is 0.6 output bits/input bit. The encoder error probability can be reduced by increasing the rate (so that more source sequences are covered). It can be reduced at the same rate by increasing the block length; for instance, source blocks of 10 encoded by the best 64-word length-6 code yields an error probability of 0.011.

An everyday example of a variable length code is the system of country codes used in international telephone dialing. Outside of North America, the area code words within countries are usually variable length as well. Table 2.1-2 reproduces a part of this system.

**Table 2.1-2.** Variable Length Code for Telephone Access. Heavily Called Countries Tend to Have Shorter Codes. More of This Code is Available in Most Telephone Directories.

| Country | Telephone Code |
|---|---|
| USA | 1 |
| Czechoslovakia | 42 |
| Austria | 43 |
| United Kingdom | 44 |
| Denmark | 45 |
| Sweden | 46 |
| Iraq | 964 |
| Kuwait | 965 |
| Oman | 968 |

If a string of block code words is received it may easily be divided into words, since every $N$-th symbol is the end of a word. If any string can be unambiguously divided into words, a code is said to be *decodable*. In some decodable variable-length codes several future words must be viewed before an earlier word can be unambiguously recognized; if words can always be recognized immediately when their last symbol appears, the code is *instantaneously* decodable. Block codes are all instantaneously decodable, but many variable length codes are not. A necessary and sufficient condition for an instantaneous variable length code is the *prefix condition*: No code word is a prefix of any other code word. The country codes in Table 2.1-2 were designed to satisfy the prefix condition so that the telephone equipment could recognize the country immediately. For example, countries 44 and 46 (United Kingdom and Sweden) cannot be confused with country 4 because no country has code word 4. Country 1 (USA) cannot be confused with other countries whose codes begin with 1 because there are none of these.

Two important results of source coding theory set the limits of decodable codes.

**The Kraft Inequality.** Suppose a code has $M$ words with word lengths $l_1, l_2 \ldots, l_M$. The symbols come from a $b$-ary alphabet. A code satisfying the prefix condition exists if and only if

$$\sum_{i=1}^{M} b^{-l_i} \leq 1. \tag{2.1-4}$$

**McMillan's Inequality.** A decodable code exists if and only if the lengths satisfy inequality (2.1-4).

These two inequalities show how many words can be packed into a decodable code. For economy we would like to have short words rather than long ones in a code, but short words increase the Kraft sum (2.1-4) faster. The sum can be thought of as suggesting the way that long and short words can be most favorably balanced. The McMillan result shows that the less restrictive class of decodable codes are not more efficient than the codes that satisfy the prefix condition. Block codes automatically satisfy (2.1-4) because there are $b^N$ words of length $N$ and $(b^N)b^{-N} = 1$. Proofs of these inequalities appear in the information theory texts listed at the end of Chapter 1.

## Shannon's First Theorem: The Noiseless Source Coding Theorem

The Noiseless Coding Theorem is one of the most important results of information theory because it gives the limit to the efficiency of source codes. The version we will give is for block-to-variable length codes and it shows that binary code words exist whose efficiency (2.1-2) tends to the entropy $H(Z)$ as the block length grows. The same result can be proven for other types of codes, and analogous proofs exist for sources with correlation and infinite alphabets [2-1].

**Theorem 2.1-1: The Noiseless Source Coding Theorem.** Let a source $Z$ have independent outcomes from a finite alphabet that are encoded in groups of $K$. The code word lengths are $l_i, i = 1, \ldots, M$. Then an instantaneously decodable binary code exists for which the efficiency $\mathcal{E}[l]/K$ satisfies

$$H(Z) \leq \frac{\mathcal{E}[l]}{K} \leq H(Z) + \frac{1}{K}. \tag{2.1-5}$$

*Proof*: The proof will work by giving each source outcome† $\mathbf{z}^{(K)}$ a code word whose length in bits is the first integer larger than $\log_2(1/P(\mathbf{z}^{(K)}))$. This creates a variable length code with lengths

†Boldface $\mathbf{z}^{(K)}$ represents a vector of length $K$; $z^K$ will represent $z$ to the $K$-th power.

$$\log_2\left(\frac{1}{P(\mathbf{z}^{(K)}(i))}\right) \le l_1 \le \log_2\left(\frac{1}{P(\mathbf{z}^{(K)}(i))}\right) + 1, \qquad i = 1, \ldots, M. \quad (2.1\text{-}6)$$

Taking the negative of both sides and exponentiating, we get

$$P\left(\mathbf{z}^{(K)}(i)\right) \ge 2^{-l_i}$$

for the left inequality. If we sum both sides of this over all the code words, the result is

$$1 \ge \sum_{i=1}^{M} 2^{-l_i},$$

which is the Kraft Inequality; this shows that a code must exist with these lengths.

To finish the proof, we take expectations of both sides of (2.1-6). This will give

$$H\left(\mathbf{Z}^{(K)}\right) \le \mathcal{E}[l] \le H\left(\mathbf{Z}^{(K)}\right) + 1.$$

It is easy to show (see Problem 2-3) that for independent outcomes, $H(\mathbf{Z}^{(K)}) = KH(Z)$. Substituting this and dividing by $K$, we get (2.1-5).
**QED**

We can apply the source coding theorem (in its block version) to the skewed source of Example 2.1-1. The code had rate 0.6 there, but $H(Z)$ is 0.286 and (2.1-5) implies that a code with rate this low exists and will give zero encoding error probability if we set $K$ large enough. The form of this conclusion, and the requirement that the block length and hence the coding delay be large, are typical of coding theorems in information theory.

## Code Trees

Throughout this book we will represent codes by a graphical construction called a code tree. Some examples are shown in Fig. 2.1-1. Code tree (a) is the block code from Example 2.1-1. Any path from the root node to a square block corresponds to a source code word, and the source sequence that was encoded appears in parentheses next to the block. The source coder output that represents each source sequence can be obtained by tracing through the tree from the root to a block. Taking an upward branch means a binary 0 output while a downward branch means a 1; in code (a) for example, three ups in succession are the encoding of 11111, represented by code word 000, while two ups and a down, 001, represents the sequence 01111. In the source coder output triples, the leftmost symbol is the first coder output, a convention we shall follow throughout the book. The up and down sequences are called *path maps*, because they map a path through the tree.

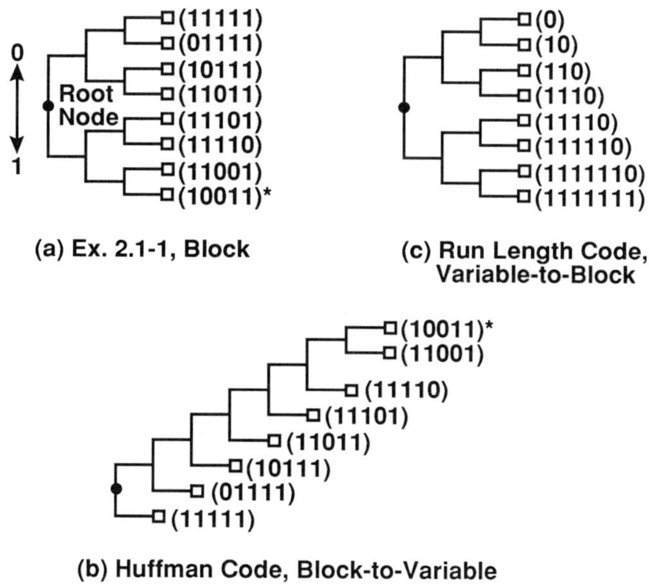


**Figure 2.1-1.** Examples of binary code trees. Code (a) is listed in Table 2.1-1 and code (c) in Table 2.2-1; code (b) is the Huffman code in Example 2.2-1. The source sequence inputs are parenthesized and the code word output is obtained by following a path to a square block, taking up=0 and down=1. *In codes (a) and (b) the code word is released if an unlisted source sequence appears.

Code (a) is a block (i.e., block-to-block) code. This is clear from the code tree since all the parenthesized words are the same length and all the tree paths are the same length. Code (b) is a block-to-variable length code, with equal length source words but variable coder outputs; the variability is clear from the different path lengths. Code (b) is a Huffman code that we will take up in Section 2.2. Code (c) is variable-to-block, the reverse of (b). Its tree paths are all the same length but the source words now vary. This code is a run length code, also discussed in

Section 2.2. All the codes in Fig. 2.1-1 are binary, by which we mean the path maps consist of binary symbols. A $b$-ary source code has $b$-ary output symbols and has up to $b$ branches out of each tree node.

Every code has a code tree. A little thought will show that the instantaneous codes are those in which all the square blocks come at the ends of paths. This follows because no source coder output sequence is the prefix of another if and only if no path is a subpart of another. Non-instantaneous codes are those that have some square blocks part way along the tree paths.

In Chapter 4 we will make use of channel code trees and a related graphical construction called a trellis. In an input-output sense, a channel code is the reverse of a source code. Now the path map sequences are the input to the channel coder and the parenthesized code words are the outputs to the channel.

Some codes are *tree codes.* This nomenclature must not be confused with a code tree, which every code has. A tree code is code whose words have a certain strict regularity: A fraction $1/b$ of the code words begin with the same group of $a$ symbols, the next fraction $1/b$ all begin with another group of $a$, and so on for all $b$ fractions; within the words of any of these fractions, the same property repeats for the second group of $a$ symbols along the code words, and for all succeeding fractions through the entire length of the code. Necessarily, tree codes are block codes and the code word length is a multiple of $a$. Figure 2.1-2 illustrates a $b = 3$, $a = 2$ tree code with code word length 4. It can be seen that $1/b$ of the words begin with the two symbols 00 while the other thirds begin with 10 and with 01, and that this property continues in the second tree level. In fact, the tree itself neatly divides the words into thirds and these thirds into thirds. As the figure shows, it is convenient to parenthesize the code words not as one word at the end of a tree path but as groups of $a$ along the path. We can do this because of the strict regularity property. We can generate code words by tracing through the tree branch by branch and collecting groups of $a$; for instance, the top most word is (00)(00), or the total word (0000).

The rate of this code is two ternary coder outputs per four source symbols. From definition (2.1-3), the rate is $(2\log 3)/4 = 0.792$. We can generalize this to any tree code by noting that $aN$ source outputs are coded into $N$ $b$-ary coder outputs, so that the tree code rate becomes

$$\frac{\log b}{a} \quad \text{bits/source symbol.} \tag{2.1-7}$$

We can apply this tree code to the source in Example 2.1-1. The code's error probability is only 0.005 instead of 0.018. This is primarily

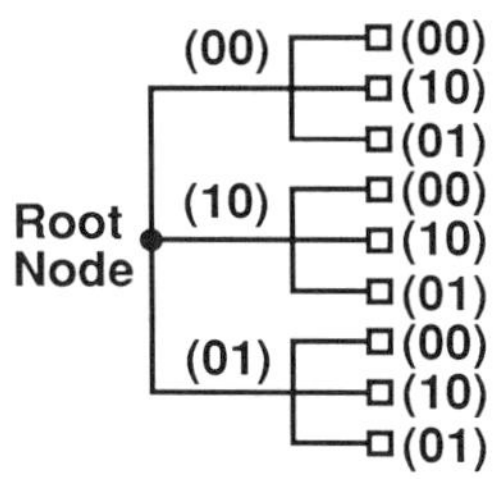


**Figure 2.1-2.** A 2-level tree code with $a = 2$ symbols/branch and branching factor $b = 3$.

because the code covers more source sequences than the rate 0.6 code in the example, or put another way, its rate is higher.

## 2.2 APPLICATIONS OF NOISELESS SOURCE CODING

Most applications of source coding are of the rate-distortion type and we will begin to discuss these in Section 2.3. Noiseless source coding applications are relatively few and simple. We will discuss three: Huffman coding, run length coding and the application of noiseless coding to PCM. The last will give us an opportunity to review quantization.

This section brings us, for the first time, to the problem of how to find *efficiently* the best code word from among the words in the code book. The source encoders here will select this word by an exhaustive search of the code book or by a similar direct approach. More sophisticated searches are far more efficient, but these find their best application with rate-distortion and channel coding. We take up code book search algorithms in Chapter 6.

### Huffman Coding

Properly speaking, the Huffman coding algorithm is a scheme for designing a code, not for searching a code book. It produces a variable-length code for either block or variable length sequences of source outputs. The procedure works best, as do noiseless coding schemes generally, when the different source sequences have dramatically varying probabilities. A Huffman code has a number of interesting properties, chief of

**The Huffman Algorithm**

**Step 1.** Assign the $b$ smallest probability sequences $z_{M-b+1}, \ldots, z_M$ to the $b$ branches coming out of a code tree node; put square blocks at the ends of these branches, as in Fig. 2.2-1, so that the tree paths ending here correspond to the code words for the $b$ sequences. The set of sequences stemming from this node is called the *union* for the node. Every tree node has a union. Assign a probability to the union that equals the sum of the probabilities of all the sequences forward of the node.

**Step 2.** Repeat **Steps 2a** and **2b** until $b$ or fewer objects remain.

> **Step 2a.** Reorder the list of remaining sequences and unions by probability, counting the newly formed union as a single object in the list.
>
> **Step 2b.** Assign the $b$ least probable objects in the list to a set of $b$ branches out of a new tree node. This set is a new union corresponding to the new node. If any of the objects is itself a union, then its branch connects to the node and subtree for that union. If the object is one of the original source sequences, put a square block at the end of its tree branch. The probability of the new union is the sum of all its object probabilities.

**Step 3.** Now $b$ or fewer sequences or unions remain. Assign these objects to branches out of the root node. Union branches connect to already formed subtrees and source sequence branches that end in square blocks. The resulting single tree is the Huffman code tree.

which is that it minimizes the efficiency expression (2.1-2) for a given set of source sequences; that is, its average coder output length is least among all codes for the given set.

We have already listed the Huffman algorithm. The following example will illustrate the procedure. The algorithm is easy to understand directly from the example, and the reader may prefer to skip to that. The idea of the procedure is to construct a variable length code tree like that of Fig. 2.1-1(b) *backwards*, beginning from the longest paths back to the root node. When the procedure terminates, we have a $b$-branching code tree, and from this we can get the source sequence – coder output pairs. To start the procedure we need a list of source sequences $\mathbf{z}_1 \ldots, \mathbf{z}_M$ ordered by probability from most to least. These can be of the same or of varying lengths. The grouping of source outputs into a sequence is called forming an extension of a source. An extension is really just another source with a longer list of outcomes. The Huffman procedure treats $\mathbf{z}_1, \ldots, \mathbf{z}_M$ as $M$ abstract objects that could just as well be the outputs of an $M$-ary source.

**Example 2.2-1.**

> We will form a Huffman code to replace the one in Example 2.1-1. The skewed source there had 0-symbol probability 0.05 and is, therefore, ideal for noiseless source coding. We will consider the length-5 sequences of source outcomes in Table 2.1-1; this means that the source coder will, again, fail with probability 0.018. (The Huffman procedure will also give a code for all 32 length-5 sequences which has zero probability of failure). The code develops as shown in Fig. 2.2-1. The probability skew in the source leads to an elongated code tree that corresponds to coder outputs with widely varying lengths. The average of these lengths is about 1.6 binary symbols, and since each coder output represents 5 source outcomes, the efficiency $\eta$ is 0.32 bits/source symbol. This is quite close to $H(Z)$ which is 0.286, and much better than the code of Example 2.1-1 whose efficiency was 0.6.
>
> Both of these codes fail with probability 0.018. The Huffman procedure applied to a full list of 32 source blocks yields a binary code with $\eta$ near 0.36 bits/source symbol (See Problem 2-4) whose failure probability is zero. Theorem 2.1-1 tells us that among the zero-failure codes, no code has $\eta$ below $H(Z) = 0.286$.

From the example we can see that the Huffman procedure can produce some very good codes. In fact it can be proven that no code with shorter average length of output symbols exists for the given set of source sequences. We can, however, drive down the average length by starting

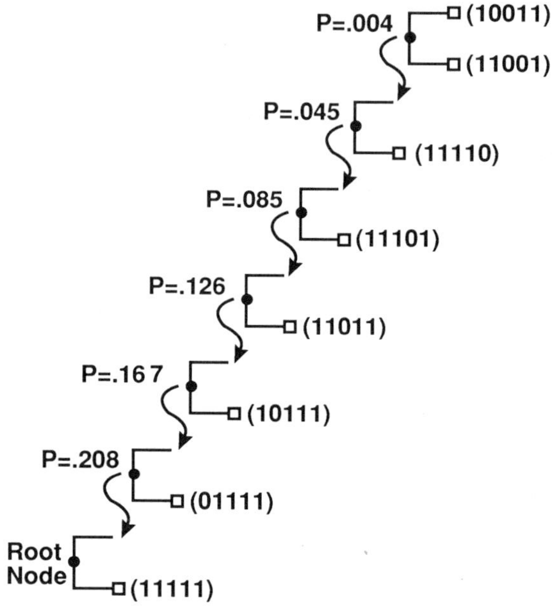


Figure 2.2-1. Progress of the Huffman code generation algorithm on the source words of Table 2.1-1. The algorithm moves from right to left. Word unions are symbolized by node dots next to the union probabilities. Source word probabilities are given in Table 2.1-1.

with a larger set of longer sequences; as the set grows the average length approaches $H(Z)$. These and other properties of Huffman codes are proven in Gallager [2-1] or Abramson [2-3].

## Run Length Coding

When a source becomes extremely skewed toward one outcome, the technique of run length coding becomes attractive. Suppose the source is binary, with the 1-symbol much more probable. The source might represent a telemetry sensor for micrometeorite hits or some other rare event, with 0 marking the hit. The idea of run length coding is to count the length of 1s until the next 0 and to send this number. The coder puts out an integer (base $b$) at each chance occurrence of a source 0.

This simple idea needs only a small repair in order to be practical. If we place a limit $J$ on the coder integer, and reserve output integer $J$

to mean "$J$ 1s have appeared and the next symbol may be another 1", then the scheme will encode any run of 1s without error. We can view the scheme now as variable-to-block. The following example illustrates these points.

Example 2.2-2.

> Returning once more to the source with 1-probability 0.95, we can set up the block length 3 run length code in Table 2.2-1. The coder output integer 0, represented by 000, means the next source symbol is a 0; integer 6, or 110, means a run of 6 is terminated by a 0; integer 7 means a run of 7 1s that may or may not be followed by a 0. Each group of three coder output bits represents on the average 6.03 source symbols (see Problem 2-6), so that the coder efficiency is 0.50 bits/source symbol. The code tree is Fig. 2.1-1(c). It not only has better efficiency than the code of Example 2.1-1, but also it is easier to use and has zero failure probability. Our run-length code has much higher rate than the Huffman code (Example 2.2-1), but it is much easier to use than the zero-failure Huffman version.

**Table 2.2-1.** A Run Length Code with Output Length 3. With the Source of Example 2.1-1, the Efficiency is 0.50 Bits/Source Symbol.

| Source Sequence | Run Length | Code Word |
|---|---|---|
| 0 | 0 | 000 |
| 10 | 1 | 001 |
| 110 | 2 | 010 |
| 1110 | 3 | 011 |
| 11110 | 4 | 100 |
| 111110 | 5 | 101 |
| 1111110 | 6 | 110 |
| 1111111 | $\geq 7$ | 111 |

## Quantizers and Noiseless Coding

Ordinary pulse code modulation (PCM) quantizers provide an interesting application for noiseless source coding because their binary output words are seldom equiprobable. By relation (1.1-3), the output word entropy $H(Z)$ of $n$-bit PCM must be less than $\log_2(2^n)$, which is $n$. By Theorem 2.1-1, there must be a noiseless coding scheme that we can attach to the PCM output that will reduce the bit stream nearly to $H(Z)$ bits per PCM input. It is important to realize that this holds true even when the analog samples at the PCM input are independent. We will take up the case of correlated samples later, in Section 2.4. The term *entropy coder* is often used for a noiseless coder that is designed to reduce the bit stream in independent-sample PCM.

Since the PCM quantizer is an important unit in many of the systems to come, we will review it now. A PCM quantizer divides the analog input range into $2^n$ segments called bins. In *linear* PCM $2^n - 2$ of these bins are identical and regions at the top and bottom of the range extend to plus and minus infinity. In other kinds of PCM, the bin sizes are scaled in an attempt to match the probability distribution of the analog samples. A sample will lie in one of the bins and the PCM output will be the $n$-bit binary code word assigned to that bin. In a 3-bit quantizer, for example, the code words $000, 001, \ldots, 111$ might be assigned to the bins from negative to positive; for an input range centered at zero volts the first bit in the word would become the sign bit, 0 for negative and 1 for positive.

Associated with each bin is a real-number value called the *reproducer level.* Most often, it lies at the center of the bin. Once an input sample is converted to PCM code its true value is lost and in all future appearances it takes on the bin reproducer level. The difference between the level and the sample is the quantizer error. These facts are illustrated for a few quantizer bins in Fig. 2.2-2.

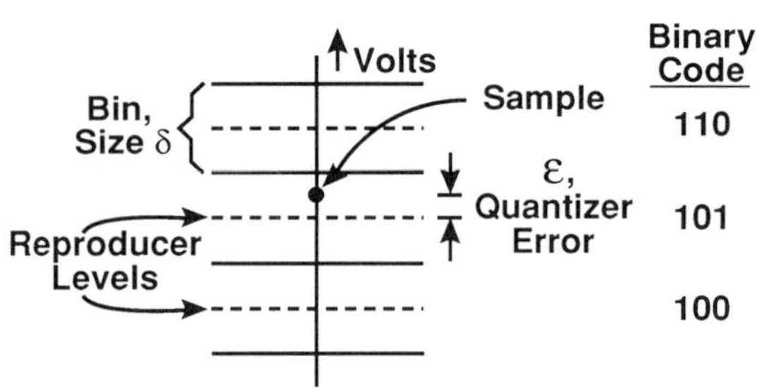


Figure 2.2-2. A few bins of a 3-bit PCM quantizer, showing reproducer levels and a quantization error.

The bin sizes and reproducer level placements that minimize squared (or other) quantizer error depend on the sample probability distribution. An algorithm for finding them was worked out by Max [2-4] and appears in Jayant and Noll [2-5]. A common assumption is that the input samples are independent Gaussian variables, and for this case Table 2.2-2 gives the least square-error quantizers for $n = 1$-5. These often serve as benchmarks in evaluating more complex digitizers; an optimized square-error quantizer like one of these is called a *Max quantizer*. Another benchmark is the Max quantizer for the uniform distribution, but it is easy to see that this consists of $2^n$ identical bins that span the uniform range and of reproducer levels that lie in the middle of the bins.

**Table 2.2-2.** MAX QUANTIZERS FOR INDEPENDENT GAUSSIAN SAMPLES WITH ZERO MEAN AND UNIT VARIANCE; FOR OTHER VARIANCES, SCALE BY THE SQUARE ROOT OF THE VARIANCE; THE HIGHEST RATE QUANTIZERS HAVE ALMOST UNIFORM BINS.

n = 1

Bins: 0
Levels: .80

n = 2

Bins: 0 .98
Levels: .45 1.51

n = 3

Bins: 0 .50. 1.05 1.75
Levels: .25 .76 1.34 2.15

n = 4

Bins: 0 .26 .52. .80 1.10 1.44 1.84 2.40
Levels: .13 .39 .66 .94 1.26 1.62 2.07 2.73

n = 5

Bins: 0 .13 .26 .40 .54 .68 .82 .97 1.13 1.30 1.48 → 1.68 1.91 2.17 2.51 2.98
Levels: .07 .20 .33 .47 .61 .75 .89 1.05 1.21 1.39 → 1.58 1.79 2.03 2.32 2.69 3.26

| Bits | Entropy | Mean Square Error |
|---|---|---|
| 1 | 1. | 0.363 |
| 2 | 1.91 | 0.118 |
| 3 | 2.83 | 0.0345 |
| 4 | 3.77 | 0.00950 |
| 5 | 4.73 | 0.00250 |

The Table gives the square error and the entropy of the Gaussian Max quantizers. The error is related to $n$ by a simple law that in fact holds for any input distribution once $n$ is larger than 3 or so. We will derive it now. The key to the law is that once the bin sizes are small then the location of a sample within a bin, given that it lies in that bin, is virtually random, even if the input distribution is not itself uniform. The mean square of the uniformly distributed error $\varepsilon$ is then

$$\mathcal{E}[\varepsilon^2] = \frac{1}{\delta}\int_{-\delta/2}^{\delta/2} \varepsilon^2 \, d\varepsilon = \frac{\delta^2}{12}$$

for a bin size of $\delta$. The mean square of the samples themselves is their square rms value $A^2_{rms}$. The ratio of the signal to the noise energies, the SNR, is then

$$\mathrm{SNR} = 12\frac{A^2_{\mathrm{rms}}}{\delta^2}. \tag{2.2-1}$$

Next we need to fix the total input range, and a standard assumption in audio work is that this should be set to $(-4A_{\mathrm{rms}}, 4A_{\mathrm{rms}})$. Samples that fall outside the range create *overload noise*, but we will ignore this now; see Problem 2-9. $2^n - 2$ bins fit into the range, so that for linear PCM we have

$$\delta = 8\frac{A_{\mathrm{rms}}}{2^n - 2}.$$

Substituting into (2.2-1), we get for large $n$ that SNR $= (3/16)2^{2n}$ and in decibel form this becomes

$$\begin{aligned}\mathrm{SNR}_{\mathrm{dB}} &= 10\big(n\log_{10}4 + \log_{10}(3/16)\big)\\ &= 6.02n - 7.3 \qquad \text{(The 6 dB rule).}\end{aligned} \tag{2.2-2}$$

Eq. (2.2-2) is called the *6-dB rule*; it states that each new PCM bit adds about 6 dB to the quantizer SNR. Although we have derived it for linear PCM, it holds for any reasonable bin spacing, input distribution, or assumption about the quantizer range. Only the constant term ("7.3" above) differs in each case. The rule also applies to predictive quantizers like those in Section 2.4, the only exception being the delta modulator. It even applies to the optimal codes of rate-distortion theory. (See Fig. 2.3-2 and Example 2.3-2). The 6-dB rule that governs the Gaussian Max quantizers in Table 2.2-2 is approximately $6n+4.1$. We can use this rule to extend the table. For instance, an 8-bit PCM digital telephone systems, modeled as carrying Gaussian-distributed calls, will have a theoretical SNR of 52.3 dB.

The subject of PCM extends well beyond the facts that we need here in a coding text. References [2-6] or [2-7] provide a wider introduction to PCM, and [2-5] is an advanced text.

The entropies in Table 2.2-2 fall steadily further from the $n$-bit output rate as $n$ grows, because the analog sample is not equally likely to fall in each bin. We see that the 5-bit PCM output has entropy of only 4.73. This implies that a source coder could reduce the PCM rate by 5.4%. Because this is so small, such a coder is not often used in practice. It is worth noting, however, the theoretical fact that the square-error minimizing quantizer does not also minimize entropy. To carry out both minimizations with one coder, we need a rate-distortion coder, the subject of the next section.

## 2.3 RATE-DISTORTION THEORY PRELIMINARIES

The framework of noiseless coding theory fails to address many natural problems. A storage medium or channel may not accept as many as $H(Z)$ bits/source output. $H(Z)$ may be infinite; this is always the case with analog data like samples of speech or pictures. Small distortions in the source may not be perceptible at all. Even if they are, high fidelity may not be worth the transmission cost. All of these situations come under the heading of rate-distortion theory. For a given coder output rate the theory analyzes what is the least obtainable distortion in the coded source outcomes, or viewed the reverse way, it finds the least coder rate required for a set distortion in $Z$. Either way, a rate-distortion code seeks to simultaneously minimize both the rate and the distortion of a coder.

A rate-distortion code would not be suitable for coding a bank transaction or a C program because these must be preserved with perfect accuracy. We should also not confuse rate-distortion coding with the removal of redundancy and correlation, since the original source data can still be perfectly recovered in these cases. True rate-distortion coding *deletes information.*

To the facts we have reviewed in Sections 1.2 and 2.1, we need to add a definition of source distortion and a new kind of source coding theorem that takes account of distortion. We will also need to develop a framework for correlated sources since most of the applications of rate-distortion coding are to this kind of source.

## Distortion Measures

Just as in Section 2.1, a rate-distortion code is a mapping $\mathcal{C}: \mathcal{Z} \to \mathcal{W}$ where $\mathcal{Z}$ is a set of source sequences and $\mathcal{W}$ is a set of code words. We will consider only block codes, in which both the words and sequences are of the same length $K$. We will call the code words by the special name *reproducer words*, because the object of the rate-distortion coder is to reproduce the source sequences by a close approximation. The reproducer words will, therefore, be made of symbols that are very much like source symbols, but there will be many fewer reproducer words than source sequences. It is usually convenient to denote the reproducer word that the coder chooses by a simpler word like a binary-symbol serial number; if there are m reproducer words the length of this output would then be $\log_2 m$. Each rate-distortion problem will have a natural form for this representation. The reproducer word terminology helps us keep the code distinct from the final form of the coder output. As in noiseless coding, the rate of a block rate-distortion coder is measured in output bits per source symbol; it is $(\log m)/K$.

Now we need to define a *distortion measure* $\rho(\mathbf{z}^{(K)}, \mathbf{w}^{(K)})$ that assigns a distortion, or a cost or an error, to encoding the source sequence $\mathbf{z}^{(K)}$ by the reproducer word $\mathbf{w}^{(K)}$. We will work only with per-letter distortion measures for which the total measure is the sum of individual nonnegative symbol distortions,

$$\rho(\mathbf{z}^{(K)}, \mathbf{w}^{(K)}) = \sum_{i=1}^{K} \rho(z_i, w_i). \tag{2.3-1}$$

Three per-letter distortion measures account for all the applications in this book. The *squared error* measure, for use with analog sources, assigns distortion

$$\rho(z_i, w_i) = (z_i - w_i)^2. \tag{2.3-2}$$

The sum of this over a word is the error energy between the reproducer word and the source sequence, and $(1/K)\rho(\mathbf{z}, \mathbf{w})$ is the mean square error. A closely related measure is the *absolute error*

$$\rho(z_i, w_i) = |z_i - w_i|. \tag{2.3-3}$$

The third measure is intended for binary sources reproduced by binary words:

$$\rho(z_i, w_i) = \begin{cases} 0, & \text{if } z_i = w_i \\ 1, & \text{otherwise.} \end{cases} \tag{2.3-4}$$

This is called the *probability of error* (PE) measure after the fact that $(1/K)\rho(\mathbf{z}, \mathbf{w})$ is the relative frequency of mismatching symbols in the reproduction. The measure assesses a unit penalty if the reproducer symbol mismatches and no penalty at all if it matches.

The rate-distortion coder chooses the closest reproducer word in the code book to the source sequence, as measured by the distortion measure. Alternately, it may save searching effort and settle for one among the closer words. In rate-distortion coding, the object is a good word, unlike channel coding where the unique transmitted word is the decoder object. Suppose that the reproducer word must be good enough that its average distortion is $D$ per letter; that is, the code mapping $\mathcal{C}$ satisfies

$$\frac{1}{K}\mathcal{E}\left[\rho\left[\mathbf{z}, \mathcal{C}(\mathbf{z})\right]\right] \leq D. \tag{2.3-5}$$

Then the *rate-distortion function* $R(D)$ is defined to be the least rate in bits per source outcome that any such coder can achieve.

$R(D)$ is a function that relates a distortion criterion $D$ to the coder output rate necessary to sustain $D$. For a discrete-valued source, $R(0)$ is $H(Z)$, the entropy of the source, since we know that a noiseless coder can code the source without distortion; for an analog source, $R(0)$ normally equals infinity, since no number of bits can specify the outcome exactly. The $D$ for which $R(D) = 0$ is the distortion from simply guessing the source outcomes without any information about them. In between these extremes $R(D)$ must be a nonincreasing function of $D$. This follows because the number of reproducer words needed to place a word $\mathcal{C}(\mathbf{z})$ closer than $D$ to the average $\mathbf{z}$ cannot increase when $D$ grows and so neither can the rate. We can think of rate-distortion coding as a covering problem in which we want to cover a set of source sequences with balls of radius $D$. The larger the balls, the fewer that are needed to cover the space.

Figure 2.3-1 shows the function $R(D)$ for the combination of the binary symmetric source and the PE measure of (2.3-4). The reproducer words are now binary sequences like the source words. Beginning at $R(0) = H(Z) = 1$, the function drops monotonically with $D$ until $R(D) = 0$ at $D = 0.5$; this is the distortion we can obtain by guessing at the source outcomes without information from the coder. When $D = 0.11$, $R(D) = 0.5$. This means that when the coder output is 0.5 bits/source outcome, a compression of 2:1 in the source information, no coder can average fewer than 11% bit errors in its reproducer word output. A second $R(D)$ function in Fig. 2.3-1 is the one for the strongly skewed binary source in Examples 2.1-1, 2.2-1, and 2.2-2, for which $P(0) = 0.05$. $R(D)$ runs from $H(Z) = 0.286$ at $D = 0$ down to zero at $D = 0.05$, a distortion we can obtain by always guessing outcome 0. The skewed-source $R(D)$ lies far below the symmetric-source $R(D)$. To achieve rate 0.286 we would reduce the skewed-source redundancy by noiseless coding, but to reduce the rate still further will require rate-distortion coding. The rate distortion functions in Fig. 2.3-1 are derived in Example 2.3-1.

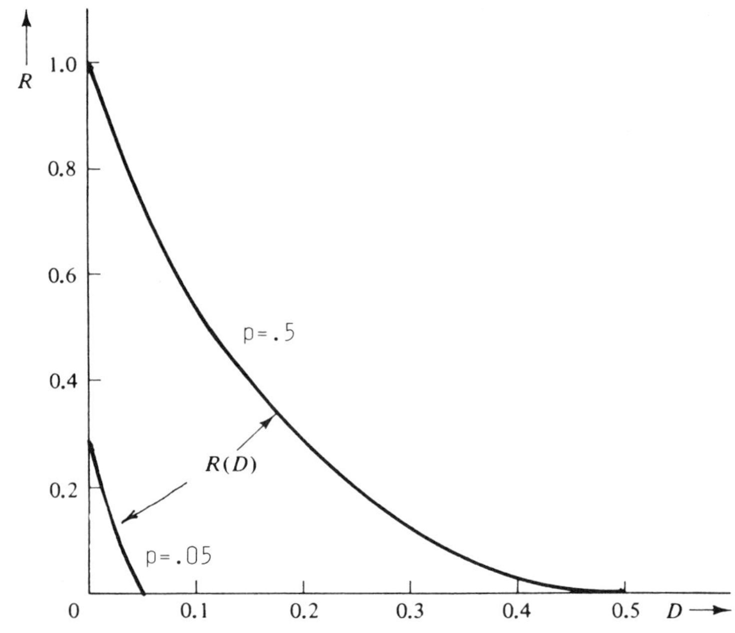


**Figure 2.3-1.** Rate-distortion functions for the binary memoryless source with PE distortion measure, for $P(0) = 0.5$ and 0.05 (the source in Ex. 2.1-1 and 2.2-1). The rate axis intercepts are $H(Z)$.

Shannon [2-8] and Kolmogoroff [2-33] first described the function $R(D)$, some 10 years after the original source and channel coding papers. Shannon [2-8] first proved that coders exist that achieve rate $R$ and distortion $D$.

## Random Codes

In proving his rate-distortion coding theorem, Shannon used once again the method of *random coding*, which he had devised for his 1948 channel coding theorem. His idea was to select the reproducer-word symbols at random, each letter of each word independently and by the same distribution. He then showed that at lease one of the $2^{KR(D)}$ words of length $K$ so chosen has distortion $\rho(\mathbf{z}^{(K)}, \mathbf{w}^{(K)})$ close to $KD$, with the probability of this tending to 1 as the length $K$ grew.

Random coding was an original and striking idea and it gives the right answer to a great many questions. It will appear several times in this book. It is used to derive channel capacity, cut-off rate, and the properties of sequential coders. Important concrete codes, like the convolutional codes, have properties in common with random codes. In a sense, Shannon information theory allows no other kind of code. We pointed out in Section 1.2 that the theory regards all information streams, including code word symbols, as probability distributions. On a deep level, information theory is a set of limit theorems about random symbol streams, that makes suggestions about real data streams; in the limit a randomly selected long code word works because no other outcome is statistically possible.

But even philosophers find random coding a frustration. Isn't there some probability that no reproducer word lies within $D$ of a source sequence? Could not the entire code be bad? The answer to these concerns lies in the probabilistic language of the coding theorem statement. The theorem claims only that source sequences will be properly encoded on the *average*; further, it claims that the random code is good only with *high probability.* In the unlikely event that our first code is poor, we can select another, and eventually we have to select a good one. Another frustration about a random coding proof is that it never actually exhibits a code, but only shows that one must exist.

We will now generate random codes whose performance lies near the $R(D)$ function. We start with a sequence of $K$ outcomes from a memoryless source, $Z_1, \ldots, Z_K$, which are distributed by the product distribution

$$p(\mathbf{z}^{(K)}) = \prod_{i=1}^{K} p_Z(z_i).$$

Hence $p_Z$ is the distribution that defines the source $Z$ and if $Z$ is analog $p_Z$ will be a density. Rate-distortion code words of length $K$, denoted $\mathbf{W}^{(K)} = W_1, \ldots, W_K$, are selected at random by

$$q(\mathbf{w}^{(K)}) = \prod_{i=1}^{K} q_W(w_i); \tag{2.3-6}$$

that is, each letter of each word is chosen independently by the distribution $q_W(w)$. A total of $m$ words are chosen, so that the rate of the code is $R = (\log_2 m)/K$ bits/source symbol.

It remains to define the distribution $q_W(w)$. What we will actually work with is the conditional distribution $q_{W|Z}(w|z)$, a perhaps counterintuitive step that needs explanation. We can always write the reproducer word distribution as the marginal distribution

$$q_W(w) = \sum_z q_{W|Z}(w|z) p_Z(z). \tag{2.3-7}$$

The formal definition of $R(D)$ includes a minimization over all the conditional distributions that lead to this same marginal; or put a little more philosophically, the rate-distortion relation between two information sources $p_Z(z)$ and $q_W(w)$ depends on the conditional distributions that can exist between them.

What conditional distribution will be used in (2.3-7)? The answer lies in the definition of $R(D)$. First, define the set

$$Q_D = \left\{ q(w|z) : \sum_w \sum_z p(z) q(w|z) \rho(z,w) \leq D \right\}; \tag{2.3-8}$$

this is the collection of all conditional densities that satisfy the criterion that reproducer-word distortion be less than $D$. As a consequence of (2.3-7) and (2.3-8), our random code will perform at distortion $D$ or less on the average. Next we define $R(D)$; this is

$$R(D) = \min_{q \text{ in } Q_D} \mathcal{E}\left[\log \frac{q(W|Z)}{q(W)}\right]. \tag{2.3-9}$$

The coding theorem will show that randomly chosen codes of this rate can indeed achieve distortion $D$, and this is the justification of this rather odd looking expression. Information theorists call the expectation over $W$ and $Z$ in (2.3-9) the mutual information

$$I(W;Z) = \sum_w \sum_z q(w|z) p(z) \log \frac{q(w|z)}{q(w)}$$

between the variables $W$ and $Z$. (See Section 3.1). $R(D)$ is the least such information when the $W$ distribution is constrained by (2.3-8).

### The Rate-Distortion Coding Theorem

In the foregoing we have defined a rate-distortion function, given a memoryless source distribution and a distortion measure. It remains to show that a random code selected by (2.3-7) and (2.3-8) can indeed have rate $R$ and distortion $D$. The following proof is based on Berger [2-2] and it does this for discrete-valued memoryless sources and a bounded, per-letter distortion measure; replacing the summations with integrals gives the proof for analog sources. Extensions of the proof to more general distortions, and a *converse*, which states that no better codes exist, may be found in [2-1] or [2-2]. Jelinek [2-19] proved the result for random tree codes.

**Theorem 2.3-1: The Rate-Distortion Coding Theorem.** Define $R(D)$ as in (2.3-9). Assume a bounded per-letter distortion measure $\rho(z, w)$ and a discrete memoryless source defined by $p_Z(z)$. Given any small $\varepsilon > 0$ and $D \geq 0$, a block length $K$ can be found such that there exists a code with

$$\frac{1}{K}\mathcal{E}\left[\rho(\mathbf{Z}^{(K)}, \mathbf{W}^{(K)})\right] < D + \varepsilon \tag{2.3-10a}$$

and rate

$$\frac{1}{K}\log_2 m < R(D) + \varepsilon. \tag{2.3-10b}$$

*Proof*: Let the source outcome be $\mathbf{z}$ and choose $m = 2^{K[R(D)+\varepsilon]}$ reproducer words $\mathbf{w}$ at random using a $q(w)$ from the set of distributions (2.3-8). From these words make up several sets:

i) $S_D(\mathbf{z}) = \{\mathbf{w} : (1/K)\rho(\mathbf{z}, \mathbf{w}) \leq D + \varepsilon/2\}$, the set of reproducer words close enough to $\mathbf{z}$;

ii) $S_R(\mathbf{z}) = \{\mathbf{w} : (1/K)\log\{q(\mathbf{w}|\mathbf{z})/q(\mathbf{w})\} \leq R(D) + \varepsilon/2\}$, the set of reproducer words with small enough information about $\mathbf{z}$;

iii) $S_{DR}(\mathbf{z}) = S_D(\mathbf{z}) \cap S_R(\mathbf{z})$, the set of reproducer words satisfying both (i) and (ii).

The proof consists of three parts that show that with certainty as $K$ grows this $m$-word code has average distortion $D + \varepsilon$.

The first part expresses the probability that the distortion criterion cannot be met in terms of the probabilities of these sets. Let $\rho_{\max}$ be the largest per-letter distortion that can occur. The probability that none of the $m$ random reproducer words lies within $D + \varepsilon/2$ of $\mathbf{z}$ is

$$\left[1 - P\{S_D(\mathbf{z})\}\right]^m. \tag{2.3-11}$$

Next we cite a series of overbounds to this probability. First, it must be true that (2.3-11) is

$$\leq \left[1 - P\{S_{DR}(\mathbf{z})\}\right]^m. \tag{2.3-12}$$

Next, since every reproducer word in $S_{DR}(\mathbf{z})$ satisfies (ii), we can write

$$q(\mathbf{w}) \geq 2^{-K[R(D)+\varepsilon/2]} q(\mathbf{w}|\mathbf{z})$$

and for the whole set $S_{DR}(\mathbf{z})$,

$$\begin{aligned} P\{S_{DR}(\mathbf{z})\} &= \sum_{\mathbf{w} \text{ in } S_{DR}(\mathbf{z})} q(\mathbf{w}) \\ &\geq 2^{-K[R(D)+\varepsilon/2]} \sum_{\mathbf{w} \text{ in } S_{DR}(\mathbf{z})} q(\mathbf{w}|\mathbf{z}). \end{aligned}$$

This means that (2.3-12) is itself overbounded by

$$\left[1 - 2^{-K[R(D)+\varepsilon/2]} \sum_{\mathbf{w} \text{ in } S_{DR}(\mathbf{z})} q(\mathbf{w}|\mathbf{z})\right]^m. \tag{2.3-13}$$

Finally, we can use the standard inequality

$$(1 - \alpha\beta)^m \leq 1 - \beta + e^{-m\alpha}, \qquad \alpha \geq 0, \beta \leq 1$$

to overbound (2.3-13) by

$$1 - \sum_{\mathbf{w} \text{ in } S_{DR}(\mathbf{z})} q(\mathbf{w}|\mathbf{z}) + e^{-m2^{-K[R(D)+\varepsilon/2]}}.$$

Since $m = 2^{K[R(D)+\varepsilon]}$, we can make the third term as small as we like by increasing $K$; for large enough $K$, (2.3-13) is

$$\leq 1 - \sum_{\mathbf{w} \text{ in } S_{DR}(\mathbf{z})} q(\mathbf{w}|\mathbf{z}) + \varepsilon/2\rho_{\max}. \tag{2.3-14}$$

In the proof's second part we bound the expected coding distortion. Given that the source puts out $\mathbf{z}$, the expected distortion is bounded by

$$\frac{1}{K}\mathcal{E}\left[\rho(\mathbf{Z}, \mathbf{W})|\mathbf{Z} = \mathbf{z}\right] \leq D + \varepsilon/2 + \left[1 - P\{S_D(\mathbf{z})\}\right]^m \rho_{\max}.$$

The bound holds because if a good reproducer word is available its distortion is better than $D + \varepsilon/2$ and if none is available the distortion is no worse than $\rho_{\max}$. We can use (2.3-11)-(2.3-14) to overbound the right-hand side; the result of substituting (2.3-14) for the bracket term is

$$\frac{1}{K}\mathcal{E}\left[\rho(\mathbf{Z},\mathbf{W})|\mathbf{Z}=\mathbf{z}\right] \le D+\varepsilon/2+\rho_{\max}\left[1-\sum_{\mathbf{w}\text{ in }S_{DR}(\mathbf{z})} q(\mathbf{w}|\mathbf{z})\right]+\varepsilon/2.$$

Now we will remove the conditioning on the expectation to get

$$\begin{aligned}\frac{1}{K}\mathcal{E}\left[\rho(\mathbf{Z},\mathbf{W})\right] &\le \sum_{\mathbf{z}} p(\mathbf{z})\left\{D+\varepsilon+\rho_{\max}\left[1-\sum_{\mathbf{w}\text{ in }S_{DR}(\mathbf{z})} q(\mathbf{w}|\mathbf{z})\right]\right\}\\ &= D+\varepsilon+\rho_{\max}\left[1-\right.\\ &\qquad \left.\sum_{\mathbf{z}}\sum_{\mathbf{w}\text{ in }S_{DR}(\mathbf{z})} q(\mathbf{w}|\mathbf{z})\,p(\mathbf{z})\right]. \end{aligned} \tag{2.3-15}$$

The third part of the proof shows that the bracket in (2.3-15) tends to 0 with $K$, so that the expected distortion is $\le D+\varepsilon$ in the limit. The double-sum probability in (2.3-15) is $P\{S_D \cap S_R\}$, where the sets $S_D$ and $S_R$ are

iv) $S_D = \{(\mathbf{z},\mathbf{w}) : \mathbf{w} \text{ in } S_D(\mathbf{z})\}$, that is, all pairs $(\mathbf{z}, \mathbf{w})$ such that $\mathbf{w}$ is close enough to $\mathbf{z}$;

v) $S_R = \{(\mathbf{z},\mathbf{w}) : \mathbf{w} \text{ in } S_R(\mathbf{z})\}$, all pairs such that w has small enough information about z.

From elementary set theory, the bracket as a whole satisfies

$$1-P\{S_D\cap S_R\} \ge 1-P\{S_D\}+1-P\{S_R\}.$$

We can show that $P\{S_D\}$ and $P\{S_R\}$ both converge to 1 by the weak law of large numbers. Looking first at $S_D$, we know from (2.3-8) that for each independent variable $(Z_i, W_i)$, $\mathcal{E}[\rho(Z_i, W_i)] \le D$; so as $K$ grows,

$$\begin{aligned}P\{S_D\} &= P\left\{\frac{1}{K}\rho(\mathbf{Z},\mathbf{W}) \le D+\varepsilon/2\right\}\\ &= P\left\{\frac{1}{K}\sum_{i=1}^{K}\rho(Z_i,W_i) \le D+\varepsilon/2\right\} \to 1.\end{aligned}$$

Looking at $S_R$, we know from the $R(D)$ definition (2.3-9) that

$$\mathcal{E}\left[\log\frac{q(W_i|Z_i)}{q(W_i)}\right] = R(D),$$

so as $K$ grows

$$P\{S_R\} = P\left\{\frac{1}{K}\log\prod_{i=1}^{K}\frac{q(W_i|Z_i)}{q(W_i)} \leq R(D) \ +\varepsilon/2\right\}$$
$$= P\left\{\frac{1}{K}\sum_{i=1}^{K}\log\frac{q(W_i|Z_i)}{q(W_i)} \leq R(D)+\varepsilon/2\right\} \rightarrow 1.$$

**QED**

We give now two important examples of the rate-distortion function. Most of the source coding in the rest of the book is based on these.

**Example 2.3-1: PE Distortion and Binary Source.**

When this memoryless source has $P(0) = P(1) = 0.5$, it is easy to derive the rate-distortion function. From the symmetry of the source, the reproducer letter distribution must also be $q(0) = q(1) = 0.5$, and the distribution $q_{W|Z}$ is

$$\begin{aligned} q(0|1) &= q(1|0) = \gamma \\ q(0|0) &= q(1|1) = 1-\gamma, \end{aligned} \tag{2.3-16}$$

with the parameter $0 \leq \gamma \leq 1$ yet to be found. Distribution (2.3-16) is a member of the $Q_D$ set (2.3-8) and from the condition there we have

$$[\rho(0,0)+\rho(1,1)]\,(1-\gamma)(1/2)+[\rho(0,1)+\rho(1,0)]\,\gamma(1/2) \leq D$$

which reduces to

$$\gamma \leq D$$

when the PE values are inserted into $\rho(z,w)$. Thus, any distribution (2.3-16) that satisfies this condition on $\gamma$ is a member of the set $Q_D$. To find $R(D)$, we need to minimize the expectation in (2.3-9); it is

$$\gamma\log 2\gamma + (1-\gamma)\log 2(1-\gamma) = 1 - h_B(\gamma).$$

This expression reaches its minimum when $\gamma$ reaches its maximum, which is $D$. Thus $R(D)$ for the BSS with PE measure is

$$R(D) = 1 - h_B(D). \tag{2.3-17}$$

To find $R(D)$ for the asymmetric binary source, we would need to optimize (2.3-9) over the constraint set (2.3-8) by means of Lagrange multipliers. This is beyond our scope (see [2-1] or [2-2]) and we will simply cite the answer:

$$R(D) = h_B(p) - h_B(D), \quad p = P(0). \tag{2.3-18}$$

The BSS/PE combination often appears in the analysis of code book search algorithms, although its practical applications are relatively few. The PE measure has the effect of counting errors. One interesting application is to the binary symmetric channel, which we will take up in Section 3.1. For now, we need only imagine a binary channel that can carry error-free only $RK$ out of the $K$ source bits presented to it, where $R < 1$. How many bit errors must occur in this overloaded channel? One estimate can be obtained by modeling the channel as a PE rate-distortion coder. $K$ binary symbols enter the coder and are rate-distortion coded down to $RK$ bits, which represent a length-$K$ reproducer word; the channel carries the $RK$ bits without error. At the rate-distortion coder, (2.3-17) tells us that the reproduction must be distorted at least $D$ on the average, where $1 - h_B(D) = R$. It turns out that this expression does, indeed, bound the average error rate of an overloaded channel. For example, a channel overloaded by 2:1 factor has average error rate at least $D$, where $1 - h_B(D) = 0.5$ implies that $D = 0.110$.

**Example 2.3-2: Squared-error Distortion and the Gaussian Source.**

Here the source is analog and puts out a sequence of independent Gaussian variables with zero mean and variance $\sigma^2$. The distortion measure is (2.3-2). This combination is an intensely practical one since it provides the most widely accepted benchmark for encoders of the analog-to-digital conversion type like those used with speech and images. For a rigorous calculation of $R(D)$ we must again refer to [2-1] or [2-2]. The reproducer word density turns out to be

$$q(w|z) = \frac{1}{\sqrt{2\pi\alpha D}}\, e^{\dfrac{-(w-\alpha z)^2}{2\alpha D}}, \quad \alpha = 1 - \frac{D}{\sigma^2}; \tag{2.3-19}$$

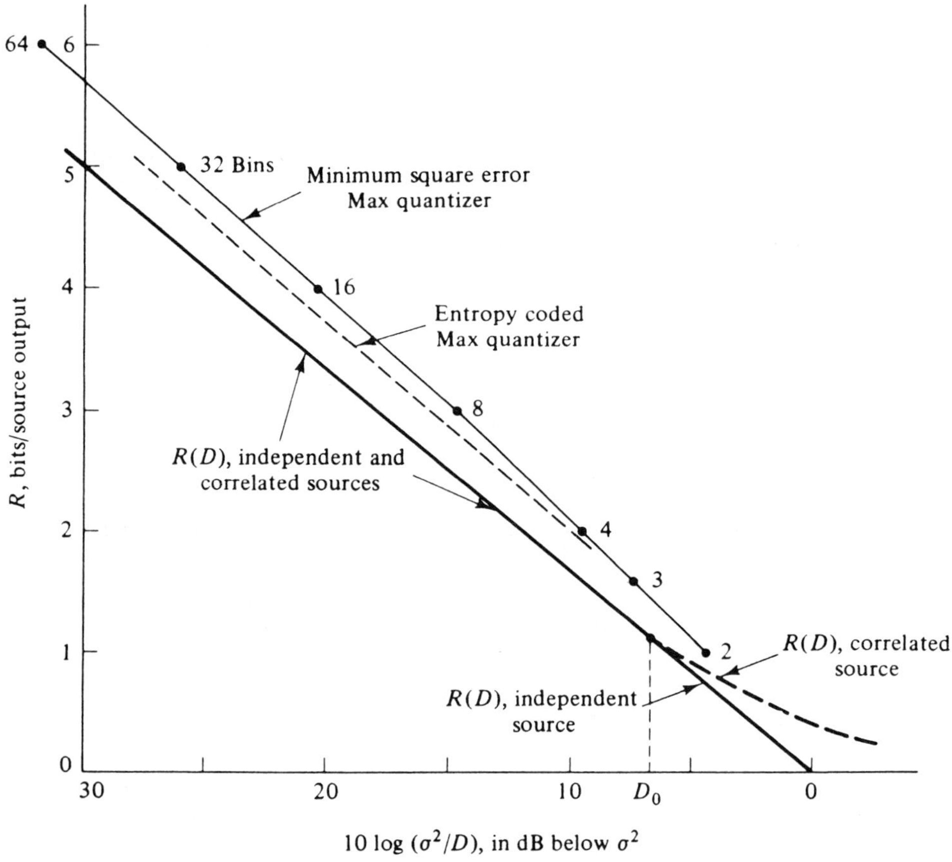


**Figure 2.3-2.** Rate-distortion functions for square error and Gaussian independent and AR correlated sources. The independent and correlated functions are identical until distortion $D_0$. Shown for comparison are the rate and distortion of the Max quantizers with and without entropy coding. (From *Modern Electrical Communications*, 2nd Ed., [2-6], ©1988 Prentice-Hall, Used by permission).

the expectation in (2.3-9) yields

$$R(D) = \frac{1}{2}\log\frac{\sigma^2}{D}, \qquad D \le \sigma^2. \tag{2.3-20}$$

We can always obtain distortion $\sigma^2$ by simply guessing that $Z$ is 0, so $R(D) = 0$ when $D \geq \sigma^2$.

Figure 2.3-2 plots this $R(D)$ function and compares it to some others. (We will take these up later). The distortion axis here is signal-to-noise ratio in $dB$: Since the signal power is $\sigma^2$ and the noise power is the distortion $D$, this is a scaling equivalent to $10 \log_{10}(\sigma^2/D)$. It can be seen that $R(D)$ obeys a 6 dB *rule* of its own (proven in Problem 2-12), with the same slope as the PCM rule (2.2-2). To explore this comparison further, we have plotted rates and distortions of some Max quantizers with and without optimal noiseless coding. The data come from Table 2.2-2; the rates after noiseless coding are the entropies in the Table. It appears that all three types of coding obey parallel 6 dB rules. The Max quantizers without noiseless coding are about 3.4 dB worse than $R(D)$ in terms of SNR at a fixed rate, and the noiseless coding reduces this gap to about 2 dB.

## Correlated Gaussian Sources

The rate-distortion theory of correlated sources is a deep subject that is the object of much present-day research. We will confine ourselves to square error and one type of correlated Gaussian source, the autoregressive one. This source is the classic model for correlated samples of speech and images because the theory for it is straightforward and fits well with the square-error performance of practical schemes. It remains true, however, that speech and image samples are often better modeled by non-Gaussian sources (for which the rate-distortion theory is much harder) and human perception cannot accurately be modeled by square-error distortion, especially when the distortion is not small.

A Gaussian source is *autoregressive of order k*, denoted AR($k$), if it satisfies

$$z_j = \sum_{i=1}^{k} a_i z_{j-i} + u_j, \qquad (2.3\text{-}21)$$

where the $u_i$ are zero-mean independent Gaussians with variance $\sigma^2$. An AR source is a filtered memoryless Gaussian source where the filter is the all-pole discrete-time one shown in Fig. 2.3-3; its $z$-transform is $1/(1 - A(z))$, where $A(z) = a_1 z^{-1} = \ldots + a_k z^{-k}$. (The $z$ here is not a source outcome!) There are good reasons to model real-life signals in this way. Speech, images, biowaveforms, and sonar signals all have a low pass spectrum. At the same time low-order all-pole filters are the simplest realizations of a low pass filter. It is natural to model these signals as the output of such a filter with an independent-sample excitation at the input. In speech modeling the model is particularly close: The excitation is white noise for unvoiced sounds and a pulse train for voiced sounds, both of which can be viewed as white in spectrum, while the all-pole filter coefficients can be related directly to vocal tract parameters such as the spacings of the lips, teeth and palate.

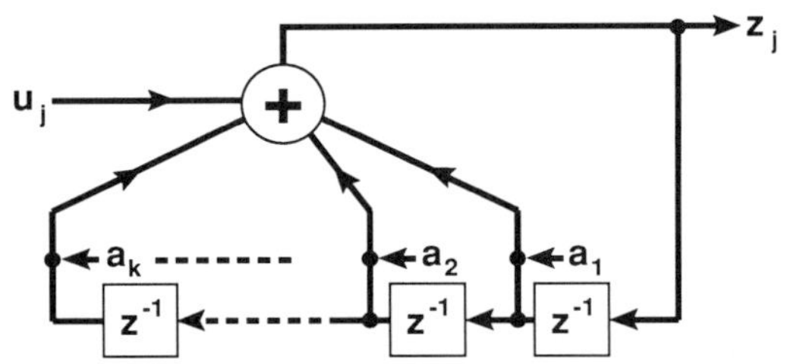


**Figure 2.3-3.** Order-$k$ all pole filter model for AR($k$) autoregressive Gaussian source. The inputs are independent zero-mean Gaussians with variance $\sigma^2$.

The structure in Fig. 2.3-3 closely resembles that of the practical differential PCM coder that would digitize this correlated source. We will take up this kind of coder in Section 2.4. Yet another resemblance exists to the structure of the random code generator that produces good reproducer words with performance close to the rate-distortion function for an AR source. We will not pursue this code generation further, but we can quote a theorem due to Berger [2-2] and Gray [2-9] which establishes the function $R(D)$ for Gaussian AR sources.

**Theorem 2.3-2: The AR Gaussian coding theorem.** Let $Z$ be an AR($k$) source as defined by (2.3-21) subject to a square-error distortion measure. Then the rate-distortion function satisfies

$$R(D)\begin{cases} = \frac{1}{2}\log\left(\frac{\sigma^2}{D}\right), & \text{if } D \le D_0, \\ \ge \frac{1}{2}\log\left(\frac{\sigma^2}{D}\right), & \text{otherwise,} \end{cases} \tag{2.3-22}$$

where

$$D_0 = \inf_{\omega}\left|H(\omega)\right|^2\sigma^2$$

and

$$H(\omega) = \frac{1}{\left[1 - \sum_{i=1}^{k} a_i e^{-ji\omega}\right]}$$

is the frequency spectrum of $1/(1 - A(z))$. Moreover, as the word length $K$ grows there exists a tree code with rate and distortion arbitrarily close to $R$ and $D$.

This is a remarkable theorem because it states that $R(D)$ for a correlated source is the same as the $R(D)$ for the *input* variates $U$ in Fig. 2.3-3. The surprise in this comes from the fact that an all-pole filter generally expands the variance of its input to a larger size at the output; in speech modeling for example, the expansion from $U$ to $Z$ is 10–20 dB. Yet the rates $R$ needed in theory to produce $D = (1/K)\mathcal{E}[(\mathbf{U} - \hat{\mathbf{U}})^2]$ and $D = (1/K)\mathcal{E}[(\mathbf{Z} - \hat{\mathbf{Z}})^2]$ are the same in both cases. $\hat{\mathbf{U}}$ and $\hat{\mathbf{Z}}$ here are the reproducer words in each case.

As $D$ grows, this trait continues until $D$ reaches $D_0$, the *critical distortion* for the AR source. $D_0$ is the infimum of the power spectrum of the filter output. Figure 2.3-2 plots the AR-source $R(D)$ along with the others and it can be seen that the correlated and uncorrelated functions are identical until $D_0$, with the distortion-axis scaling used there. After $D$ exceeds $D_0$, extra rate is required for coding the correlated source. Note that the distortion axis is no longer SNR for the correlated source because its SNR has expanded to $(\text{SNI})\sigma^2/D$; here SNI is a factor greater than 1 that we will define in eq. (2.4-5). A true rate versus SNR plot for the correlated source would typically be far to the left of an uncorrelated plot.

Modern rate-distortion theory has shown that the critical distortion controls not only the form of $R(D)$ but the structure of the encoder as well. For small distortions (high SNRs), the optimal random code generator is close in structure to differential PCM, which in turn performs very well as a practical scheme. At large distortions, as measured by $D_0$, quite different structures are required, both for practical coding and for a proof of the coding theorem; refs. [2-10] and [2-11] are typical of recent research. As a rule, it can be said that transform coding is required. Roughly speaking, $R(D_0)$ for speech and image models lies in the range 1–2 bits/sample. This is too low a rate for effective waveform speech coding, so that effective waveform coders tend to resemble differential PCM, but it is a wastefully high rate for image coding and as a consequence good image coders are transform coders.

## 2.4 PREDICTIVE SOURCE CODERS

Everyday signals like speech, picture scan lines, or electrocardiograms are heavily correlated from sample to sample because they have a low-pass signal spectrum. An effective rate-distortion coder will be one that takes advantage of the correlations. Most good coders for correlated data, whether theoretical or practical, predict future samples in some way from the data already observed and then encode the *difference* between their prediction and the sample outcome. As a generic class, we will call these predictive coders.

Historically, the first predictive quantizer to be based on sound mathematical principles was differential PCM, or DPCM, and we will start with it. Later it was shown that a structure closely related to DPCM generates the theoretically optimum rate-distortion codes. Another important fact about DPCM is that the code inherent in it is a tree code.

A code book, tree structured or not, can be searched for a good reproducer word in a simple or a complex way. We will concentrate in this section on the generation of good code books and on coders that employ very simple searching, just as we did with noiseless coders in Section 2.2; more complex code book searching is discussed in Chapter 6. As quantizers, DPCM and its relatives perform a *single path search* of their tree code books, that is, they elongate a single reproducer word in the code tree, keeping only the code tree branch at each level that gives the least square error. A quantizer that contains a predictor structure coupled to a simple quantizer like PCM is called a *predictive quantizer*. It can be thought of as a predictive coder that uses a single path search.

We will conclude with a brief review of the more recent predictive quantizers for speech. Some of these are not based on a per-letter square error distortion, but rather on peculiarities of human perception. As such, they illustrate the practical art of speech coding more than they do rate-distortion coding. The term *waveform coder* is applied to coders that attempt to find a reproducer word that is close to the source samples in the per-letter sense of Section 2.3.† Some speech coders of the non-waveform type seek only an encoding that will be perceived as similar to the speech, but may be far away from the samples in a square-error sense.

---

†Speech processing engineers consider transform coders (see Section 2.5) not to be waveform coders, even though they may put out a least square-error reproducer word.

## Differential PCM

DPCM is based on least mean-square (LMS) prediction of the next source outcome in a sequence. Figure 2.4-1 shows the basic DPCM encoder and decoder. A tapped delay line in the dashed box generates the least mean-square estimate $\tilde{z}_j$ of the outcome $z_j$, based on the previous $k$ outcomes. The difference $\Delta_j = z_j - \tilde{z}_j$ between the outcome and the estimate is quantized in the box labeled $Q$ in the encoder. $Q$ is a simple non-predictive quantizer like the one in Section 2.2. Its $n$-bit output is the encoder binary output and forms the input to the decoder. Both encoder and decoder have a copy of the delay line and of a box $Q^{-1}$ that generates reproducer levels from the binary stream.

These reproducer levels vary from the differences $\Delta_j$ by a quantization error $\varepsilon_j$, and the representation of $z_j$ that feeds the delay line is $\Delta_j + \varepsilon_j + \tilde{z}_j = z_j + \varepsilon_j$, which contains the same error. Both encoder and decoder see this error. Precisely speaking, Fig. 2.4-1 is a coder inspired by an LMS estimator, not an LMS estimator. Nonetheless, its behavior tracks that of an LMS estimator very closely when the quantizer rate is not low, and we will therefore next review LMS estimation.

Let $r(\,)$ be the discrete-time autocorrelation function of the stationary zero-mean variables $Z_j$:

$$r(\tau) = \mathcal{E}[Z_j Z_{j+\tau}] \tag{2.4-1}$$

To simplify expressions we will assume $Z$ has unit variance. From $r(\tau)$ form the matrix

$$\mathbf{R} = \begin{bmatrix} 1 & r(1) & r(2) & \dots & r(k-1) \\ r(1) & 1 & r(1) & \dots & r(k-2) \\ r(2) & r(1) & 1 & \dots & r(k-3) \\ \vdots & \vdots & \vdots & \ddots & \vdots \\ r(k-1) & r(k-2) & r(k-3) & \dots & 1 \end{bmatrix}. \tag{2.4-2}$$

Then the LMS linear estimate of $Z_j$, given $Z_{j-k}, \dots Z_{j-1}$, is [2-12]

$$\tilde{z}_j = \sum_{i=1}^{k} a_i^{\text{opt}} z_{j-i} \tag{2.4-3}$$

where

$$\mathbf{a}^{\text{opt}} = \mathbf{R}^{-1}\mathbf{r}.$$

Here $\mathbf{a}^{\text{opt}}$ and $\mathbf{r}$ denote the column vectors of coefficients and the autocorrelations $r(1), \dots, r(k)$. The error variance of the estimate $\tilde{z}_j$ is

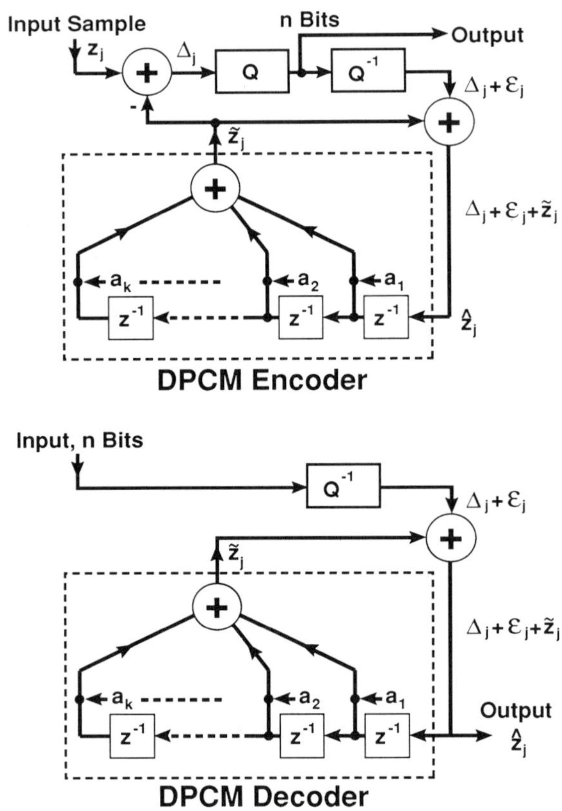


**Figure 2.4-1.** General encoder and decoder for differential PCM. The tapped delay line that realizes eq. (2.4-3) is in the dashed boxes. $Q$ denotes a quantizer (or A-to-D converter) and $Q^{-1}$ denotes the inverse function that produces the reproducer levels.

$$\sigma^2_{\text{est}} = 1 - (\mathbf{a}^{\text{opt}})^T \mathbf{r} = 1 - \mathbf{r}^T \mathbf{R}^{-1} \mathbf{r}. \tag{2.4-4}$$

Here $T$ indicates transpose.

For a general source $Z$, the estimator (2.4-3) has the least variance (2.4-4) of any linear estimator that observes $k$ outcomes. For a Gaussian source it is the best estimator of any form that observes $k$ outcomes. For the AR($k$) source of Fig. 2.3-3, we can go even further and say that (2.4-3) has the least error variance of any length estimator. It can be verified that the predictor coefficients $\mathbf{a}^{\text{opt}}$ are identical to the AR taps $\mathbf{a}$ in Fig.

2.3-3. In fact, the prediction errors $z_j - \tilde{z}_j$ are exactly the independent Gaussian inputs to the filter in Fig. 2.3-3; the AR source filter, which has transform $1/(1-A)$, is the inverse of the input subtractor/predictor combination in the DPCM circuit, which has transform $(1-A)$. Here we have a compelling reason to model practical sources like speech as independent-excitation AR sources: The source correlation mechanism and DPCM are inverses of each other.

The ratio between the DPCM coder input power and its prediction difference $\Delta_j$ power is called the *signal-to-noise improvement* ratio, or SNI, of the coder. The theoretical SNI is given by the inverse of (2.4-4), since it was assumed in the LMS derivation that the source had unit variance:

$$\text{SNI} = \frac{1}{\sigma_{\text{est}}^2} = \frac{1}{(1 - \mathbf{r}^T \mathbf{R}^{-1} \mathbf{r})}. \qquad (2.4\text{-}5)$$

The SNI is generally greater than 1, so that the sequence $\Delta_j$ applied to the quantizer in Fig. 2.4-1 is smaller in variance than $z_j$. Consequently, the quantizer error variance is scaled down by the same factor, compared to what it would have been if $z_j$ were applied directly to the quantizer.

A remarkable fact about DPCM is that this smaller error, denoted $\varepsilon_j$ in Fig. 2.4-1, propagates through the coder virtually without amplification and becomes the reproducer error of the whole coder. We saw in Section 2.2 that the SNR of a simple quantizer follows a 6 dB rule like (2.2-2) that depends on the number of bits in the output. This SNR holds between the powers of $\Delta_j$ and $\varepsilon_j$. The power ratio between $z_j$ and $\Delta_j$ is the SNI, so that the DPCM SNR, the ratio of the powers of $z_j$ and $\varepsilon_j$, is the product of the SNI and the quantizer SNR. Letting SNRD and SNRQ be the SNRs of DPCM and of the unadorned quantizer, we can write this relation in dB as

$$\text{SNRD} \approx \text{SNI} + \text{SNRQ} \qquad (\text{in dB}). \qquad (2.4\text{-}6)$$

Equation (2.4-6) is approximate because it assumes that propagating quantizer errors do not affect the estimate $\tilde{z}_j$. Still, DPCM coders with output rate $n$ above 2 bits/sample closely follow (2.4-6); that is, they obey a 6 dB rule offset by SNI. The factor SNI can be very large. For speech coding with a fixed predictor it is about 10 dB, and if the predictor is allowed to adapt to the local speech correlation, the SNI ranges over 12 to 25 dB; images and high quality speech and music have SNIs as large as 30 dB. As an example, a 6-bit fixed-predictor DPCM coder will have SNRD equal about 39 dB using (2.2-2), almost as good as an 8-bit simple quantizer, which has SNR of 41 dB.

A formal analysis that (2.4-6) applies to DPCM has proved illusive. Arnstein [2-13] derived a precise expression for SNRD for the AR(1) source. Sethia [2-14] devised a recursion that leads to an accurate SNRD for the higher-order case. Both of these results closely follow (2.4-6) when the DPCM rate is not low.

A speech coder with a fixed order-$k$ predictor implies in a sense that speech has been modeled as a fixed AR($k$) source. It is futile to set $k$ greater than 3 [2-17], because the speech correlation is widely varying. A standard fixed order-3 model for 8-kHz-sampled speech has $a_1 = 1.75$, $a_2 = -1.22$, and $a_3 = 0.30$ [2-20]. Spectrally, this creates a process that peaks in the range 100–1000 Hz and rolls off thereafter at the rate of 10–20 dB per octave. A more accurate model, of order up to 10, must vary as the speech correlation varies. We can say little in particular about this model, except that it will be bandpass with peaks that correspond to the resonances of the vocal tract. The first autocorrelation, $r(1)$, for many speech sounds is about 0.9. Images on the other hand have $r(1)$ in the range 0.95–0.99. One consequence of this is the higher SNI for images.

There are many variants of the basic DPCM scheme of Fig. 2.4-1. With nonstationary sources like speech, it is important to adapt the LMS predictor as the local source statistics change. There are two types of adaptive predictor in DPCM coding. In the adaptive filter method, the predictor coefficients continuously adapt in response to the prediction errors $\Delta_j$. Adaptation from these actual errors requires knowledge of the actual speech samples, which are unavailable at the receiver; this kind of encoder thus must send its coefficients along a separate channel, an approach called *forward adaptation.* In *backward adaptive* DPCM, changes to the coefficients are derived from the quantizer outputs, which are available to both the encoder and decoder, so that a separate channel is unnecessary. In the other type of adaptive DPCM, the encoder calculates a new set of coefficients for each frame of speech, which holds constant across the frame. These schemes are almost always forward adapting. Speech coder frames last typically 10–20 ms, which is 80–160 samples at an 8 kHz sampling rate.

Speech processing engineers refer to a tapped delay line type predictor like that in Fig. 2.4-1 as a *direct-form* predictor. The *lattice structure* predictor has largely supplanted the direct form. This structure has an equivalent $z$-transform, and even though the structure is more complicated, the coefficients are easier to compute and adapt and they resist transmission errors better.

Aside from an adaptive predictor, a DPCM coder can also have an adaptive quantizer. It can be tricky to separate these concepts. An adaptive predictor attempts to follow changes in the source correlation, while an adaptive quantizer tracks changes in the source energy. The concepts are connected, however, by the SNI formula (2.4-5); the AR($k$) source energy is the product of the SNI and the prediction error energy, in the absence of quantizer error. We can conclude that the quantizer should adapt to the variations in source energy that remain after the SNI factor is removed by prediction. Quantizer adaptivity can be forward or backward, continuous or frame-wise, and in a practical speech coder it will generally follow the style of the adaptive prediction. The idea of the adaptation is to measure or estimate the energy at the quantizer input and scale the reproducer levels by the square root of the energy. In speech coding, this more accurate scaling improves SNRQ – and through (2.4-6), SNRD – by as much as 6 dB.

The details of prediction structures are a complex subject that is beyond the scope of a coding text. From our point of view, the different structures are different code book generators, and we should note that some of them generate the same code in less computation and that some of the codes are more resistant to channel errors. Excellent texts with an emphasis on adaptive speech coding are Rabiner and Schafer [2-15] and Jayant and Noll [2-5].

DPCM without adaptation seems to have originated with a patent by Cutler [2-16] and a paper by McDonald [2-17]; the latter remains a good reference on the SNR improvement possible in nonadaptive DPCM coding of speech. A related early scheme was delta modulation, a DPCM coder with a 1-bit quantizer, a single predictor tap set to unity, and very rapid sampling rate; the scheme had the advantage of simplicity but could not equal the performance of true LMS-derived DPCM. Applications of DPCM to music and biomedical waveforms resemble speech applications because the bandwidth and correlation of all these sources are roughly similar. DPCM hardware for image coding must contend with much higher sample correlation and far wider bandwidth, so that progress has lagged 5–10 years behind speech DPCM. A method of adaptivity more favored with images is to select which among a small set of fixed image models has the best local fit. Typical of this recent image DPCM research is [2-18].

## Waveform Tree Codes

New waveform speech coders with varying degrees of adaptation and prediction appeared throughout the 1960s and 1970s. At the same time rate-distortion theorists discovered the form of optimal tree codes for the AR sources that model speech so well [2-2], [2-9]. Eventually, it was noticed that almost all the speech codes were simple tree codes and that the more successful of them resembled the optimal AR codes in their structure [2-20]. Thus, the tree coding concept provides a means to organize and understand practical coders and to relate them to theory. We will explore some of these codes now, looking first at practical tree codes and then at theoretically optimal ones.

Figure 2.4-2 is part of the tree code that is inherent in the nonadaptive DPCM of Fig. 2.4-1. The predictor taps are the standard fixed order-3 taps (1.75, -1.22, 0.30) which we recently discussed. If we denote the quantizer reproducer level output as $\hat{u}_j$ (instead of $\Delta_j + \varepsilon_j$), we find from Fig. 2.4-1 that the reproducer values $\hat{z}_j$ in the code itself satisfy

$$\hat{z}_j = \sum_{i=1}^{k} a_i \hat{z}_{j-i} + \hat{u}_j, \tag{2.4-7}$$

These values appear on the tree branches. Each value is a weighted sum of previous values plus the present $n$-bit quantizer output, and the code rate is $n$ bits/source symbol. In Fig. 2-4-2 the quantizer is a 1-bit one with outputs $\pm 1$, so that the tree is binary branching. In a specific application the speech samples would be scaled so that the average energies of $z_j$ and $\hat{z}_j$ were matched.

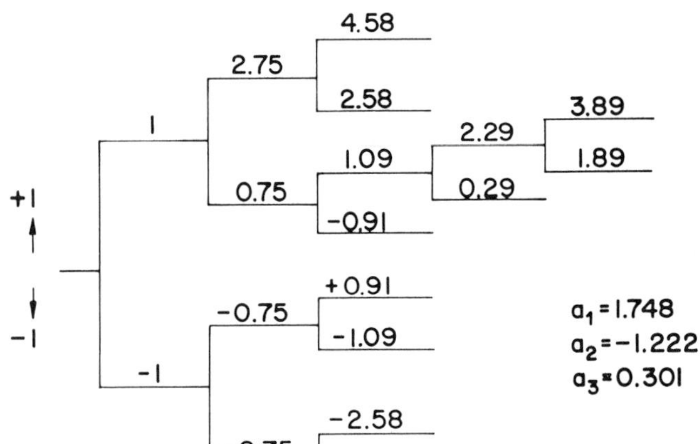


Figure 2.4-2. The tree code of the standard nonadaptive DPCM. A 1-bit quantizer is assumed with reproducer levels $\pm 1$. From Ref. [2-20] ©IEEE 1975, used by permission.

The relationship shown in (2.4-7) is identical to that in (2.3-21), which defines an autoregressive source. This means that the source creates its outputs from an excitation in the same way that (2.4-7) generates a rate-distortion code from a quantizer output sequence. If we represent that driver sequence by the $z$-transform $\hat{U}(z)$, then the code word corresponding to it has transform

$$\hat{Z}(z) = \frac{\hat{U}(z)}{1 - A(z)}, \tag{2.4-8}$$

where $A(z)$ represents the predictor taps. The form (2.4-8) also shows that in principle $\hat{\mathbf{z}}$ may be obtained directly from the sequence $\hat{\mathbf{u}}$ via the infinite impulse response filter $1/(1 - A)$. This offers a useful alternate structure for DPCM-type code generation. We can represent this in the transform domain by $\hat{Z}(z) = C(z)\hat{U}(z)$, where $C(z) = 1/(1 - A(z))$, or in the time domain by

$$\hat{z}_j = \sum_{i=0}^{\infty} c_i \hat{u}_{j-i}. \tag{2.4-9}$$

There are more tap coefficients here of course, compared to (2.4-7), but the $c_i$ diminish rapidly so that the sum can be truncated, and the quantizer outputs are low precision so that the arithmetic in (2.4-9) is easier to implement.

We can expect that the code defined by (2.4-7) will perform well if the rate is not low. The DPCM encoder performs a single-path search of this tree. For multi-path searches (as in Chapter 6) or exhaustive searches, a modified code works better, and one type is the class of *smoothed DPCM* codes. These can be defined by the $z$-transform

$$\hat{Z}(z) = \frac{F(z)\hat{U}(z)}{1 - A(z)}, \tag{2.4-10}$$

in which $F(z)$ is a lowpass filter whose stop band lies near half the sampling frequency $f_s$. $F(z)$ reduces the quantizer hunting tone at $f_s/2$ Hz that sometimes occurs in DPCM and it seems to improve the SNR of multipath searching coders [2-20].

Our second example of a waveform code tree is the code for the Jayant adaptive quantizer [2-5] [2-15] [2-21]. Unlike the previous codes which were predictive but not adaptive, the code in this scheme is adaptive but not predictive. It is a practical example of a source code that accommodates to changes in average source energy. The idea of the Jayant quantizer is to expand the bin spacing when large source outputs are encountered and reduce it when small outputs appear. If the expansion and contraction factors are close to unity the quantizer will track consistent changes in source energy but ignore short runs of unusual outputs.

A 3-bit quantizer will illustrate the principle. Figure 2.4-3 shows some parts of the tree code in this quantizer; the values on the branches are the quantizer reproducer levels $\hat{z}$ and the quantizer binary outputs follow the usual tree convention in which outputs $000, 001, 010, \ldots, 111$ correspond to the top branch through the bottom branch out of a node. The first bit functions as the sign bit. Source outcomes that are larger in absolute value lie in the top and bottom quarters of the quantizer range and cause the quantizer outputs 00X or 11X. The signal to expand the range could be agreement of the first two symbols, with contraction occurring otherwise. In the figure, this rule is illustrated with the factors 1.11 and 0.90.

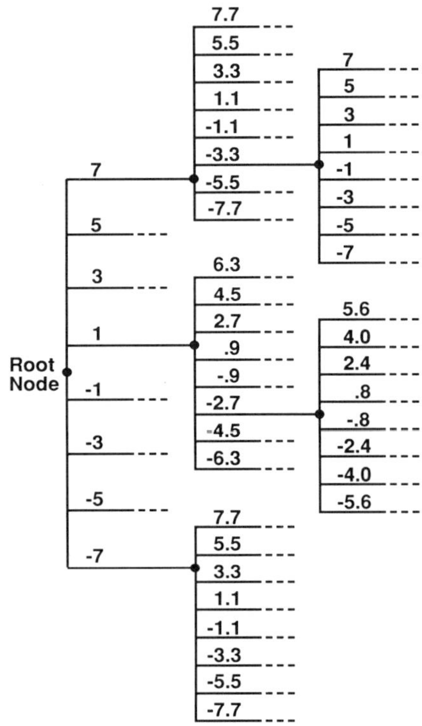


Figure 2.4-3. Parts of the tree code in the 3-bit Jayant adaptive quantizer. This code has eight branches out of each node.

## Rate-Distortion Optimal Tree Codes

A class of random tree codes for the autoregressive Gaussian source and square error whose performance achieves the rate-distortion function was found by Gray [2-9] and Berger [2-2]. They generated a random word by passing independent Gaussian variates with zero mean and variance $\sigma^2$ through a filter whose $z$-transform is $B(z)/(1 - A(z))$, where $B(z)$ is the solution of

$$|B(z)|^2 = \sigma^2 - D|A(z)|^2. \tag{2.4-11}$$

Here $\sigma^2$ is equal to the variance of the excitation $u_j$ in Eq.(2.3-21) and Fig. 2.3-3, which define the autoregressive source. A word so chosen corresponds to one path through the code tree. Another path in the tree code that branches off of this one is generated by the same excitation variates up to the branching point and thereafter by independent ones. In this way the entire tree structure is filled up. In total, the tree comprises $2^{NR(D)}$ words, with $N$ the depth of the structure, and some word in the tree has average square error close to $D$ from the source word as $N$ grows. The design target distortion $D$ must be smaller than the critical distortion $D_0$ defined in Theorem 2.3-2.

The code generation process here is closely related to practical DPCM code generation. The DPCM generator replaces the $b$ independent Gaussians that correspond to the $b$ branches out of a node with the discrete set of quantizer reproducer levels. We can view these as quantizations of the Gaussians. Pragmatically, DPCM has replaced any of these quantizations that turn out to be identical with an unused reproducer level, since otherwise some future code tree paths will be identical. A second similarity to DPCM-type codes exists because the filter $B(z)$ works out to be a very mild low pass filter. This means that spectrally $B(z)/(1-A(z))$ is almost the same as $1/(1-A(z))$, which is the DPCM generation filter. For large $D$, $B(z)$ is a little more severe as a low pass, but then the optimal code generation starts to resemble the smoothed DPCM class defined in (2.4-10). All of these similarities justify why the practical DPCM performance shown in Fig. 2.3-2 is so close to the rate-distortion function at small distortion.

## RELP, CELP, and LPC

For completeness we will review briefly some non-waveform speech coding methods. These are important in practice in the range of rates 2.4–16 kbits/s, where the performance of DPCM coding is no longer acceptable, but the influence of rate-distortion theory and of code searching algorithms is much less felt in their design. These methods produce encodings that "sound good" to human perception rather than match the speech waveform closely. Another priority in their design is an efficient algorithm to estimate the autoregressive model of the vocal tract.

Figure 2.4-4 diagrams the basic linear predictive coding (LPC) coder. LPC is as much a concept of speech analysis as it is a method of digitizing. In the LPC view, speech consists of an excitation which is applied to the vocal tract. The excitation is either voiced (produced by the vocal cords) or unvoiced (simply a rush of air) and is modulated temporally to produce explosive bursts, silent periods, or relatively extended voiced and unvoiced periods. Typical voiced and unvoiced sound waveforms are shown in Fig. 2.4-5. The vocal tract adds spectral shape to the excitation, which is approximately white in overall spectrum, and it aids in the temporal modulation. The result is a sequence of spectral shapes, energy profiles, and voiced/unvoiced switches that we call speech. Some 80 of these basic speech sounds, or *phonemes*, are recognized.

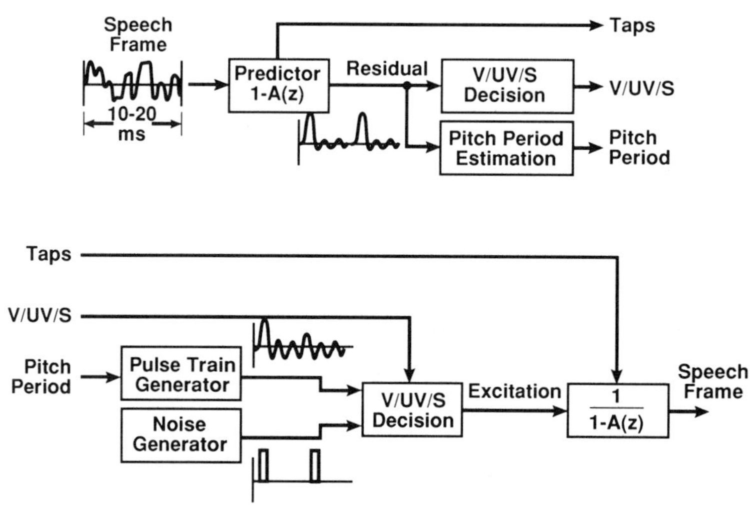


**Figure 2.4-4.** LPC encoder and decoder system diagrams, showing the coding of one speech frame. V/UV/S denotes Voiced/Unvoiced/Silent frame.

The LPC system attempts to break speech down into these components. Speech is accepted in frames of 10–20 ms, a length that includes at least one *pitch period*, that is, one period of the vocal tract vibration.

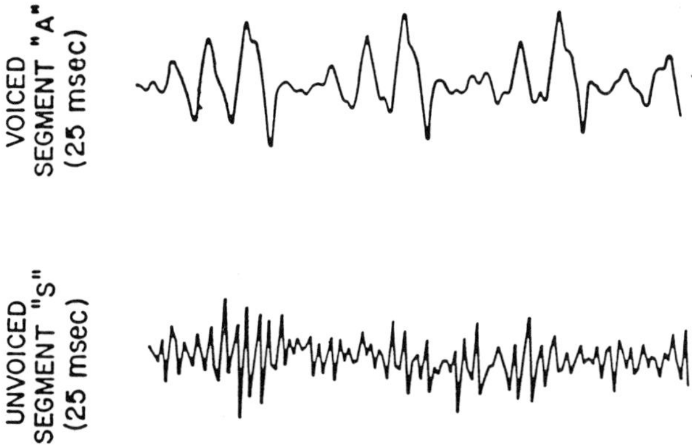


**Figure 2.4-5.** Voiced (top) and unvoiced (bottom) waveforms of speech. The scale of the unvoiced waveform is increased 10 times. From Flanagan *et al.* [2-35] ©IEEE 1979, used by permission.

The coder first extracts an 8–12 pole all-pole model from the frame expressed as either the direct form or the lattice form predictor structure referred to earlier in the section. It releases the taps and an estimate of the frame energy as output. It also filters the speech through the predictor structure to produce a decorrelated *residual* sequence. In theory, this is the speech excitation, either a pulse train for voiced sound or white noise for unvoiced sound. The coder last decides whether the residual sequence is voiced, unvoiced, or silent, and releases its decision as output; if the sound is voiced, the coder also releases an estimate of the pitch period.

The LPC decoder appears also in Fig. 2.4-4. It generates a pulse train at the pitch rate and a noise sequence and applies either one of these or silence to an inverse predictor filter. The result is a frame of synthetic speech. Since neither the pulse train nor the noise waveform is necessarily close to the original speech residual, LPC is not a waveform method and does not produce a close approximation to the speech waveform. Although LPC with a 10–12 pole model reproduces quite understandable speech, its tone quality is artificial and the speaker is difficult to recognize.

As was DPCM, LPC is based on the autoregressive source model of Section 2.3, with, of course, non-Gaussian excitation. The encoder prediction, whose $z$-transform is $(1 - A(z))$, removes the source correlation, modeled by (2.3-21). The decoder filter, whose transform is $1/(1 - A(z))$, puts it back in again. Since the residual is not encoded, but is instead characterized by a three-outcome decision, the waveform rate-distortion theory of this and the last sections has little to say about LPC. It is possible to develop a theory about coding the taps, and we will return to this briefly in Section 2.5. Practical LPC simply quantizes the taps directly, a process that requires about 40 bits per frame, or 2000 bits/s. The pitch period and the V/UV/S decision add a little to this. 2.4 kbits/s is a standard rate for LPC coding.

The quality of straight LPC is not really acceptable in commercial use. The key to improving its quality is coding at least part of the residual waveform. In this way properties of the excitation that are peculiar to the speaker can be transmitted and excitations that are not purely noise or a pulse train can be represented. If all the residual is coded, the result is simply adaptive DPCM, and so the aim now is to code only some critical part of it.

One method for this is the *sub-band coder*. This scheme presents the residual in Fig. 2.4-4 to bank of band pass filters and quantizes the outputs of the bank to differing accuracies, depending on how important the frequency ranges are to human perception. The outputs of filters in the range 100–800 Hz are most important. By more or less ignoring other frequencies in the residual, the sub-band coder can achieve reasonable quality with a total bit rate of 10–16 kbits/s, including 2 kbits/s for the predictor taps and the rest for the quantizers in the residual sub-bands. By comparison, adaptive DPCM would require 16–32 kbits/s. The concept of the sub-band coder originated with Schafer and Rabiner [2-22]. Its present day incarnation realizes the filter bank in a special way called quadrature mirror filtering [2-23].

Another technique for coding the residual has come to be called residual-excited LPC, or RELP. In this scheme, the residual waveform is low pass filtered to, say, 1 kHz, a small fraction of its original bandwidth of 3–4 kHz. The RELP coder then samples the 1 kHz waveform at its Nyquist rate of 2 kHz and quantizes these samples; in discrete time, this amounts to filtering plus decimation by 4 (assuming 4 kHz speech). At the receive end, the RELP decoder reproduces the slow speed samples and then interpolates 4-fold to produce an 8 kHz sample stream. The interpolation causes a spectral unfolding process in which the 1–4 kHz signal spectrum is filled up with replicas of the 0–1 kHz spectrum. To the ear, this partly synthetic residual sounds remarkably life-like. RELP coder overall bit rates lie in the range 8–12 kbits/s.

The older RELP schemes [2-25] have been improved by fine tuning the regeneration of the residual. One scheme [2-26], for instance, aligns the pitch harmonics that appear in the spectrum replicas, which otherwise would not all have a harmonic relation to the pitch period.

The RELP coder seeks to encode only the spectrally important parts of the residual. Other LPC-type schemes code the important parts of the residual as a time waveform. In *multi-pulse* LPC, the coder transmits the size and location of the main pitch pulse and of a few subsidiary pulses that it finds [2-24]. The source coding problem here is thus one of characterizing sizes and locations. The decoded speech loses much of the synthetic quality of LPC with a fixed pulse train. Bit rates lie in the 5–10 kbits/s range.

One can always perform traditional rate-distortion coding on the residual samples with the aid of a code book full of reproducer-sequence residuals. Good, low rate coding of the residual can be achieved, but it is difficult to make use of an effective code book without strenuous searching of the book [2-27]. Modern versions of the technique use small, carefully chosen code books [2-28], and are called code-excited LPC (CELP) coders.

A very large number of speech coding schemes exist, and we have highlighted here only some that are based at least vaguely on rate-distortion waveform coding. Good further reading may be found in the refs. [2-5], [2-15] and [2-35], and in the periodic tutorials that appear for instance in the *IEEE Communications Society Magazine* or the *IEEE ASSP Society Magazine.*

## 2.5 SOME OTHER RATE-DISTORTION APPROACHES

This section discusses transform coding and vector quantization, two further approaches to coding a waveform against a square error or similar distortion measure. This time the methods will not be inspired by the predictive coding idea. They will also tend to operate at low rates, often below 1 bit/sample. Recall that we saw in Sections 2.3 and 2.4 a connection between high rate coding and DPCM-like codes: For distortions below the critical distortion $D_0$ in Theorem 2.3-2, DPCM-like codes approach the rate-distortion function bound in theory and practical DPCM performs well, but above $D_0$ no simple form exists for the rate-distortion function or for an optimal code generator. Thus, we do not expect DPCM codes at low rates.

Another reason that DPCM codes can be unsatisfactory is that they somewhere contain an $n$-bit quantizer, which constrains them to an integer output rate. Particularly, rates below 1 bit/sample are impossible without some modification of the structure. Noiseless coding of the quantizer outputs will reduce their rate somewhat, as we saw in Section 2.2, but this strategy will not normally produce rates below 1 (see Table 2.2-2). Such "fractional" rates are the natural range for image coding and they have recently gained importance for speech coding as well.

Transform coding and vector quantization are the important methods of low rate rate-distortion coding and these form the bulk of this section. They relate to another idea, too, the empirical design of codes. In proving theorems and looking to theory for guidance, we have mostly ignored such an approach. Although transform coding is to some extent suggested by rate-distortion theory, the approach in both these methods is empirical. They are a source of ideas about how to code when the source is statistically ill-defined, when we must rely on observation of the data in order to design the code.

## Vector Quantization

Vector quantization is a name for a group of rate-distortion coding techniques that began to appear in the late 1970s. The term often refers to the simultaneous coding of a block, or "vector", of source outcomes, in the same manner as the rate-distortion coders of Section 2.3, as opposed to a non-block sample-at-a-time coder like PCM. Since we have already discussed this kind of vector coding under another name, we will not pursue it further. A more legitimate kind of vector coding occurs when the vector itself is the natural source output, that is, when the source puts out a group of related components. We have already seen the example of LPC coefficients in Section 2.4. Here 10 or so related coefficients are derived from a frame of speech to form a 10-dimensional outcome; this vector output is the model for the frame's speech, which is generated from the model by (2.4-7). How to digitize the model is the vector quantization problem now. Another vector problem of this type is the coding of a transform coder output, as we will discuss further shortly.

A simple way to digitize LPC parameters is to assign more bits to the more important parameters, fewer to the less important ones, and otherwise code the parameters independently. Lattice-form LPC parameters have an advantage here because the components in this type of vector usually appear in order of importance to the SNR, from most to least. If we concentrate on a single component, we see a more subtle phenomenon, that the component value tends to concentrate in certain ranges; the first lattice-form coefficient, for instance, is likely to lie in the range (0.7–0.9). Efficient coders will thus concentrate on these ranges (see [2-15], p. 250ff, and [2-29]). A still deeper layer of subtlety is that outcomes from component to component relate to each other, because they represent certain speech sounds, and some coder output saving must be possible by recognizing such patterns. Finally, the whole vector outcome is likely related to the previous whole outcome, and may be virtually the same if a speech sound continues across several frames.

More adventurous kinds of vector quantization have been proposed to take account of these latter subtleties. These methods are based on vector prediction, state variable descriptions of the vectors, and on generating empirical codes that recognize dependencies among components. Most of these approaches are in early stages of development; we recommend the survey article by Gray [2-30].

## Empirical Code Design

A third field that has grown up under the banner of vector quantization is the design of codes from sequences of actual source outcomes, rather than from an assumed, a priori probability distribution for the source. It may be computationally difficult or simply artificial to estimate a probability distribution for a source or the distribution may change from time to time. All three are true for speech LPC coefficients. Instead, we can fit a code book directly to the data and repeat the fitting each time the data distribution seems to change.

The most common method of empirical code book design is to iterate through successively better sets of reproducer values, testing each set against the actual source data in order to generate a better set. When applied to PCM coding of single source outcomes against square error distortion, this procedure is called *Lloyd's algorithm*, after S. P. Lloyd who first proposed it for PCM in 1957 [2-31]. The algorithm begins with an initial set of m reproducer values $\{w_j,\ j = 1, \ldots, m\}$ and a block of source outcomes called the *training sequence*; $m$ equals $2^R$, where $R$ is the PCM coder rate in bits/sample. Corresponding to the $m$ reproducer values are $m$ PCM bins $B_1, \ldots, B_m$ (see Fig. 2.2-2). $B_j$ is the subset of the real line closer to $w_j$ than to any other reproducer value. The central step in the algorithm is to find the centroid of each bin $B_j$ with respect to the training sequence, that is, to find the new reproducer value that minimizes the average distortion of training data that fall in the bin. Now the algorithm has produced a new set of reproducer values from the old set. Each iteration begins with a set of reproducer values, calculates the set of bins, and computes the new values as the centroids of the bins.

**The LBG Algorithm**

**Step 1.** Initialization:
Obtain an $N$-element training sequence; if vectors are to be coded, the elements $z$ in the sequence are vectors.
Choose a small distortion threshold $\varepsilon > 0$ to stop the algorithm.
Choose an initial set of $m$ reproducer values $w_j(1)$, $j = 1, .., m$.
Set $n = 1$; $D(0) = \infty$; error $= \infty$.

**Step 2.** Repeat Steps 2a, 2b, and 2c until error $\leq \varepsilon$.

**Step 2a.** Find the minimum distortion set of bins at the $n$-th iteration, $B_j(n)$, $j = 1, \ldots, m$:
For each training outcome $z$, locate the reproducer value $w_j(n)$ that minimizes $d(z, w_j(n))$; then $z$ belongs in set $B_j(n)$.
At the end,

$$B_j(n) = \{z : d(z, w_j(n)) \leq d\left(z, w_i(n)\right), i \neq j\}. \quad (2.5\text{-}1)$$

**Step 2b.** Find the $n$-th iteration distortion as

$$D(n) = \frac{1}{N} \sum_j \sum_{z \text{ in } B_j(n)} d(z, w_j(n)). \quad (2.5\text{-}2)$$

Set error $= |D(n) - D(n-1)|$.

**Step 2c.** $\{w_j(n)\}$ are the final reproducer letters and $\{B_j(n)\}$ are the final bins.
Find the reproducer values for iteration $n+1$. These are the optimal values for the set of $n$-th bins $\{B_j(n)\}$: $w_j(n+1)$ is the $w$ that minimizes

$$\sum_{z \text{ in } B_j(n)} d(z, w). \quad (2.5\text{-}3)$$

Set $n = n + 1$.

With a long enough training sequence, the Lloyd algorithm converges quickly, to at least a locally optimum set of reproducer values. With 10000 training data, for example, the procedure will produce the reproducer values in Table 2.2-2 in 5–10 iterations. Modifications to the algorithm have been proposed that help ensure that it finds a globally optimum design [2-32].

The modern version of Lloyd's algorithm is the LBG algorithm, named after Linde, Buzo and Gray [2-32]. We give a formal listing for this on p.66. The source outcomes now can be vectors, such as LPC coefficient vectors, and any reasonable distortion measure is allowed.

For square-error distortion, the solution to the minimization in Step 2c is the centroid of the training outcomes in bin $B_j(n)$,

$$\frac{1}{\mathrm{card}(B_j(n))} \sum_{z \text{ in } B_j(n)} z, \tag{2.5-4}$$

which is an easy calculation. For other measures, a variety of computational techniques exist [2-32]. It is important to remember that the training data and reproducer values may be vectors. In case these vectors are simply blocks of $K$ outcomes from a scalar source, the rate of the empirical code will be $\log_2 m$ bits/source block, but $(1/K)\log_2 m$ bits/source symbol. In case the source truly puts out vectors, the rate is $\log_2 m$ bits/source vector. For vector sources like the LPC coefficient source, the LBG algorithm is a powerful and important technique, since it is difficult to imagine how a good code might be designed from a theoretical principle. For LPC coefficients, special distortion measures have been proposed [2-32].

**Example 2.5-1.**

We will apply the LBG algorithm to a small collection of 2-dimensional vectors subject to square-error distortion, in an example adapted from [2-32].

Step 1: The vectors are

$$\begin{aligned} \mathbf{x}_1 &= (-.37, .99) & \mathbf{x}_7 &= (-.59, .18) \\ \mathbf{x}_2 &= (.64, -.12) & \mathbf{x}_8 &= (.14, 1.76) \\ \mathbf{x}_3 &= (-.83, .61) & \mathbf{x}_9 &= (.71, -.35) \\ \mathbf{x}_4 &= (-.71, -1.22) & \mathbf{x}_{10} &= (.30, .80) \\ \mathbf{x}_5 &= (-.29, -.95) & \mathbf{x}_{11} &= (.30, 1.07) \\ \mathbf{x}_6 &= (1.10, .52) & \mathbf{x}_{12} &= (.38, -.33) \end{aligned}$$

If $D(n) - D(n-1) = \varepsilon = 0.001$, the algorithm stops.

The initial set of reproducer values $\{w_1(1), \ldots, w_4(1)\}$ is the vertices of the 2-cube, $\{(2,2), (2,-2), (-2,2), (-2,-2)\}$.

*First Iteration:*

Step 2a: The minimum distortion set of bins for the first reproducer value set is

$$\begin{aligned} B_1(1) &= \{\mathbf{x}_6, \mathbf{x}_8, \mathbf{x}_{10}, \mathbf{x}_{11}\} \\ B_2(1) &= \{\mathbf{x}_2, \mathbf{x}_9\} \\ B_3(1) &= \{\mathbf{x}_1, \mathbf{x}_3, \mathbf{x}_7\} \\ B_4(1) &= \{\mathbf{x}_4, \mathbf{x}_5, \mathbf{x}_{12}\} \end{aligned} \tag{2.5-5}$$

Step 2b: The square error with these bins is 2.02, from (2.5-2). The stopping criterion is not yet satisfied.

Step 2c: The new reproducer values are the centroids of the bins above, which are

$$\begin{aligned} \mathbf{w}_1(2) &= (\mathbf{x}_6 + \mathbf{x}_8 + \mathbf{x}_{10} + \mathbf{x}_{11})/4 \\ &= (.46, 1.04) \\ \mathbf{w}_2(2) &= (.67, -.23) \\ \mathbf{w}_3(2) &= (-.60, .59) \\ \mathbf{w}_4(2) &= (-.46, -.83) \end{aligned} \tag{2.5-6}$$

*Second Iteration:*

Step 2a: The minimum distortion set of bins for the second reproducer value set is the same as the first set (2.5-5).

Step 2b: The square error is now .10. The stopping rule is not yet satisfied.

Step 2c: The third set of reproducer values is the same as the second, (2.5-6), because the set of bins has not changed.

*Third Iteration:*

The entire iteration is the same as the second, because the bins and reproducer values are the same; thus $D(3) - D(2) = 0$ and the algorithm stops.

In this example, the algorithm has reached a fixed point, as it is called in optimization theory, at which the bins and values are each optimal for the other. Part of vector quantization theory investigates whether this fixed point represents a globally optimum quantizer.

The rate of our quantizer is 2 bits/source vector, or 1 bit/source outcome. Because of the short training sequence, we cannot be very sure that the code (2.5-6) will perform well with new source outcomes.

## Transform Coding

The idea of transform coding is to take a transform of a block of source data and encode the transform coefficients. The code's reproducer values approximate the coefficients. An inverse transformation of the values yields an approximation to the source data. Transform coding often provides a way to code at a rate less than 1 bit/sample. A block of source data may transform to a shorter block of coefficients, so that a code rate of 1 (or more) bits per coefficient corresponds to a considerably smaller rate per source outcome. An example would be Fourier transform coding of sinusoidal waveforms that are known to have a limited number of different frequencies. Even when the transform coefficient block is long, many transforms have the property that certain coefficients are more important than others to the overall expected distortion. This provides a natural way to code at successively lower rates by ignoring more and more of the coefficients.

Among the favored transforms are the Karhunen-Loeve, Fourier, cosine, Hadamard, and Walsh transforms. The Karhunen-Loeve (or KL) transform plays a special role because it gives the minimum reconstruction error of any $M$-coefficient transform; in stating this we ignore the rate-distortion coding of the coefficients and measure only the effect of approximating the source block with the $M$ basis functions of the transform. The KL transform is derived directly from the statistical model of the source and it is computationally more difficult than the other transforms. These two facts are disadvantages in practice, where computing is expensive and models hard to define, so the KL transform is seldom used except to establish the theoretical limits to transform coding. More often than not, little loss comes of using the $M$ basis functions of another transform. Tests have shown that the Fourier approaches the KL transform performance for both one- and two-dimensional coding, as $M$ grows. The cosine transform often does this well. All the transforms perform about the same when $M$ is very small.

The Fourier and cosine transforms are of particular value in voice coding, since the sinusoidal basis functions closely relate to the way speech is perceived. It is often effective to code certain spectral ranges either coarsely or not at all because they do not contribute to the perception of a particular sound. This technique is called *zonal sampling.*

We lack the space to delve further into this large subject. Jayant and Noll [2-5] provide an excellent, modern chapter on the subject. We will conclude with a review of the rate-distortion theory that surrounds transform coding.

As early as 1956, Kolmogorov [2-33] calculated some rate-distortion functions in the large distortion ($D > D_0$) region, but he did not prove the existence of codes that perform as well as the function. This proof appears for the Gaussian source case for the first time in Gallager [2-1], who made use of the KL transform in the large distortion region and showed that the coefficients could be coded so that distortion in the source outcomes approached the rate-distortion function. His proof applied to continuous functions of time but can easily be extended to the discrete-time case in this chapter. His proof is repeated in Berger [2-2]. Berger also contributed a proof that optimal *tree* codes exist for Gaussian sources if $D \leq D_0$. Mazor and Pearlman [2-34] extended this tree coding result to the full range of distortion. Both of these results apply to coding in the time domain, not the transform domain. Pearlman and Jakatdar [2-10] showed that optimal tree codes also exist for coding of the transform coefficients, again for Gaussian sources.

These results all deal with the existence of codes and assume that large numbers of transform coefficients are grouped into blocks for coding. Practical transform coders, on the other hand, simply quantize the coefficients one by one, perhaps in a zonal strategy. More sophisticated approaches are a present-day research topic.

## Two Part Source Modeling

We have seen that it can be difficult to find an accurate probabilistic model for practical sources. Sometimes the source is not stationary over time; in other cases the excitation of a correlated source is not close to a convenient distribution like the Gaussian one. There are some practical sources that must really be viewed as a superposition of two or more quite distinct sources. An important approach to coding these is to split the source outcomes into streams that come from each distinct source. We can often model each of these accurately.

Images are effectively modeled as a superposition of two sources, a gray-scale source that is slowly varying and an edge source that is best coded according to the techniques used for line drawings [2-36]. A typical image scan line is shown in Fig. 2.5-1(a). The intensity plot jumps at edges as the gray scale changes discontinuously. By differentiating the scan line, we can easily locate the edges; after subtracting out the edge we can code the remaining gray scale variations by a method like DPCM.

Another example of a two-part source is the electrocardiogram shown in Fig. 2.5-1(b). It consists of a small region of jagged peaks, called the QRS complex, superposed on a slowly varying waveform. Once again, it is easy to remove the QRS complex and code it separately, while coding the remaining waveform by a predictive method like DPCM. If fixed DPCM is applied to the whole waveform, it will perform poorly

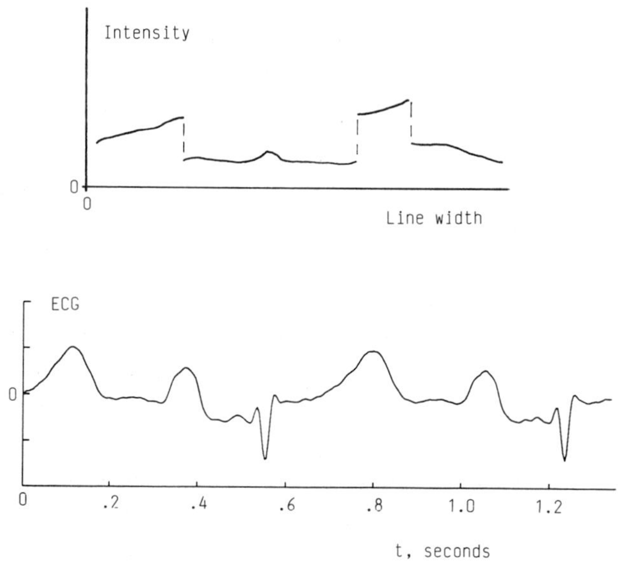


**Figure 2.5-1.** a) (Top) The intensity of an image scan line; b) (bottom) An electrocardiagram trace; Courtesy D. Ritscher, Rensselaer Polytechnic Institute.

at the QRS complex and somewhat less well in the smooth parts in the waveform because the QRS activity confuses the design of the predictor.

# PROBLEMS

2-1 Find the Kraft sums for the telephone code in Table 2.1-2 and for the codes in Fig. 2.1-1.

2-2 Rewrite the proof of Theorem 2.1-1 for the case where the code words have non-binary symbols.

2-3 For a block of $K$ independent and identically distributed outcomes $\mathbf{Z}^{(K)}$ show that $H(\mathbf{Z}^{(K)}) = KH(Z)$, where $Z$ is a single outcome.

2-4 A fair die is tossed until a "6" appears. If $T$ is the number of tosses, find the entropy of this variable.

2-5 Construct the binary Huffman code for all 32 length-5 source sequences in Ex. 2.2-1. Find its rate in average coder output bits per source symbol.

2-6 Construct the ternary Huffman code for the 32 sequences in Problem 2-5, and find the code's rate.

2-7 a) Show that the average output rate of the run length code in Ex. 2.2-2 is 0.50 bits/source symbol.
b) Repeat for a code with length-4 output blocks.

2-8 Consider a binary symmetric source whose 1-probability is 0.99. Construct a run length code with eight 3-bit output words and find the code's rate. Compare to $H(Z)$ for the source. Compare the $H(Z)$ and code rate for this source to the $H(Z)$ and rate in Ex. 2.2-2.

2-9 Re-derive Eq. (2.2-2), taking into account the overload noise caused by quantizer input samples that fall outside the quantizer range $(-4A_{\rm rms}, 4A_{\rm rms})$. Assume the inputs are zero-mean unit-variance Gaussians.

2-10 Derive the 6dB rule for a sinusoid quantized by linear PCM. Assume the quantizer range exactly fits the sinusoid's amplitude.

2-11 a) Under what conditions on the source distribution will the standard 1-bit Max quantizer have an output entropy of unity?
b) Show that with a Gaussian input, the 2-bit Max quantizer has entropy 1.91.

2-12 Suppose one guesses that (2.3-19) is the correct random code letter distribution for the independent-letter Gaussian source with square-error distortion. Show that this guess is correct by verifying that $q(w|z)$ is in the set (2.3-8) and that the expectation in (2.3-9) is $R(D)$ given by (2.3-20).

2-13 Show mathematically that (2.3-20) obeys a 6dB rule, and calculate the exact rule in terms of the code rate and the encoding SNR.

2-14 Consider the binary symmetric source/PE distortion combination in Ex. (2.3-1). It is desired to reduce the rate of the source by a percentage that lies in the range 50-99%. From rate-distortion theory, plot the distortion of the best possible codes over this range of rate reduction.

2-15 Repeat Problem 2-14 for the case where the binary source is asymmetric with 1-probability equal 0.9.

2-16 Suppose speech is modeled as an AR(2) Gaussian source with taps $a_1$ and $a_2$ as shown in Fig. 2.3-3. Find the SNI, eq. (2.4-5), in terms of $a_1$ and $a_2$ only.

2-17 Verify that the optimal LMS predictor taps $\mathbf{a}^{opt}$ in (2.4-4) for an AR($k$) Gaussian source $Z$ are the same as the taps $\mathbf{a}$ in Fig. 2.3-3 that generate $Z$.

2-18 By solving (2.4-11), find the filter $B(z)$ in the rate-distortion optimal code generation filter for the McDonald standard speech AR model ($a_1 = 1.75, a_2 = -1.22, a_3 = 0.30$). Plot some points of the impulse response of $B(z)/(1 - A(z))$.

2-19 Using a random number table, generate 20 outcomes of a uniformly distributed source whose outcomes lie in the range (-1, +1). Use the LBG algorithm to design a 2-bit quantizer based on this training sequence. Compare to the optimal square-error 2-bit quantizer.

# *REFERENCES*

[2-1] R. G. Gallager, *Information Theory and Reliable Communication*, Wiley, New York, 1968.

[2-2] T. Berger, *Rate-Distortion Theory*, Prentice-Hall, Englewood Cliffs, N.J., 1971.

[2-3] N. Abramson, *Information Theory and Coding*, McGraw-Hill, New York, 1963.

[2-4] J. Max, "Quantizing for minimum distortion," *IRE Trans. Information Theory*, vol. IT-6, pp. 7–12, March 1960.

[2-5] N. S. Jayant and P. Noll, *Digital Coding of Waveforms*, Prentice-Hall, Englewood Cliffs, N.J., 1984.

[2-6] H. Stark, F. B. Tuteur, and J. B. Anderson, *Modern Electrical Communications*, 2nd Ed., Prentice-Hall, Englewood Cliffs, N.J., 1988.

[2-7] P. Z. Peebles, Jr., *Digital Communications Systems*, Prentice-Hall, Englewood Cliffs, N.J., 1987.

[2-8] C. E. Shannon, "Coding theorems for a discrete source with a fidelity criterion," *IRE Natl. Conv. Record*, Part 4, pp. 142–163, 1959.

[2-9] R. M. Gray, "Information rates of autoregressive processes," *IEEE Trans. Information Theory*, vol. IT-16, pp. 412--421, July 1970.

[2-10] W. A. Pearlman and P. Jakatdar, "Transform tree code for stationary Gaussian sources," *IEEE Trans. Information Theory*, vol. IT- 31, pp. 761–768, Nov. 1985.

[2-11] M. L. Sethia and J. B. Anderson, "Low rate tree coding of autoregressive sources," *IEEE Trans. Information Theory*, vol. IT-30, pp. 279–284, March 1983.

[2-12] J. L. Melsa and D. L. Cohn, *Detection and Estimation Theory*, McGraw-Hill, New York, 1978.

[2-13] D. S. Arnstein, "Quantization error in predictive coders," *IEEE Trans. Communications*, vol. COM-23, pp. 423–429, April 1975.

[2-14] M. L. Sethia, "Digitization of analog sources," Ph.D. dissertation, ECSE Dept., Rensselaer Poly. Inst., Troy, N.Y., 1983.

[2-15] L. R. Rabiner and R. W. Schafer, *Digital Processing of Speech Signals*, Prentice-Hall, Englewood Cliffs, N.J., 1978.

[2-16] C. C. Cutler, "Differential quantization of communications," U.S. Patent, 2 605 361, July 1952.

[2-17] R. A. McDonald, "Signal to noise and idle channel performance of DPCM systems – Particular applications to voice signals," *Bell System Tech. J.*, vol. 45, pp. 1123–1151, Sept. 1966.

[2-18] J. W. Woods and I. Paul, "Adaptive prediction DPCM coding of images," *Conf. Record*, Intern. Conf. on Communications, Seattle, Wash., pp. 31.8.1–5, June 1980.

[2-19] F. Jelinek, "Tree encoding of memoryless time-discrete sources with a fidelity criterion." *IEEE Trans. Information Theory*, vol. IT-15, pp. 584–590, Sept. 1969.

[2-20] J. B. Anderson and J. B. Bodie, "Tree encoding of speech," *IEEE Trans. Information Theory*, vol. IT-21, pp. 379–387, July 1975.

[2-21] N. S. Jayant, "Adaptive quantization with a one word memory," *Bell System Tech. J.*, vol. 52, pp. 1119–1144, Sept. 1973.

[2-22] R. E. Crochiere, S. H. Webber, and J. L. Flanagan, "Digital coding of speech in sub-bands," *Bell System Tech. J.*, vol. 55, pp. 1069–1085, Oct. 1976.

[2-23] D. Esteban and C. Galand, "Application of quadrature mirror filters to split band voice coding schemes," *Proc.*, Intern. Conf. on Acoustics, Speech and Signal Processing, Hartford, Conn., pp. 191–195, May 1977.

[2-24] B. S. Atal and J. R. Remde, "A new model of LPC excitation for producing natural-sounding speech at low bit rates," *Proc.*, Intern. Conf. on Acoustics, Speech and Signal Processing, Paris, France, pp. 614–617, 1982.

[2-25] C. K. Un and D. T. Magill, "The residual-excited linear prediction vocoder with transmission rate below 9.6 kbits/s," *IEEE Trans. Communications*, vol. COM-23, pp. 1466–1473, Dec. 1975.

[2-26] R. L. Zinser, "An efficient, pitch-aligned high-frequency regeneration technique for RELP vocoders," *Proc.*, Intern. Conf. on Acoustics, Speech and Signal Processing, Tampa, Fla., pp. 969–972, March 1985.

[2-27] B. S. Atal and M. R. Schroeder, "Stochastic coding of speech signals at very low bit rates," *Proc.*, Intern. Conf. on Communications, Amsterdam, Netherlands, pp. 1610–1613, May 1984 (published by North-Holland).

[2-28] P. Kroon and B. S. Atal, "Strategies for improving the performance of CELP coders at low bit rates," *Proc.*, Intern. Conf. on Acoustics, Speech and Signal Processing, New York, pp. 151–154, April 1988.

[2-29] R. Viswanathan and J. Makhoul, "Quantization properties of transmission parameters in linear predictive systems," *IEEE Trans. Acoustics, Speech, and Signal Proc.*, vol. ASSP-23, pp. 309–321, June 1975.

[2-30] R. M. Gray, "Vector quantization," *ASSP Magazine*, vol. 1, pp. 4–29, April 1984.

[2-31] S. P. Lloyd, "Least squares quantization in PCM's," Bell Telephone Labs. paper, Murray Hill, N.J., 1957.

[2-32] Y. Linde, A. Buzo, and R. M. Gray, "An algorithm for vector quantizer design," *IEEE Trans. Communications*, vol. COM-28, pp. 84–95, Jan. 1980.

[2-33] A. N. Kolmogorov, "On the Shannon theory of information transmission in the case of continuous signals," *IRE Trans. Information Theory*, vol. IT-2, pp. 102–108, Sept. 1956.

[2-34] B. Mazor and W. A. Pearlman, "A tree coding theorem for stationary Gaussian sources and the squared-error distortion measure," *IEEE Trans. Information Theory*, vol. IT-32, pp. 156–165, March 1986.

[2-35] J. L. Flanagan, et. al., "Speech coding," *IEEE Trans. Communications*, vol. COM-27, pp. 710–737, April 1979.

[2-36] D. Anastassiou and D. J. Sakrison, "New bounds to R(D) for additive sources and applications to image encoding," *IEEE Trans. Information Theory*, vol. IT-25, pp. 145–155, March 1979.

# CHAPTER 3

# Error Control Coding

In this chapter, we discuss a number of codes for error control. Only block codes are treated here. Discussion on convolutional codes will be deferred until next chapter. After reviewing some information theoretic foundations of coding in the first section, *linear block codes* are treated in Section 3.2. The concepts of parity-check and generator matrices to represent linear block codes are discussed. Several examples of block codes are given, including the important class of *Hamming codes.* Principles behind syndrome decoding and decoding using a standard array are treated in Section 3.3. Section 3.4 provides some useful bounds on coding and introduces the concept of coding gain. Section 3.5 discusses the principles behind *cyclic codes.* Some important decoding techniques for these codes are treated in Section 3.6. These include the Meggitt and error-trapping decoders. After introducing some algebra in Section 3.7, in the next three sections that follow, we treat the most important and practical of all cyclic codes, the *Bose-Chaudhuri-Hocquenghem* (BCH) codes and *Reed-Solomon* codes. The treatment includes the Massey-Berlekamp algorithm for decoding these codes. In Section 3.11, we turn to coding for burst error control, which has been successfully applied to storage media such as magnetic tapes and compact disc. *Automatic-repeat-request* (ARQ) schemes find wide applicability in computer networks and these schemes are treated in the last section.

Clearly, it is difficult to cover the vast subject of error control coding in one chapter. The reader desiring for more should select from the text books [3-1] – [3-11]. In addition to these texts, [3-12] – [3-14] devote a chapter or more to the subject of coding.

## 3.1 FUNDAMENTALS OF CODING

For a century prior to the work of Shannon [3-15], communications engineers believed that information transmitted over a noisy channel could only be inaccurately received. Shannon shattered this belief with his remarkable work in which he showed that the limitation of a noisy channel is not its ability to accurately transmit information but the rate at which it can carry information. This means that the signal-to-noise ratio on the channel does not limit the reliability of the transmitted information but only its rate. Shannon showed the existence of *channel codes* which can encode the information and transmit at the rate of $R$ bits/channel use so that, as long as $R$ is less than a number $C$ called *channel capacity*, the received information can be decoded with arbitrarily small probability of decoding error. To understand this statement, more formally stated later as the channel coding theorem, we need to examine the concepts of channel capacity and channel codes.

### The Channel

Before we discuss channel capacity, we define first the notion of a *channel* in mathematical terms. Referring to Fig. 1.2-2, we observe that a digital channel has input $X$ and output $Y$, $X$ and $Y$ being *random variables* taken from input and output alphabets, respectively. The channel is mathematically characterized by a conditional probability distribution $p_{Y|X}(y|x)$, the probability that the channel output is $y$ given that its input is $x$. The channel may also be defined by the following transition probability matrix, where the input and output alphabet sizes are $I$ and $J$, respectively.

$$[p(y|x)] = \begin{bmatrix} p(y_1|x_1) & p(y_2|x_1) & \dots & p(y_J|x_1) \\ p(y_1|x_2) & p(y_2|x_2) & \dots & p(y_J|x_2) \\ \vdots & \vdots & \ddots & \vdots \\ p(y_1|x_I) & p(y_2|x_I) & \dots & p(y_J|x_I) \end{bmatrix}.$$

Observe that

$$\sum_{j=1}^{J} p(y_j|x_i) = 1, \qquad 1 \le i \le I,$$

since given input $x_i, 1 \le i \le I$, the channel output must be some $y_j, 1 \le j \le J$. Knowing the input distribution $p_X(x)$, the output distribution may be calculated using

$$p(y_j) = \sum_{i=1}^{I} p(y_j|x_i)p(x_i), \qquad 1 \le j \le J.$$

The entropy of $Y$, $H(Y)$, may then be calculated using (1.1-1).

**Example 3.1-1: The Binary Symmetric Channel.**

The channel model that we use in most of this chapter is the *binary symmetric channel* (BSC). The input of this channel is the binary symbols 0 and 1. Each binary symbol presented to the channel is received correctly with probability $(1-p)$ and incorrectly with probability $p$; $p$ is called the *crossover* probability. Figure 3.1-1 shows the BSC.

Let the input distribution be given by $p_X(0) = p_0$ and $p_X(1) = p_1 = 1 - p_0$. Then

$$\begin{aligned} p_Y(0) &= p_{Y|X}(0|0)p_X(0) + p_{Y|X}(0|1)p_X(1) \\ &= (1-p)p_0 + p(1-p_0), \end{aligned}$$

and

$$\begin{aligned} p_Y(1) &= p_{Y|X}(1|0)p_X(0) + p_{Y|X}(1|1)p_X(1) \\ &= pp_0 + (1-p)(1-p_0). \end{aligned}$$

Using (1.1-1) and (1.1-2) to calculate $H(Y)$ yields

$$H(Y) = h_B\big(pp_0 + (1-p)(1-p_0)\big).$$

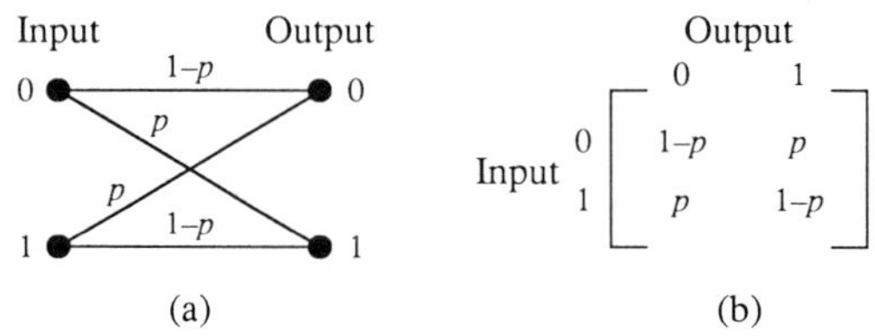


**Figure 3.1-1.** (a) The binary symmetric channel; (b) the transition probability matrix.

A channel is said to be *memoryless* if the successive outputs of a channel due to successive transmissions or uses of the channel are independent. For a memoryless channel, probability of receiving a block of $n$ symbols $\mathbf{y} = (y_0, y_1, \ldots, y_{n-1})$ due to the transmission of a block of $n$ symbols $\mathbf{x} = (x_0, x_1, \ldots, x_{n-1})$ is given by

$$p_{Y|X}(\mathbf{y}|\mathbf{x}) = \prod_{i=0}^{n-1} p_{Y|X}(y_i|x_i).$$

A *discrete* channel is one for which the sizes of the input and output alphabets are finite. A channel that is both discrete and memoryless is called a *discrete memoryless channel* (DMC).

The BSC is an appropriate channel model for the transmission of binary data such as in binary PSK or FSK. (Modulations like these are reviewed in Appendix A). A more appropriate model for BPSK in the white Gaussian noise of Section 1.3 is the *binary erasure channel* (BEC). A BEC is a channel that has two inputs 0 and 1 and three outputs 0, 1, and $e$, $e$ representing erasure. An erasure occurs when the channel output lies very close to the decision threshold. The BEC enables us to perform *soft decision decoding* instead of the *hard decision decoding* as in the case of the BSC. An extension of the BEC idea is to quantize the channel output to more than three levels. In practice, eight output levels are found to approach the performance attainable with infinite number of levels.

## Mutual Information and Channel Capacity

We now introduce the concept of *mutual information* which will be needed to define channel capacity. Prior to the transmission of information over the channel, the uncertainty present in $X$ is its entropy $H(X)$. The uncertainty present in $X$ after the reception of $Y$ is given by the conditional entropy $H(X|Y)$. (Refer to Section 1.2 for discussion on entropy and conditional entropy.) The difference between these quantities, called the *mutual information* between $X$ and $Y$, must be the amount of information conveyed over the channel. The mutual information between $X$ and $Y$ is the amount of information known about $X$ by the reception of $Y$ and is defined as

$$I(X;Y) = H(X) - H(X|Y). \tag{3.1-1}$$

It is possible to show that mutual information is symmetrical in $X$ and $Y$, *i.e.*,

$$I(X;Y) = H(Y) - H(Y|X), \tag{3.1-2}$$

and that either (3.1-1) or (3.1-2) leads to (see Problem 3-1)

$$I(X;Y) = \sum_x \sum_y P(x,y) \log \frac{P(x,y)}{P(x)P(y)}. \tag{3.1-3}$$

**Example 3.1-2: Mutual Information for the BSC.**

Let the input symbol distribution be $p(0) = p_0$ and $p(1) = p_1 = 1 - p_0$. Then using (1.1-4) and (1.1-5) in (3.1-2) yields

$$\begin{aligned} I(X;Y) &= H(Y) - H(Y|X) \\ &= H(Y) - \sum_x P(x) \sum_y P(y|x) \log \frac{1}{P(y|x)} \\ &= H(Y) - \sum_x P(x) \left( p \log \frac{1}{p} + (1-p) \log \frac{1}{1-p} \right) \\ &= H(Y) - \left( p \log \frac{1}{p} + (1-p) \log \frac{1}{1-p} \right) \\ &= H(Y) - h_B(p). \end{aligned} \tag{3.1-4}$$

The previous example gives $H(Y)$, with which $I(X;Y)$ may be calculated:

$$\begin{aligned} I(X;Y) &= H(Y) - h_B(p) \\ &= h_B\big((1-p_0)(1-p) + p_0 p\big) - h_B(p). \end{aligned} \tag{3.1-5}$$

> Given a BSC for which $p$ is fixed, we may vary the input distribution to maximize the mutual information in (3.1-5). We find that only the first term on the right side of (3.1-5) is dependent on $p_0$ and hence the value of $p_0$ that maximizes this term also maximizes $I(X;Y)$. This term is maximized if the argument $(1-p_0)(1-p)+p_0p=\frac{1}{2}$, so that the maximum of $I(X;Y)$ is $h_B(\frac{1}{2})-h_B(p)=1-h_B(p)$. Note that the maximum mutual information is dependent only on the channel transition probability distribution. Fig. 3.1-2 provides a graphical interpretation of the mutual information for the BSC.

We now formally define the capacity of a channel as the maximum mutual information that it can carry, the maximization being carried out over all possible source distributions:

$$C=\max_{P(x)} I(X;Y). \tag{3.1-6}$$

Capacity as defined above is independent of the input statistics and depends only on the channel transition probabilities. For the BSC, $C=C(p)=1-h_B(p)$. If $p=0$, the BSC becomes error free and carries 1 bit of information each time the channel is used. For $p=0.5$, the BSC channel capacity is 0. Fig. 3.1-3 plots the capacity of the BSC.

## Channel Coding

*Channel coding* is the act of transforming a length-$K$ data sequence into a length-$N$ code word. The set of rules that specify the transformation forms a *channel code*. We may define channel code as the mapping

$$\mathcal{C}:\mathcal{U}\rightarrow\mathcal{X},$$

where $\mathcal{C}$ is the channel code, $\mathcal{U}$ the set of information sequences of length $K$, and $\mathcal{X}$ the set of code words of length $N$. The *rate* of the code is $K/N$ symbols per channel use. If the channel input is taken from a $q$-ary alphabet, there are $q^K$ input sequences, each of length $K$, and hence $q^K$ code words, each of length $N$. For the channel code to be effective $q^K$ must be a small fraction of $q^N$, the total number of $N$-tuples.

We introduce the *repetition* code as an example of a channel code and demonstrate how it can reduce channel noise. The repetition code transmits $N$ repetitions of the same symbol. Thus, for the BSC with $q=2$, the code has two code words $00\cdots0$, and $11\cdots1$. Each code word carries one data bit and the rate of the code is $1/N$ bits per channel use. For example, when $N=3$, the two code words are 000 and 111 and

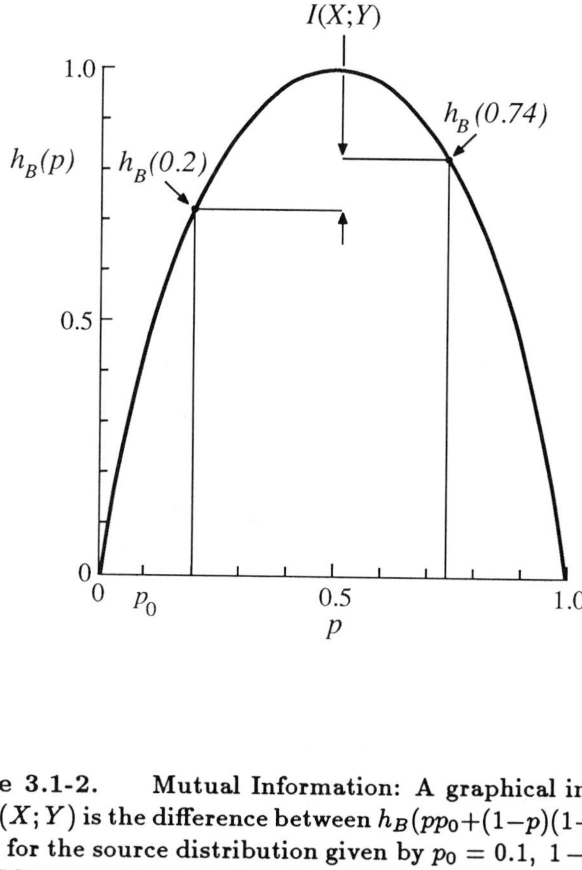


**Figure 3.1-2.** Mutual Information: A graphical interpretation; $I(X;Y)$ is the difference between $h_B(pp_0+(1-p)(1-p_0))$ and $h_B(p)$; for the source distribution given by $p_0 = 0.1$, $1-p_0 = 0.9$, and BSC crossover probability $p = 0.2$, the figure gives $I(X;Y) = h_B(0.74) - h_B(0.2)$.

the rate of the code is 1/3 bits per channel use. When a code word is transmitted over the BSC, it is possible to receive any of the $2^N$ $(= 8)$ possible sequences because of the channel noise. The decoder chooses that code word that has the smallest *Hamming distance* to the received word, Hamming distance between two sequences being defined as the number of positions in which they differ. Using this minimum Hamming distance decoder to decode the repetition code with $N = 3$ results in

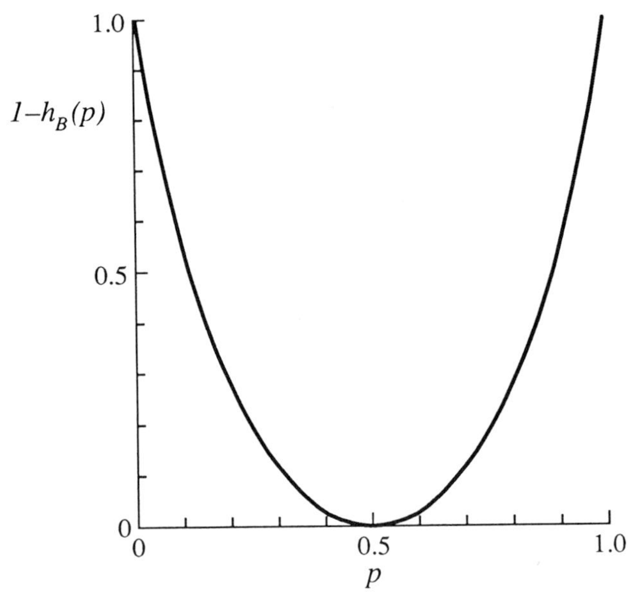


**Figure 3.1-3.** Capacity of the BSC.

decoding error whenever 2 or 3 errors occur in the transmitted code word. In this case the decoder will choose the code word that was not transmitted. The probability of error is therefore

$$P_e = \binom{3}{2}(1-p)p^2 + p^3 = 3(1-p)p^2 + p^3.$$

For $p = 0.01$ (one in one hundred uses of the channel results in error), $P_e \approx 3 \times 10^{-4}$, which is considerably less than the error rate without coding.

We can show that such a decoder is a maximum-likelihood† decoder. Let $\mathbf{x}_i$ be the closest code word to the received word $\mathbf{y}$, and the Hamming

†The concept of maximum-likelihood decision making is reviewed in Appendix A.

distance between $\mathbf{y}$ and $\mathbf{x}_i$ be $n_i$. Then, for the BSC with crossover probability $p$

$$P(\mathbf{y}|\mathbf{x}_i) = (1-p)^{N-n_i}p^{n_i}, \qquad p < \frac{1}{2}.$$

For any other code word $\mathbf{x}_j$, $j \neq i$,

$$P(\mathbf{y}|\mathbf{x}_j) = (1-p)^{N-n_j}p^{n_j},$$

where $n_j$ is the Hamming distance between $\mathbf{y}$ and $n_j$. For a maximum-likelihood decoder we require

$$P(\mathbf{y}|\mathbf{x}_i) > P(\mathbf{y}|\mathbf{x}_j), \qquad i \neq j.$$

Equivalently, taking logarithms on both sides and simplifying leads to the requirement

$$n_i \log \frac{p}{1-p} > n_j \log \frac{p}{1-p}, \qquad i \neq j.$$

Since $p/(1-p)$ is less than 1, the requirement is satisfied if $n_i$ is the least of the Hamming distances. Since $n_i$ is the distance between $\mathbf{y}$ and the closest code word $\mathbf{x}_i$, the minimum Hamming distance decoder is also the maximum-likelihood decoder.

## Random Channel Codes

Before formally stating and proving the channel coding theorem for the BSC, we need to select codes that achieve rates close to capacity, producing arbitrarily small decoding error. Shannon gave no specific code construction but instead resorted to a random selection of code words out of all the $2^N$ possible sequences. We introduced the idea of random coding in the context of rate-distortion coding in Section 2.2. Historically though, it was to prove the channel coding theorem that Shannon introduced the idea; rate-distortion random coding followed a decade later.

A random code book of $J$ code words is constructed for the BSC by tossing a $2^N$-sided fair die $J$ times. Each outcome specifies a unique code word of length $N$. Note that such a procedure may give rise to two or more code words being the same in any given code book, but the probability of this event becomes insignificant as $N$ grows.

Next we choose $J = 2^K$ such that the code rate is close to the capacity of the BSC. Recall that the code rate is

$$R = \frac{K}{N} = \frac{\log_2 J}{N}. \tag{3.1-7}$$

By choosing $R = C - \varepsilon$, where $\varepsilon > 0$ is small, we can make $R$ as close to capacity as we wish; by selecting $J = 2^{N(C-\varepsilon)}$ code words, we can achieve rates as close to capacity as possible. Since $J = 2^{NC}/2^{N\varepsilon}$, we can look upon $J$ as

$$\left(\frac{1}{2^{N\varepsilon}}\right) \text{(number of possible words of length } N)^C.$$

Thus, by taking a suitably large $N$, $(1/2^{N\varepsilon})$ can be made as small as possible or, equivalently, $J$ can be made a small fraction of the total number of possible length-$N$ words. This increases the distance between code words and reduces the risk of a received word being decoded incorrectly.

## Channel Coding Theorem for the BSC

**Theorem 3.1-1: The BSC Channel Coding Theorem.** Consider transmitting $J$ equiprobable messages through the BSC, where $J = 2^{N(C-\varepsilon)}$. Then there exists a channel code with rate $R < C$ and $J$ code words selected from all $2^N$ possible words such that the probability of error in decoding can be made arbitrarily small as $N$ grows.

*Proof*: We assume that a channel code consisting of $J$ code words is randomly selected by throwing the $2^N$-sided die $J$ times. From the preceding discussion, when $J = 2^{N(C-\varepsilon)}$, the rate of the code $R$ is less than $C$. Let us assume that $\mathbf{x}_0$ is the transmitted code word and $\mathbf{y}_j$ is the received word. To help visualize the decoding procedure, we look upon sequences of length $N$ as points in the $N$-dimensional space. We expect, on an average, $Np$ errors will occur in transmitting $\mathbf{x}_0$ through the BSC and hence $\mathbf{y}_j$ will be at a distance of $Np$ from $\mathbf{x}_0$. From the law of large numbers, we know that the probability that more than $N(p+\varepsilon)$ errors occur tends to zero as $N$ grows. So we expect $\mathbf{y}_j$ to lie within a sphere of radius $N(p+\varepsilon)$ around $\mathbf{x}_0$ in the $N$-dimensional space. On receiving $\mathbf{y}_j$, the decoder constructs a sphere of radius $N(p+\varepsilon)$ around $\mathbf{y}_j$ and looks for a unique code word within the sphere (see Fig. 3.1-4). If the decoder finds more than one code word within the sphere, we will say that a decoding error occurs. Due to the random code selection, this may happen even though the number of channel errors is less than $N(p+\varepsilon)$. If more than $N(p+\varepsilon)$ errors occur, then the transmitted code word will not lie within the sphere about $\mathbf{y}_j$ and we will again say that a decoding error occurs. If $\mathcal{S}$ represents the sphere of radius $N(p+\varepsilon)$ centerd at $\mathbf{y}_j$, then the probability of decoding error is given by

$$P_E = P\{\mathbf{x}_0 \notin \mathcal{S}\} + P\{\mathbf{x}_0 \in \mathcal{S}\}P\{\text{one or more code words other than } \mathbf{x}_0 \in \mathcal{S}\}. \tag{3.1-8}$$

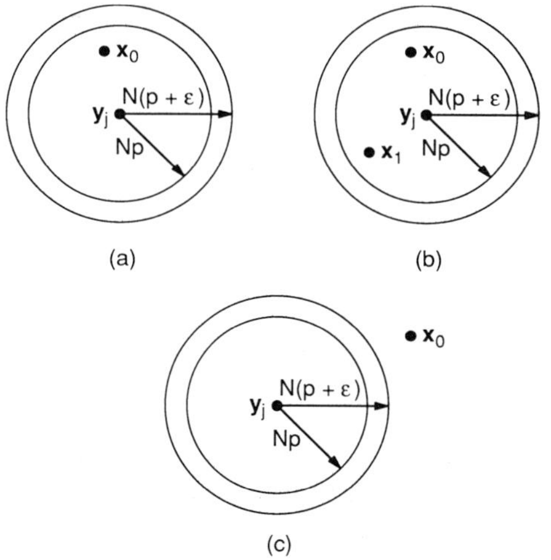


**Figure 3.1-4.** A decoder for random codes; (a) no decoder error; (b) and (c) decoder error occurs.

Here $\mathrm{x}_0 \in \mathcal{S}$ refers to the event that the code word is contained in the sphere and $\mathrm{x}_0 \notin \mathcal{S}$ to the event complement. Since $P\{\mathrm{x}_0 \in \mathcal{S}\} \leq 1$, we can write $P_E$ as

$$\begin{aligned} P_E &\leq P\{\mathrm{x}_0 \notin \mathcal{S}\} + P\left\{\bigcup_{i\neq 0}\{\mathrm{x}_i \in \mathcal{S}\}\right\} \\ &\leq P\{\mathrm{x}_0 \notin \mathcal{S}\} + \sum_{i\neq 0} P\{\mathrm{x}_i \in \mathcal{S}\}, \end{aligned} \tag{3.1-9}$$

where we have used the union bound from probability theory to upper-bound the second term. The first term in (3.1-9) denotes the event that more than $N(p+\varepsilon)$ errors occur in the channel. To show that this term can be made vanishingly small, we invoke the weak law of large numbers which states that if $z_i$, $i = 1, 2, \ldots, N$ are i.i.d random variables with mean $m$, then $P\{m - \varepsilon \leq \frac{1}{N}\sum_i z_i \leq m + \varepsilon\} \to 1$ for any $\varepsilon > 0$. We may look upon $z_i$ as taking values 1 when a channel error occurs and 0 otherwise and conclude that the probability of the total number of channel errors, $\sum_i z_i$, exceeding $N(p+\varepsilon)$ tends to zero as $N$ grows.

To bound the second term in (3.1-9), we average it over all possible randomly chosen codes. Since each code word is chosen by throwing a

$2^N$-sided fair die there are $2^N$ possible choices for each code word, and since there are $J$ code words in each code, there are a total of $2^{NJ}$ codes. Each such code occurs with probability $2^{-NJ}$. Averaging (3.1-9) over all $2^{NJ}$ codes, weighted by their probability $2^{-NJ}$, and recognizing that there are $(J-1)$ terms of the form $P\{\mathbf{x}_i \in \mathcal{S}\}$ in the second term in (3.1-9) and that these terms are all identical, we obtain

$$\begin{aligned}\overline{P_E} &\leq \delta + (J-1)\overline{P\{\mathbf{x}_i \in \mathcal{S}\}} \\ &\leq \delta + J\overline{P\{\mathbf{x}_i \in \mathcal{S}\}}. \end{aligned} \tag{3.1-10}$$

Here $\delta$ represents the first term in (3.1-9), which can be made as small as we wish by increasing $N$ and the overbar denotes averaging over the random codes.

Now consider the term $\overline{P\{\mathbf{x}_i \in \mathcal{S}\}}$, the average probability that the outcome of throwing the $2^N$-faced die lies within the sphere of radius $N(p+\varepsilon)$ centered at the received word $\mathbf{y}_j$. This must be equal to the ratio of the total number of outcomes that lie within this sphere, $\mathcal{N}(\mathcal{S})$, to the total of all possible outcomes, $2^N$. That is,

$$\overline{P\{\mathbf{x}_i \in \mathcal{S}\}} = \frac{\mathcal{N}(\mathcal{S})}{2^N}. \tag{3.1-11}$$

For the BSC, $\mathcal{N}(\mathcal{S})$ is given by

$$\mathcal{N}(\mathcal{S}) = \sum_{i=0}^{N(p+\varepsilon)} \binom{N}{i}. \tag{3.1-12}$$

If $N(p+\varepsilon)$ is not an integer, it may be replaced by the largest integer less than $N(p+\varepsilon)$. What follows does not change as a result of this. We now use a result from combinatorics (see, for example, Hamming [3-16]) to bound (3.1-11):

$$\sum_{i=0}^{N\alpha} \binom{N}{i} \leq 2^{Nh_B(\alpha)}, \qquad \alpha < \frac{1}{2}. \tag{3.1-13}$$

Using (3.1-11), (3.1-12), and (3.1-13) in (3.1-10), we get

$$\overline{P_E} \leq \delta + J2^{-N[1-h_B(p+\varepsilon)]}. \tag{3.1-14}$$

To ensure that $\overline{P_E}$ tends to zero as $N$ increases, we must ensure that

$$\log_2 J < N[1 - h_B(p+\varepsilon)]. \tag{3.1-15}$$

Since $\varepsilon > 0$ is an arbitrarily small quantity, (3.1-15) essentially states that

$$R = \frac{\log_2 J}{N} < \left[1 - h_B(p)\right] = C, \tag{3.1-16}$$

where $C = 1 - h_B(p)$ is apparently the capacity of the BSC. Since, for $R < C$, the average error probability over all possible random codes tends to zero as $N$ becomes arbitrarily large, there must exist one code for which this holds. This proves the channel coding theorem for the BSC.

**QED**

We do not pursue the proof to the converse of the channel coding theorem, which states that, for rates above capacity, the probability of error will not go to zero. We refer the reader to the texts [1-2] – [1-6] for its proof as well as for proofs of the channel coding theorem for the more general cases of the discrete memoryless channel, the antipodal modulation channel, and the Gaussian noise channel. In the rest of the chapter, we will simply state and use these results as needed.

## 3.2 LINEAR BLOCK CODES

Recall that in the random coding argument of the previous section, each random code consisted of $J = 2^K$ code words, each of length $N$ bits. This is an $(N, K)$ *block code.* One problem with random block codes like this is that the encoder will need to store all the $2^K$ code words in memory. Each block of $K$ information or message or data bits specifies a unique address to the memory. The encoder retrieves the code word stored at this address. We see that encoder's storage increases exponentially with increasing $K$. Practical encoders impose a certain structure on the code, thereby greatly simplifying the encoding process but at the expense of decreased performance compared to the random codes of the previous section. Though random codes have little practical value, we will use the performance promised by these codes as a yard stick to measure the performance of practical codes. In what follows, we study such practical codes, with emphasis given to the use of these codes rather than to the underlying mathematical structure.

A characteristic that most practical codes possess is the property of *linearity.* We say that a block code is linear if the sum of any two code words $\mathbf{x}_1$ and $\mathbf{x}_2$, denoted as $\mathbf{x}_1 + \mathbf{x}_2$, is another code word, where the operation + denotes the bit-by-bit exclusive-or operation, or bit-by-bit modulo-2 addition.

**Example 3.2-1: The $(N, 1)$ Repetition Code.**

> Consider the $(N, 1)$ repetition code introduced in Section 3.1. This code has two code words $\mathbf{x}_0 = (0, 0, \ldots, 0)$ and $\mathbf{x}_1 = (1, 1, \ldots, 1)$. Clearly $\mathbf{x}_0 + \mathbf{x}_0 = \mathbf{x}_0$, $\mathbf{x}_0 + \mathbf{x}_1 = \mathbf{x}_1 = \mathbf{x}_1 + \mathbf{x}_0$, and $\mathbf{x}_1 + \mathbf{x}_1 = \mathbf{x}_0$. The $(N, 1)$ repetition code is, therefore, a linear block code.

Note that the set of code words from a linear block code forms a *group*† under the binary operation +, since the all-zeros code word serves the role of the identity element and each code word is its own inverse. For this reason a linear block code is also called a *group code.*

One may also view an $(N, K)$ linear block code as a $K$-dimensional subspace of the vector space of all the $2^N$ $N$-tuples over the binary field GF(2).‡ To see this, we first show that the set of all $N$-tuples over GF(2) forms a vector space. Consider two $N$-tuples or vectors $\mathbf{a} = (a_0, a_1, \ldots, a_{N-1})$ and $\mathbf{b} = (b_0, b_1, \ldots, b_{N-1})$. Their sum $\mathbf{c} = \mathbf{a} + \mathbf{b} = (a_0 + b_0, a_1 + b_1, \ldots, a_{N-1} + b_{N-1})$, where + denotes modulo-2 addition, is clearly an $N$-tuple. Denoting the all-zeros vector $(0, 0, \ldots, 0)$ by $\mathbf{0}$, we find that, for any vector $\mathbf{a}$, $\mathbf{a} + \mathbf{0} = \mathbf{0} + \mathbf{a} = \mathbf{a}$. Hence $\mathbf{0}$ is the identity element. The inverse of each vector $\mathbf{a}$ is itself since

$$\begin{aligned}\mathbf{a} + \mathbf{a} &= (a_0 + a_0, a_1 + a_1, \ldots, a_{N-1} + a_{N-1})\\ &= (0, 0, \ldots, 0) = \mathbf{0}.\end{aligned}$$

With these three properties and the fact that $\mathbf{a} + \mathbf{b} = \mathbf{b} + \mathbf{a}$, the set of all vectors of $N$-tuples over GF(2) forms a commutative group under modulo-2 addition. Define scalar multiplication of a vector $\mathbf{a}$ by a field element $\alpha = 0$ or 1 as

$$\alpha\mathbf{a} = (\alpha a_0, \alpha a_1, \ldots, \alpha a_{N-1}).$$

Clearly, $\alpha\mathbf{a}$ is also an $N$-tuple. For the multiplicative identity element 1 from GF(2)

$$1 \cdot \mathbf{a} = \mathbf{a}.$$

---

†Those readers not familiar with the concepts of groups, fields and vector spaces are urged to read Appendix B.

‡GF(2) is the binary field consisting of field elements 0 and 1, with addition, "+", and multiplication, "·", operations defined by the following tables.

| + | 0 1 |
|---|---|
| 0 | 0 1 |
| 1 | 1 0 |

| · | 0 1 |
|---|---|
| 0 | 0 0 |
| 1 | 0 1 |

The reader may verify that for any two elements $\alpha$ and $\beta$ from GF(2), and for any two $N$-tuples $\mathbf{a}$ and $\mathbf{b}$ the following laws are easily satisfied.

*Distributive laws:*

$$\alpha(\mathbf{a}+\mathbf{b}) = \alpha\mathbf{a} + \alpha\mathbf{b}$$
$$(\alpha+\beta)\mathbf{a} = \alpha\mathbf{a} + \beta\mathbf{a}.$$

*Associative law:*

$$(\alpha \cdot \beta)\mathbf{a} = \alpha \cdot (\beta\mathbf{a}).$$

The set of all $N$-tuples is therefore a vector space over GF(2).

We have noted earlier, that the set of all code words from an $(N, K)$ linear block code forms a commutative group. We can proceed further to show, as we did in the previous paragraph for the set of all $N$-tuples, that the set of all code words of an $(N, K)$ linear block code forms a vector space. We leave out the arguments to show that the code words form a $K$-dimensional vector space, *i.e.*, there exist $K$ code words (or *basis vectors*) such that every code word is a linear combination of these $K$ code words.

**Example 3.2-2: The $(N, 1)$ Repetition Code.**

Consider the $(N, 1)$ repetition code, $\mathcal{C} = \{(0, 0, \ldots, 0), (1, 1, \ldots, 1)\}$. The two code vectors in $\mathcal{C}$ can be expressed as linear combinations of the all-ones vector as shown below:

$$(1, 1, \ldots, 1) = 1 \cdot (1, 1, \ldots, 1),$$
$$(0, 0, \ldots, 0) = 1 \cdot (1, 1, \ldots, 1) + 1 \cdot (1, 1, \ldots, 1).$$

The all-ones vector is therefore the basis vector of $\mathcal{C}$ and $\mathcal{C}$ is a one dimensional subspace of the $N$-dimensional space of all $N$-tuples.

**Example 3.2-3: A (6,3) Code.**

Consider the following (6,3) code:

$$\mathcal{C} = \{(000000), (001011), (010101), (011110), (100111), (101100), (110010), (111001)\}.$$

$\mathcal{C}$ is easily shown to be a 3-dimensional vector space over GF(2), with basis vectors given by $\mathbf{g}_0 = (100111)$, $\mathbf{g}_1 = (010101)$, and $\mathbf{g}_2 = (001011)$. Table 3.2-1 expresses each of the eight code vectors in $\mathcal{C}$ as linear combinations of $\mathbf{g}_0$, $\mathbf{g}_1$, and $\mathbf{g}_2$.

**Table 3.2-1.** CODE VECTORS EXPRESSED AS LINEAR COMBINATIONS OF BASIS VECTORS.

| $u_0$ | $u_1$ | $u_2$ | $u_0\mathbf{g}_0 + u_1\mathbf{g}_1 + u_2\mathbf{g}_2$ |
|---|---|---|---|
| 0 | 0 | 0 | (0 0 0 0 0 0) |
| 0 | 0 | 1 | (0 0 1 0 1 1) |
| 0 | 1 | 0 | (0 1 0 1 0 1) |
| 0 | 1 | 1 | (0 1 1 1 1 0) |
| 1 | 0 | 0 | (1 0 0 1 1 1) |
| 1 | 0 | 1 | (1 0 1 1 0 0) |
| 1 | 1 | 0 | (1 1 0 0 1 0) |
| 1 | 1 | 1 | (1 1 1 0 0 1) |

## Generator Matrix For Linear Block Codes

Generalizing the preceding two examples to any $(N, K)$ linear block code, we see that any code word may be expressed as a linear combination of $K$ basis vectors $\mathbf{g}_i$, $i = 0, 1, \ldots, K-1$. If $\mathbf{x}$ is any code word, then

$$\begin{aligned}\mathbf{x} &= u_0\mathbf{g}_0 + u_1\mathbf{g}_1 + u_2\mathbf{g}_2 + \cdots + u_{K-1}\mathbf{g}_{K-1} \\ &= \mathbf{u}\begin{bmatrix}\mathbf{g}_0 \\ \mathbf{g}_1 \\ \vdots \\ \mathbf{g}_{K-1}\end{bmatrix} = \mathbf{uG},\end{aligned} \tag{3.2-1}$$

where $\mathbf{u} = (u_0, u_1, \ldots, u_{K-1})$ is a $K$-bit message, or information, or data vector and

$$\mathbf{G} = \begin{bmatrix}\mathbf{g}_0 \\ \mathbf{g}_1 \\ \vdots \\ \mathbf{g}_{K-1}\end{bmatrix} \tag{3.2-2}$$

is a $(K \times N)$ matrix of binary field elements from GF(2). By letting $\mathbf{u}$ take on all the $2^K$ binary $K$-tuple message vectors, one may generate all the $2^K$ code words given by (3.2-1). Observe that for $\mathbf{u} = (00\ldots1\ldots0)$, with 1 in the $i$th place and zeros elsewhere, $\mathbf{x}$ is $\mathbf{g}_i$, the $i$th row of the $\mathbf{G}$ matrix. Thus each row of the $\mathbf{G}$ matrix is itself a code vector. The vector space consisting of all the $2^K$ code words, generated by the $K$ rows of $\mathbf{G}$, is called the *rowspace* of $\mathbf{G}$. Since $\mathbf{G}$ serves as the generator of all the code words, it is called the *generator* matrix.

**Example 3.2-4: Generator Matrices for Some Codes.**

The generator matrix for the $(N, 1)$ repetition code is

$$\mathbf{G} = [1\ 1\ \dots\ 1],$$

a $1 \times N$ matrix of all 1s.

For the (6,3) code introduced earlier, the generator matrix is

$$\mathbf{G} = \begin{bmatrix} 1 & 0 & 0 & 1 & 1 & 1 \\ 0 & 1 & 0 & 1 & 0 & 1 \\ 0 & 0 & 1 & 0 & 1 & 1 \end{bmatrix}.$$

A *single-parity-check* code is an $(N, N-1)$ code that has $(N-1)$ information bits and a *parity check* bit given by the modulo-2 sum of the information bits, $u_0 + u_1 + \cdots + u_{N-1}$. The effect of adding the extra check bit is to make an even number of 1's in any code word. The generator matrix for this code is

$$\mathbf{G} = \begin{bmatrix} 1 & 0 & 0 & \dots & 0 & 1 \\ 0 & 1 & 0 & \dots & 0 & 1 \\ \vdots & \vdots & \vdots & \ddots & \vdots & \vdots \\ 0 & 0 & 0 & \dots & 1 & 1 \end{bmatrix}.$$

In all these examples, observe that the $\mathbf{G}$ matrix is of the form

$$\mathbf{G} = [\mathbf{I}_K; \mathbf{P}], \tag{3.2-3}$$

where $\mathbf{I}_K$ is the $K \times K$ identity matrix and $\mathbf{P}$ is a $K \times (N-K)$ matrix given by

$$\mathbf{P} = \begin{bmatrix} p_{00} & p_{01} & \cdots & p_{0,N-K-1} \\ p_{10} & p_{11} & \cdots & p_{1,N-K-1} \\ \vdots & \vdots & \ddots & \vdots \\ p_{K-1,0} & p_{K-1,1} & \cdots & p_{K-1,N-K-1} \end{bmatrix}. \tag{3.2-4}$$

Each code word generated by matrices of the form given in (3.2-3) has for its first $K$ bits the message bits and for the remaining $(N-K)$ bits some linear combinations of the message bits, called the parity-check bits. Such codes are called *systematic codes.* The $\mathbf{G}$ matrix is said to be in *systematic form*, or *reduced echelon* form. *Nonsystematic* codes may be reduced to systematic codes by performing elementary row operations and column interchanges on the $\mathbf{G}$ matrix. Hence systematic codes are not any less powerful than nonsystematic codes. Moreover, systematic codes do not require any further transformation of the received sequence to recover the information bits as nonsystematic codes do.

## Parity-Check Matrix for Linear Block Codes

Another useful matrix associated with linear block codes is the parity-check matrix **H**. To understand its significance, we focus on the case of the systematic codes just introduced. Expand out the matrix equation in (3.2-1), which is

$$\mathbf{x} = \mathbf{u}\mathbf{G},$$

to get

$$\begin{aligned}
x_0 &= u_0\\
x_1 &= u_1\\
&\vdots\\
x_{K-1} &= u_{K-1}\\
x_K &= u_0p_{00} + u_1p_{10} + \cdots + u_{K-1}p_{K-1,0}\\
x_{K+1} &= u_0p_{01} + u_1p_{11} + \cdots + u_{K-1}p_{K-1,1}\\
&\vdots\\
x_{N-1} &= u_0p_{0,N-K-1} + u_1p_{1,N-K-1} + \cdots + u_{K-1}p_{K-1,N-K-1}.
\end{aligned} \tag{3.2-5}$$

As noted earlier, the above equation states that the first $K$ code bits are the information bits and the last $(N-K)$ code bits are the redundant parity-check bits, expressed as some linear combinations of the information bits. The last $r = (N-K)$ parity-check equations may be rewritten as

$$\begin{matrix}
x_0p_{00} + x_1p_{10} + \ldots + x_{K-1}p_{K-1,0} + x_K & & & = 0\\
x_0p_{01} + x_1p_{11} + \ldots + x_{K-1}p_{K-1,1} & +x_{K+1} & & = 0\\
\vdots & & \ddots & \vdots\\
x_0p_{0,r-1} + x_1p_{1,r-1} + \ldots + x_{K-1}p_{K-1,r-1} & & +x_{N-1} & = 0.
\end{matrix} \tag{3.2-6a}$$

Rewriting the above equation in matrix form, we get

$$[x_0\ x_1\ \ldots\ x_{N-1}] \begin{bmatrix}
p_{00} & p_{10} & \cdots & p_{r-1,0} & 1 & 0 & \ldots & 0\\
p_{01} & p_{11} & \cdots & p_{r-1,1} & 0 & 1 & \ldots & 0\\
\vdots & \vdots & \ddots & \vdots & \vdots & \vdots & \ddots & \vdots\\
p_{0,r-1} & p_{1,r-1} & \cdots & p_{K-1,r-1} & 0 & 0 & \ldots & 1
\end{bmatrix}^T = \mathbf{0}, \tag{3.2-6b}$$

or, equivalently,

$$\mathbf{x}\mathbf{H}^T = \mathbf{0}, \tag{3.2-7}$$

where **H** is an $(N-K) \times N$ matrix given by

$$\mathbf{H} = [\mathbf{P}^T; \mathbf{I}_{N-K}], \tag{3.2-8}$$

and $\mathbf{0}$ is the all-zero row vector of size $N$. In the language of vector spaces, (3.2-7) states that the set of all code vectors $\mathbf{x}$ is in the *null space* of $\mathbf{H}$. As noted earlier, since each row of the $\mathbf{G}$ matrix is itself a code vector, we have

$$\mathbf{G}\mathbf{H}^T = \mathbf{0}, \tag{3.2-9}$$

where $\mathbf{0}$ here represents the all-zero $K \times (N-K)$ matrix. Equation (3.2-9) defines $\mathbf{H}$ for any code, whether systematic or not. For systematic codes it also follows from

$$\mathbf{G}\mathbf{H}^T = [\mathbf{I}_K; \mathbf{P}] \begin{bmatrix} \mathbf{P} \\ \mathbf{I}_{N-K} \end{bmatrix} = \mathbf{P} + \mathbf{P} = \mathbf{0}.$$

**Example 3.2-5: Parity-Check Matrices for Some Codes.**

For the $(N, 1)$ repetition code, the parity-check matrix is given by the following $(N-1) \times N$ matrix

$$\mathbf{H} = \begin{bmatrix} 1 & 1 & 0 & \dots & 0 \\ 1 & 0 & 1 & \dots & 0 \\ \vdots & \vdots & & \ddots & \vdots \\ 1 & 0 & 0 & \dots & 1 \end{bmatrix}.$$

For the (6,3) code, the $\mathbf{H}$ matrix is given by

$$\mathbf{H} = \begin{bmatrix} 1 & 1 & 0 & 1 & 0 & 0 \\ 1 & 0 & 1 & 0 & 1 & 0 \\ 1 & 1 & 1 & 0 & 0 & 1 \end{bmatrix},$$

and for the single-parity-check code

$$\mathbf{H} = [1\ 1\ 1\ \dots\ 1\ 1],$$

the all-ones row vector of length $N$. Observe the interesting fact that the $\mathbf{H}$ matrix for the single-parity-check code is the same as the generator matrix for the $(N, 1)$ repetition code. For all these three examples, the reader is urged to verify that (3.2-9) holds (see Problem 3-6).

### Dual Codes

Rewrite (3.2-9) to obtain

$$\mathbf{H}\mathbf{G}^T = \mathbf{0}^T. \tag{3.2-10}$$

Comparing this equation with (3.2-9), we see that the roles of the G and H matrices are reversed in (3.2-10). We may look upon the H matrix as the generator matrix for a code with $2^{N-K}$ code words and G as the parity-check matrix for this code. The set of all code words $\mathbf{x}^d$ generated by H is said to be in the null space of G. It is possible to show that this set forms a vector space of dimension $(N-K)$ by a theorem from linear algebra; it also states that if a subspace of all $N$-tuples is of dimension $K$ then its null space has dimension $(N-K)$. The set of code words generated by H is said to be the *dual* of that generated by G, and *vice versa.* Example 3.2-5 pointed out that the H matrix for the $(N, N-1)$ single-parity-check code was identical to the G matrix for the $(N, 1)$ repetition code. These codes are therefore *dual codes.*

## 3.3 SYNDROME DECODING

### The Syndrome

Consider transmitting an arbitrary code word x over the BSC. The received word y may be written as $\mathbf{y} = \mathbf{x} + \mathbf{e}$, where $\mathbf{e} = (e_0, e_1, \ldots, e_{N-1})$ denotes the error vector, with $e_i = 1$ if an error occurred in the $i$th position and $e_i = 0$ otherwise. Define the *syndrome* corresponding to the received word y as $\mathbf{S} = \mathbf{y}\mathbf{H}^T$. It is easy to see that S depends only on the error pattern and not on the particular code word transmitted, since

$$\mathbf{S} = \mathbf{y}\mathbf{H}^T = (\mathbf{x} + \mathbf{e})\mathbf{H}^T = \mathbf{0} + \mathbf{e}\mathbf{H}^T = \mathbf{e}\mathbf{H}^T.$$

The term syndrome is used in the medical world to denote symptoms that characterize a particular abnormality or disease. In the coding world, it denotes what abnormality (error pattern) occurs in x. Unfortunately, the syndrome does not identify the error pattern uniquely. To show this, form another error pattern $\mathbf{e}_i$ by adding the actual error e to $\mathbf{x}_i$, the $i$th code word in the code book, *i.e.,* $\mathbf{e}_i = \mathbf{e} + \mathbf{x}_i$. Then

$$\mathbf{e}_i\mathbf{H}^T = \mathbf{e}\mathbf{H}^T + \mathbf{0} = \mathbf{e}\mathbf{H}^T.$$

Thus all the $2^K$ distinct error vectors $\mathbf{e}_i$, where $\mathbf{e}_i = \mathbf{e} + \mathbf{x}_i$, have the same syndrome. A maximum-likelihood decoder is one that chooses the least-weight error pattern from among the $2^K$ error patterns.

Let us suppose that a single error occurs in the $l$th position of the transmitted code word $\mathbf{x}$. Then the error vector $\mathbf{e}$ is given by the $N$-tuple $(0, 0, \ldots, 1, \ldots, 0)$, with 1 in the $l$th position. The syndrome is therefore $\mathbf{S} = \mathbf{e}\mathbf{H}^T = \sum_{i=0}^{N-1} e_i \mathbf{h}_i = \mathbf{h}_l$, where $\mathbf{h}_i$ denotes the $i$th row of $\mathbf{H}^T$. Similarly, the syndrome due to two BSC channel errors in positions $l$ and $m$ is $\mathbf{S} = \mathbf{h}_l + \mathbf{h}_m$. In general, the syndrome is obtained by adding those rows of $\mathbf{H}^T$ corresponding to the error positions. A decoding rule therefore corresponds to inverting the received bits in positions corresponding to the rows of $\mathbf{H}^T$ whose sum equals the syndrome. A maximum-likelihood decoder must choose the minimum number of such rows of $\mathbf{H}^T$ and, for unambiguous decoding, such a choice of rows must be unique. This implies that for a code to be able to correct all single errors, no two rows of $\mathbf{H}^T$ (equivalently, no two columns of $\mathbf{H}$) can be the same. To be able to correct single and double errors, all $N$ rows of $\mathbf{H}^T$ and all possible $\binom{N}{2} = N(N-1)/2$ sums of two rows must be distinct, and so on, for higher orders of error correction. It is not a trivial matter, however, to determine which choice of rows yields the syndrome. We will investigate decoding algorithms based on syndromes later.

**Example 3.3-1: Syndrome Calculations for the (6,3) Code.**

Consider the (6,3) code of our previous examples. Let the received vector $\mathbf{y}$ be $(1\ 1\ 0\ 1\ 1\ 1)$. Then the syndrome is

$$\begin{aligned}\mathbf{S} &= \mathbf{y}\mathbf{H}^T \\ &= (1\ 1\ 0\ 1\ 1\ 1)\begin{bmatrix}1 & 1 & 1\\ 1 & 0 & 1\\ 0 & 1 & 1\\ 1 & 0 & 0\\ 0 & 1 & 0\\ 0 & 0 & 1\end{bmatrix} \\ &= (1\ 0\ 1).\end{aligned}$$

Since

$$\begin{aligned}\mathbf{S} &= \mathbf{e}\mathbf{H}^T \\ &= (e_0\ e_1\ e_2\ e_3\ e_4\ e_5)\mathbf{H}^T \\ &= (1\ 0\ 1),\end{aligned}$$

we have the following set of simultaneous equations:

$$\begin{aligned} e_0 + e_1 + e_3 &= 1 \\ e_0 + e_2 + e_4 &= 0 \\ e_0 + e_1 + e_2 + e_5 &= 1. \end{aligned}$$

There exist $2^3 = 8$ error vectors that satisfy the above set of equations and they are

| $e_0$ | $e_1$ | $e_2$ | $e_3$ | $e_4$ | $e_5$ |
|---|---|---|---|---|---|
| 0 | 1 | 0 | 0 | 0 | 0 |
| 0 | 1 | 1 | 0 | 1 | 1 |
| 0 | 0 | 0 | 1 | 0 | 1 |
| 0 | 0 | 1 | 1 | 1 | 0 |
| 1 | 1 | 0 | 1 | 1 | 1 |
| 1 | 1 | 1 | 1 | 0 | 0 |
| 1 | 0 | 0 | 0 | 1 | 0 |
| 1 | 0 | 1 | 0 | 0 | 1 |

Of these eight error vectors, (0 1 0 0 0 0) has the least weight. The transmitted code vector is therefore

$$\begin{aligned}\mathbf{x} &= \mathbf{y} + \mathbf{e}\\ &= (1\ 1\ 0\ 1\ 1\ 1) + (0\ 1\ 0\ 0\ 0\ 0)\\ &= (1\ 0\ 0\ 1\ 1\ 1).\end{aligned}$$

## Hamming Codes

Hamming codes form a class of codes that correct all single errors and they have found wide applications in practical systems. They were the first class of linear block codes to be discovered [3-16].

We have seen that for single error correction capacity all $N$ columns of $\mathbf{H}$ must be distinct, where each column has $(N - K)$ elements. Since there are $2^{N-K} - 1$ distinct nonzero vectors of length $(N - K)$, we can form a single-error correcting code whose $\mathbf{H}$ matrix consists of these nonzero vectors. The block length of this code is $N = 2^{N-K} - 1$. This is a Hamming code. With $N = 3$ and $K = 1$ $(N - K = 2)$, we obtain the (3,1) repetition code, whose $\mathbf{H}$ matrix consists of the 3 nonzero vectors of length 2. With $N = 7$ and $K = 4$ $(N - K = 3)$, we obtain the first interesting (7,4) Hamming code. With $N = 15$ and $K = 11$ $(N - K = 4)$, we get the (15,11) Hamming code, and so on. Two possible $\mathbf{H}$ matrices for the (7,4) Hamming code are given below. They are both equivalent, but the one on the right is in systematic form:

$$\mathbf{H} = \begin{bmatrix} 0\,0\,0\,1\,1\,1\,1 \\ 0\,1\,1\,0\,0\,1\,1 \\ 1\,0\,1\,0\,1\,0\,1 \end{bmatrix}; \qquad \mathbf{H} = \begin{bmatrix} 0\,1\,1\,1\,1\,0\,0 \\ 1\,0\,1\,1\,0\,1\,0 \\ 1\,1\,0\,1\,0\,0\,1 \end{bmatrix}.$$

**Example 3.3-2: Syndrome Decoding for the (7,4) Hamming Code.**

For the (7,4) systematic Hamming code, we may construct a syndrome decoding table as shown in Table 3.3-1.

Each nonzero syndrome denotes a single error in the position of the column of **H** that equals the transpose of the syndrome. Suppose the received word is $\mathbf{y} = (0\,0\,1\,1\,1\,0\,0)$. Then the syndrome $\mathbf{S} = \mathbf{y}\mathbf{H}^T = (1\;0\;1)$. Performing a table lookup to retrieve the error pattern from the syndrome decoding table yields $\mathbf{e} = (0\;1\;0\;0\;0\;0\;0)$. Adding **e** and **y** yields $(0\;1\;1\;1\;1\;0\;0\;)$ for the transmitted code vector. If a double error occurs in transmission over the BSC, say $\mathbf{e} = (0\;0\;1\;0\;1\;0\;0)$, then the syndrome is the sum of the second and fourth columns of **H**. In this case $\mathbf{S} = (0\;1\;0)$ and the corresponding error pattern is $(0\;0\;0\;0\;0\;1\;0)$. Adding this error vector to the received word is equivalent to inverting the bit in the fifth position of the received word, which results in incorrect decoding. Note that double error in the channel has actually caused a triple error at the output of the decoder.

**Table 3.3-1.** SYNDROME DECODING TABLE FOR THE (7,4) HAMMING CODE. THE WORDS OF THE CODE APPEAR IN TABLES 3.5-1 AND 3.5-3.

| Syndrome | Error Pattern |
|---|---|
| 0 0 0 | 0 0 0 0 0 0 0 |
| 0 0 1 | 0 0 0 0 0 0 1 |
| 0 1 0 | 0 0 0 0 0 1 0 |
| 0 1 1 | 1 0 0 0 0 0 0 |
| 1 0 0 | 0 0 0 0 1 0 0 |
| 1 0 1 | 0 1 0 0 0 0 0 |
| 1 1 0 | 0 0 1 0 0 0 0 |
| 1 1 1 | 0 0 0 1 0 0 0 |

## The Minimum Distance

We have seen that the (3,1) repetition code can successfully correct a single error. In general, the $(N, 1)$ repetition code can correct $\lfloor (N-1)/2 \rfloor$ or fewer errors, where $\lfloor x \rfloor$ denotes the integer part of $x$. The (6,3) code in the examples can correct a single error. The reader may verify that the code cannot correct double errors by observing that a double error in the transmitted code word makes the received word closer to some other code word. The $(N, N-1)$ single-parity-check code can only detect an odd number of errors and cannot correct any errors. With any even number of errors in transmission, the parity will check at the receiver and a decoder error will occur. Thus codes have varying degrees of error correcting or error detecting capabilities.

The ability of a code to correct $t$ or fewer errors is related to a quantity called the *minimum distance* of that code. We define the minimum distance, $d_{\min}$, of a code as the smallest Hamming distance between any two code words in that code. If $\lfloor d_{\min} - 1\rfloor/2$ or fewer errors occur, then, by the definition of $d_{\min}$, the received word is closer to the transmitted code word than to any other code word. Hence the received word will be decoded correctly. If more than $\lfloor d_{\min} - 1\rfloor/2$ errors occur, then the received word will be closer to a code word different from the transmitted word and a decoder error will occur. The code can therefore correct $t$ or fewer errors as long as

$$t \leq \frac{\lfloor d_{\min} - 1\rfloor}{2},$$

or

$$d_{\min} \geq 2t + 1.$$

Equivalently, every code word must differ from every other code word in at least $2t + 1$ positions for the code to correct $t$ or fewer errors. For the $(N, 1)$ repetition code, $d_{\min} = N$ and hence the code corrects $\lfloor (N-1)\rfloor/2$ or fewer errors in transmission. The reader may verify that $d_{\min} = 3$ for the (6,3) code in our examples. It can therefore correct at most a single error.

To determine the $d_{\min}$ of a code, it is not necessary to determine the distances between all possible pairs of code words, as we will now show. Observe that

$$d_H(\mathbf{x}_1, \mathbf{x}_2) = w(\mathbf{x}_1 + \mathbf{x}_2),$$

where $d_H(\mathbf{x}_1, \mathbf{x}_2)$ refers to the Hamming distance between two code words $\mathbf{x}_1$ and $\mathbf{x}_2$, $+$ denotes bit-by-bit modulo-2 addition, and $w$, called the *Hamming weight* or simply *weight*, is the number of ones in $\mathbf{x}_1 + \mathbf{x}_2$. For example, for the (6,3) code, $d_H(0\,0\,1\,0\,1\,1,\ 0\,1\,0\,1\,0\,1) = w(0\,1\,1\,1\,1\,0) = 4$. Since in a linear block code the sum of any two code words is another code word, we have $d_H(\mathbf{x}_1, \mathbf{x}_2) = w(\mathbf{x}_1 + \mathbf{x}_2) = w(\mathbf{x}_3)$, where $\mathbf{x}_3$ is some other code word. From the definition of minimum distance, we have

$$\begin{aligned} d_{\min} &= \min\{d_H(\mathbf{x}_1, \mathbf{x}_2) \,|\, \mathbf{x}_1 \neq \mathbf{x}_2\} \\ &= \min\{w(\mathbf{x}_1 + \mathbf{x}_2) \,|\, \mathbf{x}_1 \neq \mathbf{x}_2\} \\ &= \min\{w(\mathbf{x}_3) \,|\, \mathbf{x}_3 \neq \mathbf{0}\}. \end{aligned}$$

Hence we see that the minimum distance is the minimum of the weights of all nonzero code words.

From (3.2-7), we have

$$\begin{aligned}\mathbf{x}\mathbf{H}^T &= \mathbf{x}[h_0, h_1, \ldots, h_{N-1}]^T \\ &= x_0 h_0 + x_1 h_1 + \ldots + x_{N-1} h_{N-1} \\ &= \mathbf{0}, \end{aligned} \tag{3.3-1}$$

where $h_i$ refers to the $i$th column of $\mathbf{H}$. If $\mathbf{x}$ is a code word with minimum weight, then there are $d_{\min}$ nonzero $x_i$s in $\mathbf{x}$ and consequently (3.3-1) states that $d_{\min}$ columns of $\mathbf{H}$ sum to $\mathbf{0}$. Thus $d_{\min}$ for a code may also be determined as the minimum number of columns of $\mathbf{H}$ that sum to $\mathbf{0}$.

**Example 3.3-3: $d_{\min}$ for the (6,3) Code.**

For the (6,3) code of our previous examples,

$$\mathbf{H} = \begin{bmatrix} 1&1&0&1&0&0 \\ 1&0&1&0&1&0 \\ 1&1&1&0&0&1 \end{bmatrix}.$$

The reader may verify that no two columns of $\mathbf{H}$ sum to $\mathbf{0}$. $d_{\min}$ is therefore greater than 2. The zeroth, first, and fourth columns do sum to $\mathbf{0}$ so that $d_{\min} = 3$.

Figure 3.3-1 serves to illustrate further the concept of minimum distance. Here the $\times$'s indicate code words and the bullets ($\bullet$) indicate received words. Fig. 3.3-1(a) illustrates the code geometry of a single-parity-check code. Here two code words and a received word lying "between" them are shown. The received word cannot be uniquely decoded. However, the code can detect an odd number of errors in the received word. Fig. 3.3-1(b) illustrates the structure of the (5,1) repetition code. The code has a minimum distance of 5 and can therefore correct at most two errors. A single error due to the transmission of a code word over the BSC results in $\binom{5}{1} = 5$ possible received words, and these five received words are shown lying on the inner circle around each code word. Similarly, a double error results in $\binom{5}{2} = 10$ possible received words. These words are shown lying on the outer circle around each code word. Decoding is the act of determining the code word lying at the center of the circle on which the received word lies. If more than two BSC transmission errors occur, the received word will lie on a circle associated with a code word different from the transmitted code word, resulting in decoding error.

Figure 3.3-1(c) shows the code geometry of the (7,4) Hamming code. Associated with each code word is a set of seven received words at distance 1 from it. As in the case of Fig. 3.3-1(b), any received word is decoded into the code word at the center of the circle on which the received word lies. Clearly, more than one transmission error results in incorrect decoding.

The decoders of the type shown in Fig. 3.3-1 are capable of correcting $t$ or fewer errors and no others, where $t = \lfloor d_{\min} - 1 \rfloor / 2$. Such decoders are called *bounded-distance* decoders.

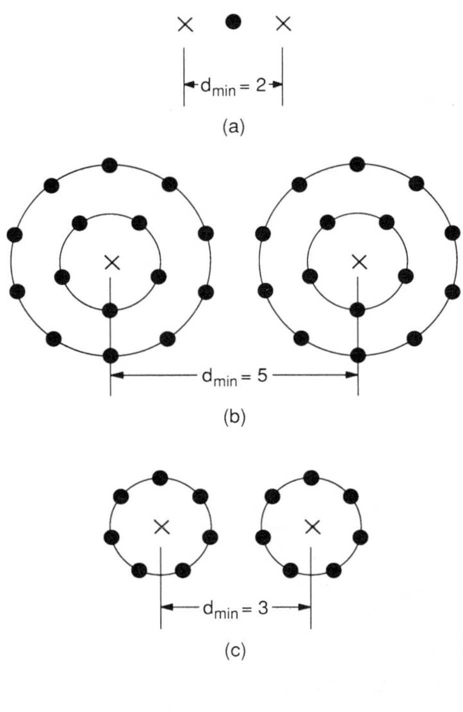


Figure 3.3-1. Code geometry of (a) the single-parity-check code word; (b) the (5,1) repetition code; and (c) the (7,4) Hamming code.

## The Standard Array

The discussion on code geometry has revealed the fact that the $2^K$ code words partition the space of all received words into $2^K$ disjoint subsets. Any received word in a subset is decoded as the unique code word within that subset. A standard array provides a technique by which such a partitioning can be achieved.

The algorithm for a standard array construction of an $(N, K)$ linear block code follows.

**Standard Array Construction**

**Step 1.** Write down the $2^K$ code words as elements of the first row, with the all-zeros code word as the leading (left-most) element.

**Step 2.** Repeat Steps 2(a) and 2(b) until all the $2^N$ words are exhausted.

**Step 2(a).** Out of the remaining unused $N$-tuples, select one with the least weight for the leading element of the next row. Call this the current row.

**Step 2(b).** Complete the current row by adding the leading element to each nonzero code word appearing in the first row and writing down the resulting sum in the corresponding column.

The rows of a standard array are called *cosets* and the leading element of a coset the *coset leader*. Several properties of the standard array follow immediately (see also Problem 3-10).

1. All $N$-tuples of a row are distinct.
2. Each $N$-tuple appears exactly once in the standard array.
3. There are exactly $2^{N-K}$ rows in the standard array.
4. For perfect codes (those that satisfy (3.4-1) in the next section), all $N$-tuples of weight $t = \lfloor (d_{\min} - 1)/2 \rfloor$ or less appear as coset leaders.

5. For *quasi-perfect* codes, in addition to all $N$-tuples of weight $t$ or less, some but not all $N$-tuples of weight $t+1$ appear as coset leaders.
6. All elements in a row (coset) have the same syndrome.
7. Elements in different rows have different syndromes.
8. There are $2^{N-K}$ different syndromes corresponding to $2^{N-K}$ rows.

The key to decoding using a standard array is to view the coset leaders as error patterns caused by the BSC. If $\mathbf{e}_i$ refers to the coset leader of the $i$th row, any element $\mathbf{y}$ in that row is expressed as $\mathbf{y} = \mathbf{x} + \mathbf{e}_i$, where $\mathbf{x}$ is the code word lying at the top of the column in which $\mathbf{y}$ lies. If $\mathbf{y}$ is the received word, the closest code word to $\mathbf{y}$ is easily seen to be $\mathbf{x}$ and hence $\mathbf{y}$ is decoded into $\mathbf{x}$. To see this, observe that

$$d_H(\mathbf{y},\mathbf{x}) = w(\mathbf{y}+\mathbf{x}) = w(\mathbf{e}_i+\mathbf{x}+\mathbf{x}) = w(\mathbf{e}_i).$$

The distance between $\mathbf{y}$ and any other code word $\mathbf{x}_1$ is

$$d_H(\mathbf{y},\mathbf{x}_1) = w(\mathbf{y}+\mathbf{x}_1) = w(\mathbf{e}_i+\mathbf{x}+\mathbf{x}_1) = w(\mathbf{e}_i+\mathbf{x}_2)$$

since by the property of linearity $\mathbf{x}+\mathbf{x}_1$ is another code word $\mathbf{x}_2$. Note that $\mathbf{e}_i+\mathbf{x}_2$ is in the same coset as $\mathbf{e}_i$ and $\mathbf{y}$. Since by construction a coset leader is chosen to be a minimum weight word from the remaining words, $w(\mathbf{e}_i) \leq w(\mathbf{e}_i+\mathbf{x}_2)$. Hence $d_H(\mathbf{y},\mathbf{x}) = w(\mathbf{e}_i) \leq w(\mathbf{e}_i+\mathbf{x}_2) \leq d_H(\mathbf{y},\mathbf{x}_1)$.

The reader can easily see that any error pattern caused by the BSC that is not a coset leader will result in a decoding error. Let $w_i$, $i = 0,1,\ldots,N$, denote the number of coset leaders of weight $i$. $w_i$, $i = 0,1,\ldots,N$, are called the *weight distribution* of the coset leaders. If $p$ denotes the crossover probability of the BSC, the probability of word error is given by

$$P_w(e) = 1 - \sum_{i=0}^{N} w_i p^i (1-p)^{N-i}. \tag{3.3-2}$$

We illustrate the technique of decoding using a standard array by an example.

**Example 3.3-4: Standard Array for the (6,3) Code.**

Consider the (6,3) code of our previous examples. There are $2^6 = 64$ possible 6-tuples, $2^3 = 8$ code words, and $2^{6-3} = 8$ rows in a standard array. Fig. 3.3-2 shows a standard array for this code obtained by following the algorithm described earlier.

The weight distribution of the coset leaders is $w_0 = 1$, $w_1 = 6$, and $w_2 = 1$; $w_i = 0$, $i = 3, 4, 5$ and 6. The error probability for this code may be calculated using (3.3-2):

$$P_w(e) = 1 - (1-p)^6 - 6p(1-p)^5 - p^2(1-p)^4.$$

Since all single errors and one double error are correctable, the fourteen remaining double error patterns dominate $P_w(e)$. Hence $P_w(e) \approx 14p^2(1-p)^4$. Using $p = 10^{-3}$ for the BSC yields $P_w(e) \approx 1.4 \times 10^{-5}$.

If the sequence 111110 is received, it is decoded into the code word 011110 appearing at the top of the column in which 111110 lies. In this instance the decoder corrects the most likely error in the zeroth position given by the coset leader 100000. In addition to correcting all single error patterns, the code corrects one double error pattern given by the coset leader 100001 appearing in the last row. For example, if 110100 is received it is decoded into $110100 + 100001 = 010101$, the code word at the top of the column in which the received sequence lies.

After constructing the first seven rows of the standard array shown in Fig. 3.3-2, three sequences of weight 2 remain. They are 100001, 011000, 000110. We have arbitrarily chosen one of these, the sequence 100001, as the coset leader for the last row. We could have instead chosen 011000 as the coset leader. In this case, the last row will be

011000 010011 001101 000110 111111 110100 101010 100001.

If the same sequence 110100 considered earlier is now received, it will be decoded into 1001100.

We now make use of Properties 6–8 to simplify the decoding procedure. To show Property 6, observe that any element in a coset, whose coset leader is $\mathbf{e}_i$, may be expressed as $\mathbf{y} = \mathbf{e}_i + \mathbf{x}_j$, for some code word $\mathbf{x}_j$. The syndrome of $\mathbf{y}$ is $\mathbf{S} = \mathbf{y}\mathbf{H}^T = \mathbf{e}_i\mathbf{H}^T + \mathbf{0} = \mathbf{e}_i\mathbf{H}^T$, the same as the syndrome of $\mathbf{e}_i$. Making use of the one-to-one correspondence between the syndromes and error patterns (Property 7), we form a lookup table containing the $2^{N-K}$ error patterns. The syndrome of the received sequence serves as the address of the location in the lookup table that contains the error pattern. The following example illustrates this simplified procedure, called *syndrome decoding.*

| coset leader | code words | | | | | | |
|---|---|---|---|---|---|---|---|
| 000000 | 001011 | 010101 | 011110 | 100111 | 101100 | 110010 | 111001 |
| 000001 | 001010 | 010100 | 011111 | 100110 | 101101 | 110011 | 111000 |
| 000010 | 001001 | 010111 | 011100 | 100101 | 101110 | 110000 | 111011 |
| 000100 | 001111 | 010001 | 011010 | 100011 | 101000 | 110110 | 111101 |
| 001000 | 000011 | 011101 | 010110 | 101111 | 100100 | 111010 | 110001 |
| 010000 | 011011 | 000101 | 001110 | 110111 | 111100 | 100010 | 101001 |
| 100000 | 101011 | 110101 | 111110 | 000111 | 001100 | 010010 | 011001 |
| 100001 | 101010 | 110100 | 111111 | 000110 | 001101 | 010011 | 011000 |

**Figure 3.3-2.** A standard array for the (6,3) code.

**Example 3.3-5: Syndrome Decoding for the (6,3) Code.**

Consider the (6,3) code of our examples for which the **H** matrix is

$$\mathbf{H} = \begin{bmatrix} 1 & 1 & 0 & 1 & 0 & 0 \\ 1 & 0 & 1 & 0 & 1 & 0 \\ 1 & 1 & 1 & 0 & 0 & 1 \end{bmatrix}.$$

Table 3.3-2 gives a lookup table for decoding this code.

Assume that the same sequence $\mathbf{y} = (111110)$ as in the preceding example is received. The following steps are performed in decoding the sequence.

Step 1. Compute the syndrome $S = \mathbf{y}\mathbf{H}^T$.
Step 2. Use the syndrome as address to retrieve the contents (error pattern) at that location.
Step 3. Decode the received sequence into the code vector $\mathbf{x} = \mathbf{y} + \mathbf{e}$.

Performing the above three steps using $\mathbf{y}$ yields $S = (1\ 1\ 1)$ after Step 1, $\mathbf{e} = (1\ 0\ 0\ 0\ 0\ 0)$ after Step 2, and $\mathbf{x} = (0\ 1\ 1\ 1\ 1\ 0)$ at Step 3.

Using $\mathbf{y} = (1\ 1\ 0\ 1\ 0\ 0)$ as the received sequence to perform the above three steps yields $S = (1\ 1\ 0)$, $\mathbf{e} = (1\ 0\ 0\ 0\ 0\ 1)$, and $\mathbf{x} = (0\ 1\ 0\ 1\ 0\ 1)$. $\mathbf{e} = (1\ 0\ 0\ 0\ 0\ 1)$ is the only double error pattern that the code corrects. Any other pattern of two or more errors will cause a decoding error. For example, if $\mathbf{x} = (1\ 1\ 1\ 0\ 0\ 1)$ is transmitted and the double error pattern $\mathbf{e} = (0\ 0\ 0\ 0\ 1\ 1)$ occurs in the BSC, the sequence $\mathbf{y} = (1\ 1\ 1\ 0\ 1\ 0)$ will be received. Performing the three steps given above to decode this sequence results in $S =$

(0 1 1), $\mathbf{e}' = (0\ 0\ 1\ 0\ 0\ 0)$, and $\mathbf{x}' = (1\ 1\ 0\ 0\ 1\ 0)$. Observe that $\mathbf{x}'$ is at a Hamming distance of three from the transmitted code word $\mathbf{x}$. Thus a double error in the channel has resulted in a triple error because of the decoding procedure employed.

**Table 3.3-2.** Syndrome Decoding Table for the (6,3) Code.

| Syndrome (address to the lookup table) | Error Pattern (contents of lookup table) |
|---|---|
| 0 0 0 | 0 0 0 0 0 0 |
| 0 0 1 | 0 0 0 0 0 1 |
| 0 1 0 | 0 0 0 0 1 0 |
| 0 1 1 | 0 0 1 0 0 0 |
| 1 0 0 | 0 0 0 1 0 0 |
| 1 0 1 | 0 1 0 0 0 0 |
| 1 1 0 | 1 0 0 0 0 1 |
| 1 1 1 | 1 0 0 0 0 0 |

# 3.4 BOUNDS ON PERFORMANCE

## Some Useful Coding Bounds

An interesting observation can be made from Fig. 3.3-1. For the (5,1) repetition code, there are $2^5 = 32$ words of block length 5 that can be received and each such word falls within one of the two spheres (or circles) associated with the two code words. For the (7,4) Hamming code there are $2^4 = 16$ code words and each sphere associated with a code word contains 8 words of length 7. Together, the sixteen spheres account for all the $2^7 = 128$ possible words of length 7. Codes in which any of the $2^N$ received word falls within one of the $2^K$ spheres of radius $t = \lfloor (d_{\min} - 1)/2 \rfloor$ are called *perfect* codes. Such $t$-error-correcting perfect binary codes obey the equality

$$\left[1 + \binom{N}{1} + \binom{N}{2} + \cdots + \binom{N}{t}\right] 2^K = 2^N. \qquad (3.4\text{-}1)$$

Within any sphere associated with a code word, there are $N$ words at distance one from that code word, $\binom{N}{2}$ words at distance two, and so on. $[1+\binom{N}{1}+\cdots+\binom{N}{t}]$ is therefore the number of words within any one sphere of radius $t$. The expression on the left in (3.4-1) gives the total number of words associated with all the $2^K$ spheres, which, for perfect codes, must equal $2^N$. For an $(N,K)$ Hamming code, we have pointed out that the block length $N$ satisfies the equation

$$N = 2^{N-K} - 1,$$

which, after rearranging, yields

$$(1+N)2^K = 2^N.$$

Since $t = 1$ for Hamming codes, the above equation shows that these codes satisfy (3.4-1) and hence are perfect codes. The proof that all $(N,1)$ repetition codes are perfect is left to the reader. One can easily show that the (6,3) code considered earlier is not a perfect code. Apart from the two code types just mentioned, the only other binary perfect code is the 3-error correcting (23,12) Golay code, which satisfies

$$\left[1+\binom{23}{1}+\binom{23}{2}+\binom{23}{3}\right]2^{12} = 2^{23}.$$

For a general $q$-ary $(N,K)$ linear block code, (3.4-1) may be generalized to

$$\left[1+(q-1)\binom{N}{1}+(q-1)^2\binom{N}{2}+\cdots+(q-1)^t\binom{N}{t}\right]q^K \le q^N. \qquad (3.4\text{-}2)$$

The term $(q-1)^i\binom{N}{i}$ in (3.4-2) gives the number of received words that differ from a given code word in $i$ symbols. There are $\binom{N}{i}$ ways in which $i$ symbols can be chosen out of $N$ and there are $(q-1)^i$ possible choices for these $i$ symbols. The term within square brackets therefore represents the number of possible received words within the sphere of radius $t$ surrounding any code word. This term multiplied by the number of spheres (or code words), $q^K$, must be less than or equal to the total number of $q$-ary $N$-tuples, $q^N$. The bound implied by (3.4-2) is called the *Hamming bound* or the *sphere-packing bound.*

The Hamming bound may be expressed as a bound on the maximum number of code words that the code can have for a given block length $N$ and error correction capacity $t$:

$$q^K \le \frac{q^N}{\sum_{i=0}^{t}\binom{N}{i}(q-1)^i}. \qquad (3.4\text{-}3)$$

One may also express this bound as a bound on the minimum redundancy required for a given $K$ and $t$:

$$N - K \geq \log_q \left[ \sum_{i=0}^{t} \binom{N}{i} (q-1)^i \right]. \tag{3.4-4}$$

To obtain the *Plotkin bound*, note that the minimum among the weights of all nonzero code words in a code $C$ must be less than or equal to the average of the weights of all nonzero code words. If all the $2^K$ code words are laid out as rows of a $2^K \times N$ matrix, one can show that each column contains exactly $2^{K-1}$ 0s and $2^{K-1}$ 1s. Thus the total weight of all the code words is $N2^{K-1}$ and the average weight of a nonzero code word $N2^{K-1}/(2^K - 1)$. Hence

$$\begin{aligned} d_{\min} &= \min_{\mathbf{x} \in C} w(\mathbf{x}) \\ &\leq \frac{N2^{K-1}}{2^K - 1}. \end{aligned} \tag{3.4-5}$$

The bound given by (3.4-5) on the minimum distance of a code is known as the Plotkin bound.

The *Gilbert-Varsharmov* bound is based on the property that for a linear $(N, K)$ block code with minimum distance $d_{\min}$, there exist $d_{\min}$ columns of the $\mathbf{H}$ matrix that sum to zero but no combinations of $d_{\min} - 1$ or fewer columns exist that sum to zero. In the language of vector spaces, any $d_{\min} - 1$ or fewer columns of the $\mathbf{H}$ matrix are linearly independent. Based on the above observation we may proceed as follows to construct a $(N, K)$ linear block code with minimum distance $d_{\min}$. Choose any nonzero $(N-K)$-tuple for the first column of the $\mathbf{H}$ matrix and any other nonzero $(N-K)$-tuple for the second column. Any nonzero $(N-K)$-tuple may be chosen for the third column that is linearly independent of the first two columns. One may proceed in this manner, adding a column at a time to the $\mathbf{H}$ matrix, making sure that the new column added is linearly independent of any $d_{\min} - 2$ or fewer preceding columns. After choosing the $(N-1)$th column of the $\mathbf{H}$ matrix, it must be true that all $d_{\min} - 2$ or fewer combinations from the $(N-1)$ columns do not exhaust all the nonzero $(N-K)$-tuples. That is

$$\binom{N-1}{1} + \binom{N-1}{2} + \cdots + \binom{N-1}{d_{\min} - 2} < 2^{N-K} - 1,$$

or,

$$\sum_{i=0}^{d_{\min}-2} \binom{N-1}{i} < 2^{N-K} \tag{3.4-6}$$

must be true before we can add the $N$th column. For $q$-ary codes, the bound in (3.4-6), known as the Gilbert-Varsharmov bound, generalizes to

$$\sum_{i=0}^{d_{\min}-2} (q-1)^i \binom{N-1}{i} < q^{N-K}. \tag{3.4-7}$$

## Coding Gain

*Coding gain* refers to the savings attainable in the energy per information bit to noise spectral density ratio ($E_b/N_0$) required to achieve a given bit error probability when coding is used compared to that with no coding. We assume that the bits to be transmitted are modulated using binary coherent PSK for transmission over the AWGN channel. (These are discussed in Chapter 4 and Appendix A).

Let $E_c$ refer to the transmitted bit energy. When coding is used, the total transmitted energy per code word is $NE_c$. Since each information bit has energy $E_b$ and there are $K$ information bits carried in a code word, the total energy is also $KE_b$. Thus a simple relation is obtained between $E_c$ and $E_b$ by equating $NE_c$ and $KE_b$,

$$E_c = \frac{K}{N} E_b = RE_b. \tag{3.4-8}$$

When no coding is used ($R = 1$), $E_c = E_b$.

The PSK modulator, the AWGN channel, and the coherent demodulator that makes bit-by-bit (hard) decisions may be looked upon as a BSC with crossover probability $p$ given by

$$\begin{aligned} p &= Q\left(\sqrt{\frac{2E_c}{N_0}}\right) \\ &= Q\left(\sqrt{\frac{2RE_b}{N_0}}\right), \end{aligned} \tag{3.4-9}$$

where the function Q is defined in (1.3-2). At high signal-to-noise ratio ($E_b/N_0$), the event that dominates the probability of incorrectly decoding a word, $P_w(e)$, for a $t$-error-correcting code is a $t+1$ error event. Hence a good approximation to $P_w(e)$ is

$$\begin{aligned} P_w(e) &\approx \binom{N}{t+1} p^{t+1}(1-p)^{N-t-1} \\ &\approx \binom{N}{t+1} p^{t+1} \end{aligned} \tag{3.4-10}$$

for small $p$. When $t+1$ errors occur in the transmitted code word, the received word will be decoded into another code word at a distance greater or equal to $2t+1$ from the transmitted code word. This was pointed out in the examples on syndrome decoding using the (7,4) Hamming code and the (6,3) linear block code, both of which have $t = 1$; in these cases, two BSC channel errors were transformed into three errors by the decoder. Since these $2t+1$ or more errors can occur anywhere in the $N$ positions of the received word, a good estimate for the bit error probability $P_b$ is

$$P_b \approx \frac{2t+1}{N} P_w(e). \tag{3.4-11}$$

Combining (3.4-10) and (3.4-11), we obtain

$$P_b \approx cp^{t+1}, \tag{3.4-12}$$

where $c$ is some constant dependent on the code parameters $N$ and $t$. Using (3.4-9) in (3.4-12) and applying the tight upper bound on $Q(x)$,

$$Q(x) \leq \frac{1}{2} \exp\left(-\frac{x^2}{2}\right), \tag{3.4-13}$$

we obtain

$$P_b \approx \frac{c}{2}\left[\exp\left(-\frac{RE_b}{N_0}\right)\right]^{t+1}. \tag{3.4-14}$$

For the same $E_b/N_0$ ratio, the probability of bit error obtained without coding is

$$P_b \approx \frac{1}{2} \exp\left(-\frac{E_b}{N_0}\right). \tag{3.4-15}$$

Taking logarithms on both sides of (3.4-14) and (3.4-15), and ignoring $\log c$ at large $E_b/N_0$, we observe that the ratio of $E_b/N_0$ for the uncoded case to that for the coded case to achieve the same bit error rate is

$$\frac{(E_b/N_0)_{uc}}{(E_b/N_0)_c} \approx R(t+1). \tag{3.4-16}$$

The above ratio, expressed in dB, is called the asymptotic coding gain $G_h$. It is given by

$$G_h = 10\log_{10}\left[R(t+1)\right] \tag{3.4-17}$$

If, instead of making bit-by-bit hard decisions at the demodulator, the $N$ unquantized outputs of the demodulator are taken together in making a soft decision on the transmitted code word, the asymptotic coding gain becomes (see Section 4.2)

$$G_s = 10\log_{10}\left[Rd_{\min}\right], \tag{3.4-18}$$

where $d_{\min}$ is the minimum distance of the code. Comparing the expressions for $G_s$ and $G_h$ indicates that soft-decision receivers provide an improvement of 3 db over hard-decision receivers. This improvement is achievable only for high signal-to-noise ratios. In practice, the improvement will be somewhat less.

We illustrate the concept of coding gain with the following example.

**Example 3.4-1: Coding Gain for the (7,4) Hamming Code.**

Consider the (7,4) Hamming code of our examples. Since this is a perfect code that corrects all single errors, error patterns of weight 2 or more result in a decoding error. So the probability of word error is given by

$$\begin{aligned} P_w(e) &= 1-(1-p)^7-7p(1-p)^6 \\ &= \binom{7}{2}p^2(1-p)^5+\binom{7}{3}p^3(1-p)^4 \\ &\quad+\binom{7}{4}p^4(1-p)^3+\binom{7}{5}p^5(1-p)^2+\binom{7}{6}p^6(1-p)+p^7 \\ &\approx 21p^2. \end{aligned}$$

Observe that the above result follows from (3.4-10) since double errors are the events that dominate $P_w(e)$ most. Since double errors in the channel result in triple errors at the decoder (see examples on syndrome decoding), an estimate of the bit error rate is

$$P_b \approx \frac{3}{7}P_w(e).$$

This expression may also be obtained by applying (3.4-11) to the code.

Given a certain bit error rate at the decoder's output, say $\alpha$, we see that needed word error rate is $P_w(e) = \frac{7}{3}\alpha$. Since $P_w(e) \approx 21p^2$, we compute the BSC crossover probability $p$ as $\sqrt{\alpha}/3$. Equating $p = \sqrt{\alpha}/3$ to $Q\left(\sqrt{\frac{2RE_b}{N_0}}\right)$ (see (3.4-9)), we can determine the ratio $E_b/N_0$ required to achieve the given post-decoding BER (or delivered BER) $\alpha$. The reader can plot BER versus $E_b/N_0$ for BPSK with and without coding and find that the (7,4) Hamming code achieves only a modest coding gain at $P_b = 10^{-6}$. We will consider more powerful codes in the following sections, yielding greater coding gain.

## 3.5 CYCLIC CODES

*Cyclic codes* are obtained by imposing additional structural properties on top of linearity required by linear codes and form an important subclass of linear codes. These codes were first studied by Prange [3-18] and since then many developments have taken place in the study of cyclic codes. The underlying structure, the Galois field and the polynomial description of finite fields, all lead to simplified encoding and decoding techniques that are algorithmic in nature. Computationally efficient techniques have been proposed in recent years for the decoding of a subclass of cyclic codes, the BCH codes. The structure of BCH codes and the algorithmic techniques for decoding them will be described in Section 3.8. References [3-1]–[3-10] provide in-depth treatment of cyclic codes.

First we provide an intuitive understanding of the workings of a cyclic code before considering their precise structure. Consider the (7,4) Hamming code with its generator matrix given in systematic form by

$$\mathrm{G} = \begin{bmatrix} 1 & 0 & 0 & 0 & 1 & 1 & 0 \\ 0 & 1 & 0 & 0 & 0 & 1 & 1 \\ 0 & 0 & 1 & 0 & 1 & 1 & 1 \\ 0 & 0 & 0 & 1 & 1 & 0 & 1 \end{bmatrix} = \begin{bmatrix} \mathbf{g}_0 \\ \mathbf{g}_1 \\ \mathbf{g}_2 \\ \mathbf{g}_3 \end{bmatrix}.$$

Table 3.5-1 lists all the 16 code words generated by this G matrix.

The reader can easily verify that the code words of Table 3.5-1 satisfy the following two properties:

1. The sum of any two code words is again a code word, which is the linearity property.
2. The result of cyclically shifting a code word is another code word. (Cyclically shifting a code word one position left is shifting its bits left one bit and carrying the leftmost bit shifted out to the rightmost position).

Codes satisfying these two properties are called cyclic codes.

To understand why the (7,4) Hamming code satisfies these two properties, consider its G matrix just given. All the rows of G may be generated by $\mathbf{g}_3$ alone, as will be shown now. $\mathbf{g}_2$ can be obtained by shifting $\mathbf{g}_3$ left by one bit and adding the result to $\mathbf{g}_3$, that is,

$$\mathbf{g}_2 = (0011010) + (0001101) = (0010111).$$

Shifting $\mathbf{g}_2$ left by one position and adding the resulting vector to $\mathbf{g}_3$ produces $\mathbf{g}_1$, that is,

$$\mathbf{g}_1 = (0101110) + (0001101) = (0100011).$$

**Table 3.5-1.** CODE WORDS OF THE SYSTEMATIC (7,4) HAMMING CODE.

| Information Vector | Code Word |
|---|---|
| 0 0 0 0 | 0 0 0 0 0 0 0 |
| 0 0 0 1 | 0 0 0 1 1 0 1 |
| 0 0 1 0 | 0 0 1 0 1 1 1 |
| 0 0 1 1 | 0 0 1 1 0 1 0 |
| 0 1 0 0 | 0 1 0 0 0 1 1 |
| 0 1 0 1 | 0 1 0 1 1 1 0 |
| 0 1 1 0 | 0 1 1 0 1 0 0 |
| 0 1 1 1 | 0 1 1 1 0 0 1 |
| 1 0 0 0 | 1 0 0 0 1 1 0 |
| 1 0 0 1 | 1 0 0 1 0 1 1 |
| 1 0 1 0 | 1 0 1 0 0 0 1 |
| 1 0 1 1 | 1 0 1 1 1 0 0 |
| 1 1 0 0 | 1 1 0 0 1 0 1 |
| 1 1 0 1 | 1 1 0 1 0 0 0 |
| 1 1 1 0 | 1 1 1 0 0 1 0 |
| 1 1 1 1 | 1 1 1 1 1 1 1 |

$\mathbf{g}_0$ is the result of shifting $\mathbf{g}_1$ left by one bit. Since any code word is a linear combination of the rows of $\mathbf{G}$ and since all rows of $\mathbf{G}$ can be obtained as a sum of $\mathbf{g}_3$ and its shifts, all code words satisfy the linearity and cyclic shift properties. By performing some elementary row operations on $\mathbf{G}$, we can, in fact, produce another generator matrix in which every row is a shifted version of any other row. To see this, add $\mathbf{g}_3$ to $\mathbf{g}_2$ to get $\mathbf{g}_2' = (0011010)$. Next add $\mathbf{g}_2$ to $\mathbf{g}_1$ to get $\mathbf{g}_1' = (0110100)$. Finally, add $\mathbf{g}_3 + \mathbf{g}_1$ to $\mathbf{g}_0$ to get $\mathbf{g}_0' = (1101000)$. The resulting generator matrix $\mathbf{G}'$ is given by

$$\mathbf{G}' = \begin{bmatrix} 1 & 1 & 0 & 1 & 0 & 0 & 0 \\ 0 & 1 & 1 & 0 & 1 & 0 & 0 \\ 0 & 0 & 1 & 1 & 0 & 1 & 0 \\ 0 & 0 & 0 & 1 & 1 & 0 & 1 \end{bmatrix} = \begin{bmatrix} \mathbf{g}_0' \\ \mathbf{g}_1' \\ \mathbf{g}_2' \\ \mathbf{g}_3' \end{bmatrix}. \tag{3.5-1}$$

We see that every row of $\mathbf{G}'$ is a shifted version of $\mathbf{g}_3'$, where $\mathbf{g}_3' = \mathbf{g}_3$. Since any code word is expressible as a linear combination of the rows of $\mathbf{G}$, which are themselves some shifted version of the last row, it is possible to express each code word in terms of the last row alone.

## The Generator Polynomial

At this stage we introduce some polynomial algebra to facilitate explaining the properties of cyclic codes. We denote an $N$-tuple $\mathbf{a}$ as

$$\mathbf{a} = (a_{N-1}, a_{N-2}, \ldots, a_0).$$

In the discussion of cyclic codes, we find it convenient to let the subscripts run in the reverse order, from $N-1$ down to 0, instead of in the forward order. We define the polynomial representation $a(D)$ of $\mathbf{a}$ as

$$a(D) \triangleq \sum_{i=0}^{N-1} a_i D^i. \tag{3.5-2}$$

Using the polynomial notation in (3.5-2), we denote the polynomial corresponding to $\mathbf{g}'_3 = (0\ 0\ 0\ 1\ 1\ 0\ 1)$ in (3.5-1) as $g'_3(D) = g(D) = D^3 + D^2 + 1$. We see that $g'_2(D) = Dg(D)$, $g'_1(D) = D^2 g(D)$, and $g'_0(D) = D^3 g(D)$. We can now represent the matrix operation $\mathbf{x} = \mathbf{u}\mathbf{G}'$ in terms of polynomial multiplication. To see this, consider generating the code word corresponding to $\mathbf{u} = (1\ 0\ 1\ 1)$, which is $\mathbf{x} = (1\ 0\ 1\ 1)\mathbf{G}' = (1\ 1\ 1\ 1\ 1\ 1\ 1)$. Representing the rows of $\mathbf{G}'$ in terms of $g(D)$, we get the polynomial corresponding to $\mathbf{x}$ as

$$\begin{aligned} x(D) &= (1\ 0\ 1\ 1)\Big(D^3 g(D)\ D^2 g(D)\ Dg(D)\ g(D)\Big)^T \\ &= (1\cdot D^3 + 0\cdot D^2 + 1\cdot D + 1)g(D) \\ &= u(D)g(D), \end{aligned} \tag{3.5-3}$$

where $u(D) = D^3 + D + 1$ is the information polynomial corresponding to input $\mathbf{u} = (1\ 0\ 1\ 1)$. From (3.5-3), we observe that every code polynomial $x(D)$ contains $g(D)$ as a factor. We carry out the operation in (3.5-3) to get

$$\begin{aligned} x(D) &= u(D)g(D) \\ &= (D^3 + D + 1)(D^3 + D^2 + 1) \\ &= (D^6 + D^5 + D^4 + D^3 + D^2 + D + 1), \end{aligned}$$

which corresponds to the all-ones code word. We obtained the same result using the matrix operation $\mathbf{x} = \mathbf{u}\mathbf{G}'$.

In (3.5-3), $u(D)$ is, in general, a polynomial of degree $K-1$ whose coefficients form the information vector and $g(D)$ is a polynomial of degree $N-K$ that multiplies the information polynomial to generate the degree-$(N-1)$ code polynomial $x(D)$. Since $g(D)$ generates all the code polynomials in the code, it is called the *generator polynomial.* Since there are $2^K$ distinct information vectors giving rise to $2^K$ information polynomials, there are $2^K$ distinct code polynomials and, therefore, $2^K$ distinct code words.

Equation (3.5-3) expresses the linearity property of code words and states further that every code polynomial contains the generator polynomial as a factor. How is one to represent the second property of cyclic codes, the cyclic shift property? Recall that the cyclic property states that any cyclic shift of a code word produces another code word. Consider the (7,4) cyclic Hamming code for which the generator polynomial was given earlier as $g(D) = D^3 + D^2 + 1$. The code word corresponding to the information polynomial $u(D) = D^3 + 1$ (information vector $\mathbf{u} = (1\ 0\ 0\ 1)$) is

$$\begin{aligned} x(D) &= u(D)g(D) \\ &= (D^3+1)(D^3+D^2+1) \\ &= D^6+D^3+D^5+D^2+D^3+1 \\ &= D^6+D^5+D^2+1, \end{aligned}$$

which corresponds to the code word $\mathbf{x} = (1\ 1\ 0\ 0\ 1\ 0\ 1)$. Shifting $\mathbf{x}$ cyclically one position left produces the code word $\mathbf{x}' = (1\ 0\ 0\ 1\ 0\ 1\ 1)$, which corresponds to the polynomial $x'(D) = D^6+D^3+D+1$. Multiplying $x(D)$ by $D$ to emulate shifting $\mathbf{x}$ one position left produces

$$Dx(D) = D^7 + D^6 + D^3 + D.$$

We observe that equating $D^7$ to 1 in $Dx(D)$ yields $x'(D)$. In the terminology of polynomial algebra, this observation can be expressed as

$$x'(D) = Dx(D) \bmod D^7 - 1.$$

In words, $x'(D)$ is the remainder polynomial obtained by dividing $Dx(D)$ by $D^7 - 1$ (or $D^7 + 1$). The long division that follows illustrates this.

$$
\begin{array}{r|l}
 & \quad 1 \\ \hline
D^7 - 1 & D^7 + D^6 + D^3 + D \\
 & D^7 \qquad\qquad\quad - 1 \\ \cline{2-2}
 & \underbrace{D^6 + D^3 + D + 1}_{\text{Remainder} \,=\, x'(D)}
\end{array}
$$

What is the property of $g(D)$ that makes every cyclic shift of a code word another code word? Multiplying both sides of (3.5-3) by $D^i$ and taking $\bmod D^N - 1$ to emulate the cyclic shift by $i$ positions to the left, we get

$$D^i x(D) \bmod D^N - 1 = D^i u(D) g(D) \bmod D^N - 1. \tag{3.5-4}$$

We must show that the term $D^i x(D) \bmod D^N - 1$ in (3.5-4) is a code word; that is, it must be divisible by $g(D)$. Hence we must show that

$$\left[D^i x(D) \bmod D^N - 1\right] \bmod g(D) = \left[D^i u(D) g(D) \bmod D^N - 1\right] \bmod g(D)$$
$$= 0.$$

If we assume that $g(D)$ divides $D^N - 1$, or $D^N - 1 = g(D)h(D)$, then clearly, for any $a(D)$,

$$[a(D) \bmod D^N - 1] \bmod g(D) = a(D) \bmod g(D).$$

Therefore

$$\left[D^i u(D) g(D) \bmod D^N - 1\right] \bmod g(D) = D^i u(D) g(D) \bmod g(D)$$
$$= 0,$$

since $D^i u(D) g(D)$ contains $g(D)$ as a factor and is therefore divisible by $g(D)$. Thus the desired condition is that $g(D)$ divides $D^N - 1$.

We state below, without proof, several properties of the generator polynomial $g(D)$.

1. $g(D)$ is the unique nonzero code polynomial of least degree in the code.
2. $g(D)$ is of the form $D^{N-K} + g_{N-K-1} D^{N-K-1} + \cdots + g_1 D + 1$.
3. Every code word is a multiple of $g(D)$, and there are $2^K$ distinct code words in the code.
4. $g(D)$ is a factor of $D^N - 1$.
5. $g(D)$ generates an $(N, K)$ cyclic code.

**Example 3.5-1: The Cyclic Binary (7,4) Hamming Code.**

Consider the (7,4) cyclic Hamming code with $g(D) = D^3 + D^2 + 1$. The degree of $g(D)$ is $N - K = 7 - 4 = 3$. $g(D)$ divides $D^7 - 1$ as verified below by long division.

$$
\begin{array}{r|llllll}
 & D^4 + D^3 + D^2 + 1 & & & & & \\
\hline
D^3 + D^2 + 1 & D^7 & & & & & -1 \\
 & D^7 + D^6 & & +D^4 & & & \\
\hline
 & & D^6 & +D^4 & & & -1 \\
 & & D^6 + D^5 & & +D^3 & & \\
\hline
 & & & D^5 + D^4 + D^3 & & & -1 \\
 & & & D^5 + D^4 & & +D^2 & \\
\hline
 & & & & D^3 + D^2 & & -1 \\
 & & & & D^3 + D^2 & & +1 \\
\hline
 & & & & & & 0 \\
\hline
\end{array}
$$

## Encoders for Cyclic Codes

Figure 3.5-1 shows a circuit for the (7,4) cyclic Hamming code that produces $x(D)$ by multiplying $u(D)$ by $g(D)$. We see that multiplying $u(D)$ by $g(D)$ produces

$$
\begin{aligned}
x(D) &= (u_3 D^3 + u_2 D^2 + u_1 D + u_0)(D^3 + D^2 + 1) \\
&= u_3 D^6 + (u_3 + u_2) D^5 + (u_2 + u_1) D^4 + (u_3 + u_1 + u_0) D^3 \\
&\quad + (u_2 + u_0) D^2 + u_1 D + u_0.
\end{aligned}
$$

Figure 3.5-2 verifies that the circuit in Fig. 3.5-1 does indeed produce the required code word. After all the four information bits have been shifted in, the register contains the code word, which may then be shifted out.

Figure 3.5-3 generalizes the circuit in Fig. 3.5-1, where $g(D)$ denotes the polynomial $D^{N-K} + g_{N-K-1} D^{N-K-1} + \cdots + g_1 D + 1$. After all the $K$ information bits are shifted in, the shift register contains the code

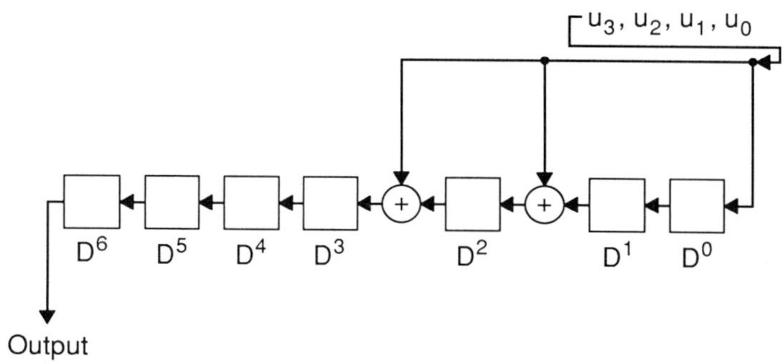


**Figure 3.5-1.** A circuit for generating $x(D) = u(D)g(D)$.

| Shift | Shift Register Contents | | | | | | |
|---|---|---|---|---|---|---|---|
| | $D^6$ | $D^5$ | $D^4$ | $D^3$ | $D^2$ | $D$ | 1 |
| 0 | 0 | 0 | 0 | 0 | 0 | 0 | 0 |
| 1 | 0 | 0 | 0 | $u_3$ | $u_3$ | 0 | $u_3$ |
| 2 | 0 | 0 | $u_3$ | $u_3 + u_2$ | $u_2$ | $u_3$ | $u_2$ |
| 3 | 0 | $u_3$ | $u_3 + u_2$ | $u_2 + u_1$ | $u_3 + u_1$ | $u_2$ | $u_1$ |
| 4 | $u_3$ | $u_3 + u_2$ | $u_2 + u_1$ | $u_3 + u_1 + u_0$ | $u_2 + u_0$ | $u_1$ | $u_0$ |

**Figure 3.5-2.** Contents of the shift register in Fig. 3.5-1.

polynomial, with the coefficient of $D^{N-1}$ in the left most stage and that of $D^0$ in the right most.

The circuit in Fig. 3.5-3 will, in general, produce code words that are not in systematic form. Systematic codes can be produced by carrying out the following steps.

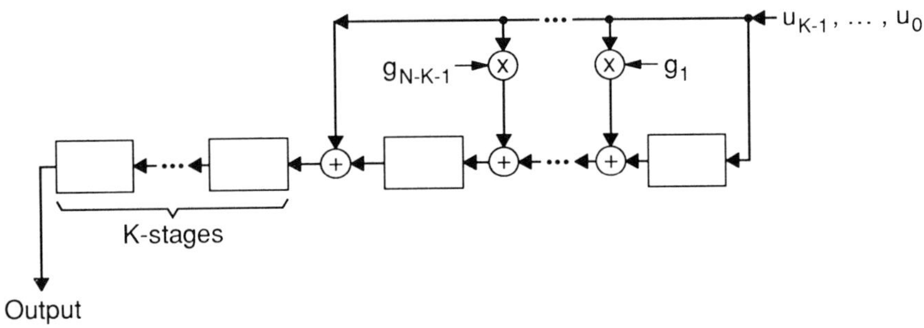


**Figure 3.5-3.** A general circuit for generating a code word via polynomial multiplication.

> **Systematic Encoder for Cyclic Codes**
>
> **Step 1.** Shift the information polynomial $u(D)$ left by $(N-K)$ places; that is, multiply $u(D)$ by $D^{N-K}$.
>
> **Step 2.** Divide $D^{N-K}u(D)$ by $g(D)$ to obtain the remainder $D^{N-K}u(D) \bmod g(D)$.
>
> **Step 3.** Add the remainder to $D^{N-K}u(D)$ to obtain the code polynomial; that is,
>
> $$x(D) = D^{N-K}u(D) + D^{N-K}u(D) \bmod g(D). \tag{3.5-5}$$

To verify that $x(D)$ is a code word, let $q(D)$ be the quotient polynomial obtained by dividing $D^{N-K}u(D)$ by $g(D)$. Then

$$\underbrace{D^{N-K}u(D)}_{\text{dividend}} = \underbrace{q(D)g(D)}_{\text{quotient} \times \text{divisor}} + \underbrace{D^{N-K}u(D) \bmod g(D)}_{\text{remainder}}.$$

Hence the code word is

$$\begin{aligned} x(D) &= D^{N-K}u(D) + D^{N-K}u(D) \bmod g(D) \\ &= q(D)g(D). \end{aligned}$$

That is, $x(D)$ is divisible by $g(D)$ and is therefore a code word. Moreover, $x(D)$ is in systematic form, since the information symbols are the coefficients of the highest $K$ terms of $x(D)$.

Systematic codes may have code words given by either

$$x(D) = u_0 + u_1 D + u_2 D^2 + \cdots + u_{K-1} D^{K-1} + p_0 D^K + \cdots + p_{N-K-1} D^{N-1}$$

or

$$x(D) = p_0 + p_1 D + p_2 D^2 + \cdots + p_{N-K-1} D^{N-K-1} + u_0 D^{N-K} + \cdots + u_{K-1} D^{N-1},$$

where $p_i, i = 0, 1, \ldots, N-K-1$ denote parity bits. The latter one in which the information bits form the coefficients of the highest $K$ terms is best since it simplifies encoding circuitry. As already pointed out, the preceding algorithm produces code words in this form and we assume this particular form for systematic codes in the rest of this chapter.

Tables 3.5-2 and 3.5-3 show the nonsystematic and systematic codes, respectively, generated by $g(D)$. Observe that $g(D)$ has the lowest degree among all nonzero code polynomials in both.

**Table 3.5-2.** NONSYSTEMATIC (7,4) BINARY HAMMING CODE GENERATED BY $g(D) = D^3 + D^2 + 1$.

| Information **u** | Information $u(D)$ | Nonsystematic Code $x(D) = u(D)g(D)$ | Codevector **x** |
|---|---|---|---|
| 0 0 0 0 | 0 | 0 | 0 0 0 0 0 0 0 |
| 0 0 0 1 | 1 | $D^3 + D^2 + 1$ | 0 0 0 1 1 0 1 |
| 0 0 1 0 | $D$ | $D^4 + D^3 + D$ | 0 0 1 1 0 1 0 |
| 0 0 1 1 | $(D+1)$ | $D^4 + D^2 + D + 1$ | 0 0 1 0 1 1 1 |
| 0 1 0 0 | $D^2$ | $D^5 + D^4 + D^2$ | 0 1 1 0 1 0 0 |
| 0 1 0 1 | $(D^2+1)$ | $D^5 + D^4 + D^3 + 1$ | 0 1 1 1 0 0 1 |
| 0 1 1 0 | $(D^2+D)$ | $D^5 + D^3 + D^2 + D$ | 0 1 0 1 1 1 0 |
| 0 1 1 1 | $(D^2+D+1)$ | $D^5 + D + 1$ | 0 1 0 0 0 1 1 |
| 1 0 0 0 | $D^3$ | $D^6 + D^5 + D^3$ | 1 1 0 1 0 0 0 |
| 1 0 0 1 | $(D^3+1)$ | $D^6 + D^5 + D^2 + 1$ | 1 1 0 0 1 0 1 |
| 1 0 1 0 | $(D^3+D)$ | $D^6 + D^5 + D^4 + D$ | 1 1 1 0 0 1 0 |
| 1 0 1 1 | $(D^3+D+1)$ | $D^6 + D^5 + D^4$ $+D^3 + D^2 + D + 1$ | 1 1 1 1 1 1 1 |
| 1 1 0 0 | $(D^3+D^2)$ | $D^6 + D^4 + D^3 + D^2$ | 1 0 1 1 1 0 0 |
| 1 1 0 1 | $(D^3+D^2+1)$ | $D^6 + D^4 + 1$ | 1 0 1 0 0 0 1 |
| 1 1 1 0 | $(D^3+D^2+D)$ | $D^6 + D^2 + D$ | 1 0 0 0 1 1 0 |
| 1 1 1 1 | $(D^3+D^2+D+1)$ | $D^6 + D^3 + D + 1$ | 1 0 0 1 0 1 1 |

**Table 3.5-3.** SYSTEMATIC (7,4) BINARY HAMMING CODE GENERATED BY $g(D) = D^3 + D^2 + 1$.

| Information $\mathbf{u}$ | Information $u(D)$ | Systematic Code $x(D) = D^3u(D) + D^3u(D) \bmod g(D)$ | Codevector $\mathbf{x}$ |
|---|---|---|---|
| 0 0 0 0 | 0 | 0 | 0 0 0 0 0 0 0 |
| 0 0 0 1 | 1 | $D^3 + D^2 + 1$ | 0 0 0 1 1 0 1 |
| 0 0 1 0 | $D$ | $D^4 + D^2 + D + 1$ | 0 0 1 0 1 1 1 |
| 0 0 1 1 | $(D + 1)$ | $D^4 + D^3 + D$ | 0 0 1 1 0 1 0 |
| 0 1 0 0 | $D^2$ | $D^5 + D + 1$ | 0 1 0 0 0 1 1 |
| 0 1 0 1 | $(D^2 + 1)$ | $D^5 + D^3 + D^2 + D$ | 0 1 0 1 1 1 0 |
| 0 1 1 0 | $(D^2 + D)$ | $D^5 + D^4 + D^2$ | 0 1 1 0 1 0 0 |
| 0 1 1 1 | $(D^2 + D + 1)$ | $D^5 + D^4 + D^3 + 1$ | 0 1 1 1 0 0 1 |
| 1 0 0 0 | $D^3$ | $D^6 + D^2 + D$ | 1 0 0 0 1 1 0 |
| 1 0 0 1 | $(D^3 + 1)$ | $D^6 + D^3 + D + 1$ | 1 0 0 1 0 1 1 |
| 1 0 1 0 | $(D^3 + D)$ | $D^6 + D^4 + 1$ | 1 0 1 0 0 0 1 |
| 1 0 1 1 | $(D^3 + D + 1)$ | $D^6 + D^4 + D^3 + D^2$ | 1 0 1 1 1 0 0 |
| 1 1 0 0 | $(D^3 + D^2)$ | $D^6 + D^5 + D^2 + 1$ | 1 1 0 0 1 0 1 |
| 1 1 0 1 | $(D^3 + D^2 + 1)$ | $D^6 + D^5 + D^3$ | 1 1 0 1 0 0 0 |
| 1 1 1 0 | $(D^3 + D^2 + D)$ | $D^6 + D^5 + D^4 + D$ | 1 1 1 0 0 1 0 |
| 1 1 1 1 | $(D^3 + D^2 + D + 1)$ | $D^6 + D^5 + D^4 + D^3 + D^2 + D + 1$ | 1 1 1 1 1 1 1 |

The circuit in Fig. 3.5-4 performs the following algorithm.

**Alternate Systematic Encoder for Cyclic Codes**

**Step 1.** With switch S1 in position 1 and S2 closed, shift the information bits into the shift register and also onto the channel.

**Step 2.** Once all the information bits are shifted into the register in $K$ shifts, now with S1 in position 2 and S2 open, the contents of the $(N - K)$-stage shift register are shifted onto the channel.

Observe that shifting the information bits directly onto the chan-

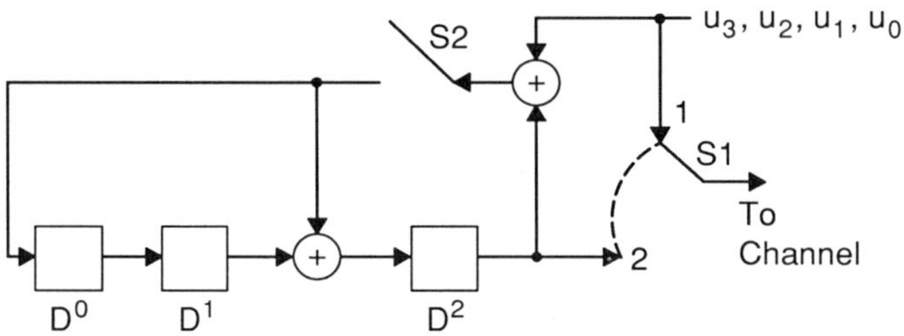


(a)

| Shift | Shift Register Contents $D^2$ | $D^1$ | $D^0$ |
|---|---|---|---|
| 0 | 0 | 0 | 0 |
| 1 | $u_3$ | 0 | $u_3$ |
| 2 | $u_3+u_2$ | $u_3$ | $u_3+u_2$ |
| 3 | $u_2+u_1$ | $u_3+u_2$ | $u_3+u_2+u_1$ |
| 4 | $u_3+u_1+u_0$ | $u_3+u_2+u_1$ | $u_2+u_1+u_0$ |

(b)

**Figure 3.5-4.** (a) Circuit for generating the systematic (7,4) binary cyclic Hamming code; (b) contents of the shift register in (a).

nel (Step 1) is equivalent to multiplying the information polynomial by $D^{N-K}$. This is step 1 of the previous algorithm. Alternatively, since the information bits are transmitted first, they form the $K$ bits $x_i, i = N-1,\ N-2,\ \ldots, N-K$, which are the coefficients of the highest $K$ terms of the code polynomial. As the information bits are shifted into the circuit of Figure 3.5-4, it performs the operation $D^{N-K}u(D) \bmod g(D)$ in $K$ steps. The reader may verify this by carrying out long division and comparing it with the contents of the shift register as given in Fig. 3.5-4(b). Steps 2 and 3 of the previous algorithm are combined into one step, Step 2, in the present one. The reader may try generalizing the (7,4) systematic encoder to any $(N, K)$ cyclic code.

One can gain additional insight into the structure of cyclic codes by studying the circuit of Fig. 3.5-4 more closely. With S1 closed and no input applied, let the circuit be initialized so that the left most stage contains 1. Then in successive shifts the contents of the shift register stages 2, 1, and 0, (the coefficients of $D^2$, $D^1$, and $D^0$) written as 3-tuples, are 001, 010, 100, 111, 011,110, 001, .... Clearly, the cycle repeats itself after seven shifts. The first seven 3-tuples represent all the nonzero 3-tuples of the parity check matrix of the (7,4) Hamming code and they form the length-$(2^3 - 1)$ *maximum length shift register sequence,* also called an *m-sequence.* This sequence of seven 3-tuples also represents all the nonzero elements of the finite field $GF(2^3)$.†

The relationship between the circuit in Fig. 3.5-4 and the generator matrix of the (7,4) Hamming code it generates will be evident from the discussion that follows. Suppose the sequence 1000 is applied to the input of the circuit and shifted into the register. As the first bit (the left most) is shifted in, the register contains 101; the next shift produces 111; the third shift 011; the fourth shift results in 110 for the parity-check bits. Thus applying 1000 to the input of this circuit results in 1000110 being transmitted. Similarly, applying 0100 results in 0100011 for the transmitted sequence; and 0010 and 0001 result in 0010111 and 0001101 being transmitted over the channel. This circuit therefore corresponds to the generator matrix

$$\mathbf{G} = \begin{bmatrix} 1 & 0 & 0 & 0 & 1 & 1 & 0 \\ 0 & 1 & 0 & 0 & 0 & 1 & 1 \\ 0 & 0 & 1 & 0 & 1 & 1 & 1 \\ 0 & 0 & 0 & 1 & 1 & 0 & 1 \end{bmatrix}. \qquad (3.5\text{-}6)$$

The parity-check matrix corresponding to $\mathbf{G}$ in (3.5-6) is

$$\mathbf{H} = \begin{bmatrix} 1 & 0 & 1 & 1 & 1 & 0 & 0 \\ 1 & 1 & 1 & 0 & 0 & 1 & 0 \\ 0 & 1 & 1 & 1 & 0 & 0 & 1 \end{bmatrix}. \qquad (3.5\text{-}7)$$

Adding the first row to second and the sum of first and second rows to third of (3.5-7) produces the parity-check matrix $\mathbf{H}'$ given by

$$\mathbf{H}' = \begin{pmatrix} u_3 & u_2 & u_1 & u_0 & p_2 & p_1 & p_0 \\ 1 & 0 & 1 & 1 & 1 & 0 & 0 \\ 0 & 1 & 0 & 1 & 1 & 1 & 0 \\ 0 & 0 & 1 & 0 & 1 & 1 & 1 \end{pmatrix}. \qquad (3.5\text{-}8)$$

In this matrix, the left most four columns correspond to information bits and the right most three to parity-check bits. The matrix $\mathbf{H}'$ given by (3.5-8) suggests that the parity bit $p_2$ is obtained by adding $u_3$, $u_1$, and $u_0$; the parity bit $p_1$ by adding $u_2$, $u_0$, and $p_2$; and $p_0$ by adding $u_1, p_2$, and $p_1$. We thus observe that each parity bit is obtained by the same linear

†The Galois field $GF(2^m)$, called the extension field of GF(2), will be discussed in the next section.

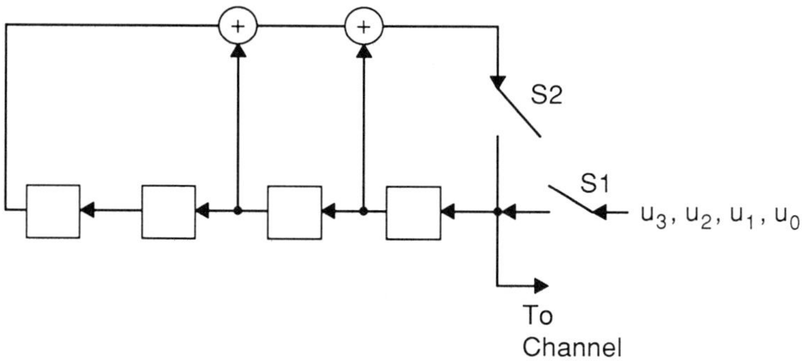


**Figure 3.5-5.** A 4-stage encoder for the (7,4) Hamming code based on $h(D) = D^4 + D^3 + D^2 + 1$.

combination of four bits to its left. This observation leads to the encoder of Fig. 3.5-5 for the (7,4) Hamming code.

The encoder of Fig. 3.5-5 works by the following algorithm:

**A $K$-stage encoder for cyclic codes**

**Step 1.** With S1 closed and S2 open, shift the information bits into the shift register and also, simultaneously, onto the channel in $K$ shifts.

**Step 2.** With S1 open and S2 closed, shift the $N - K$ parity-check bits that arrive at the input to stage 0 onto the channel in the next $N - K$ shifts.

One can easily generalize the steps taken to arrive at Fig. 3.5-5. We have pointed out earlier that $g(D)$ is a factor of $D^N - 1$ and hence

$$D^N - 1 = g(D)h(D), \tag{3.5-9}$$

where $h(D)$ is a degree-$K$ polynomial of the form

$$h(D) = h_K D^K + h_{K-1} D^{K-1} + \cdots + h_1 D + h_0.$$

If $x(D)$ is any code word generated by $g(D)$, then

$$\begin{aligned} x(D)h(D) &= u(D)g(D)h(D) \\ &= u(D)(D^N - 1), \end{aligned} \tag{3.5-10}$$

and after taking $\bmod D^N - 1$ on both sides,

$$\begin{aligned} x(D)h(D) \bmod D^N - 1 &= u(D)(D^N - 1) \bmod D^N - 1 \\ &= 0. \end{aligned} \tag{3.5-11}$$

We can show that this is the relationship $\mathbf{x}\mathbf{H}^T = \mathbf{0}$ from Section 3.2. From (3.5-10)

$$x(D)h(D) = u(D) + D^N u(D), \tag{3.5-12}$$

where we have identified 1 with $-1$ in GF(2). Since $u(D)$ contains terms with powers of $D$ less than or equal to $K-1$, and $D^N u(D)$ contains terms $D^N$ or greater, the coefficients of $D^K, D^{K+1}, \ldots, D^{N-1}$ of the left side in (3.5-12) are zeros. This observation leads to the following set of $(N-K)$ equations:

$$\sum_{i=0}^{K} h_i x_{N-i-j} = 0, \qquad \text{for } 1 \le j \le N-K. \tag{3.5-13}$$

Expressing (3.5-13) in matrix form, we get

$$\mathbf{x}\mathbf{H}^T = \mathbf{0}, \tag{3.5-14}$$

where $\mathbf{x} = (x_{N-1}\ x_{N-2} \ldots x_0)$ and

$$\mathbf{H} = \begin{bmatrix} h_0 & h_1 & \ldots & \ldots & h_K & 0 & \ldots & 0 \\ 0 & h_0 & h_1 & \ldots & h_{K-1} & h_K & \ldots & 0 \\ \vdots & & \ddots & & & \ddots & & \vdots \\ 0 & 0 & \ldots & h_0 & \ldots & \ldots & h_{K-1} & h_K \end{bmatrix}. \tag{3.5-15}$$

Since $h(D)$ can be used to construct the parity-check matrix, $h(D)$ is called the parity-check polynomial. Figure 3.5-6 illustrates a circuit for generating a cyclic code based on its parity-check polynomial.

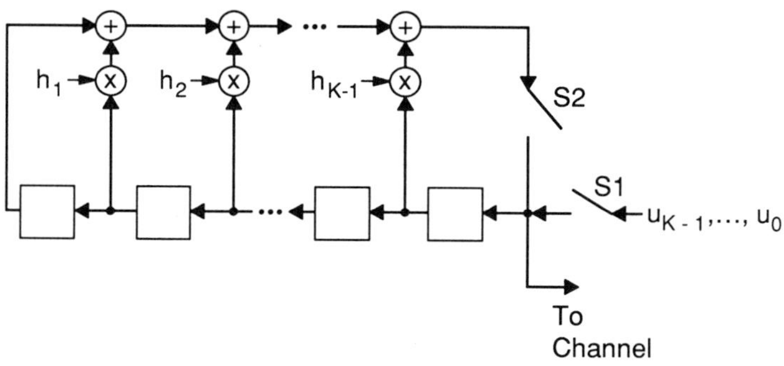


Figure 3.5-6. Generating a cyclic code by its parity-check polynomial $h(D)$, where $h(D) = h_K D^K + h_{K-1}D^{K-1} + \cdots + h_1 D + h_0$; $h_0 = h_K = 1$.

**Example 3.5-2: Parity-check Polynomial for the (7,4) Code.**

Consider the (7,4) cyclic Hamming code with $g(D) = D^3 + D^2 + 1$. One may easily verify that $D^7 - 1$ has factors

$$D^7 - 1 = (D+1)(D^3 + D + 1)(D^3 + D^2 + 1).$$

Hence

$$\begin{aligned} h(D) &= \frac{D^7 - 1}{D^3 + D^2 + 1} \\ &= D^4 + D^3 + D^2 + 1, \end{aligned}$$

from which $h_4 = h_3 = h_2 = h_0 = 1$ and $h_1 = 0$. The circuit in Fig. 3.5-6 reduces to that in Fig. 3.5-5 for these $h_i$. The parity-check matrix is given by $\mathbf{H}'$ in (3.5-8). Let the input $\mathbf{u} = (1\ 0\ 1\ 1)$ be applied to the circuit in Fig. 3.5-5. After all the four information bits are shifted into the register, S2 closes and the fifth, sixth, and seventh shifts produce, respectively, the sequence 1 0 0 at the channel input. The code word transmitted over the channel is therefore $\mathbf{x} = (1\ 0\ 1\ 1\ 1\ 0\ 0)$. It is easily verified that $\mathbf{x}\left[\mathbf{H}'\right]^T = \mathbf{0}$.

## 3.6 DECODING CYCLIC CODES

### Syndrome Computations for Cyclic Codes

Recall from Section 3.3 that we defined the syndrome S of a block code as $\mathbf{S} = \mathbf{y}\mathbf{H}^T$, where y is the received word and H is the parity-check matrix. If the received word is a code word, then $\mathbf{S} = \mathbf{0}$ and a nonzero syndrome indicates that one or more errors occurred on the channel. The idea of syndrome is similar for cyclic codes, but now we resort to polynomial notation in defining it. We have pointed out in the preceding section that every code polynomial is divisible by the generator polynomial $g(D)$. If there are no errors in the received sequence, the polynomial corresponding to it must be divisible by $g(D)$. Otherwise the remainder is nonzero and is symptomatic of the errors in the received sequence. We therefore define the syndrome as the remainder obtained by dividing the received polynomial by $g(D)$ and let the received word be y, where $\mathbf{y} = (y_{N-1}\ y_{N-2} \cdots y_0)$. Let $y(D)$ be the corresponding polynomial given by $y(D) = \sum_{i=0}^{N-1} y_i D^i$. Let $q(D)$ be the quotient obtained by dividing $y(D)$ by $g(D)$ and $s(D)$ be the remainder. Then

$$y(D) = q(D)g(D) + s(D), \tag{3.6-1}$$

where the syndrome polynomial $s(D)$, or simply the syndrome, is a nonzero polynomial of degree at most $(N - K - 1)$ if errors occurred in the channel and $s(D)$ is zero if the received word is a code word.

Example 3.6-1: Syndrome for the (7,4) Hamming Code.

The syndrome computation is accomplished by using the division circuit of Fig. 3.5-4, which performs division of $D^{N-K}u(D)$ by $g(D)$. Observe that feeding the information bits at the right end of the shift register is equivalent to premultiplying the information polynomial $u(D)$ by $D^{N-K}$ before carrying out division by $g(D)$. To compute the syndrome from the received polynomial, we need to divide $y(D)$ by $g(D)$. Since premultiplication is not required now, we feed $y(D)$ at the left end of the shift register. Figure 3.6-1 illustrates this.

Let $\mathbf{y} = (1\ 0\ 0\ 1\ 1\ 1\ 1)$ be the received word. Then the corresponding polynomial is $y(D) = D^6 + D^3 + D^2 + D + 1$. Dividing $y(D)$ by $g(D$, where $g(D) = D^3 + D^2 + 1$, yields $D^2$ for the remainder and so $s(D)$ equals $D^2$. The computations carried out in Fig. 3.6-1 confirm that after seven shifts the circuit contains (1 0 0). This sequence corresponds to the syndrome $s(D) = D^2$.

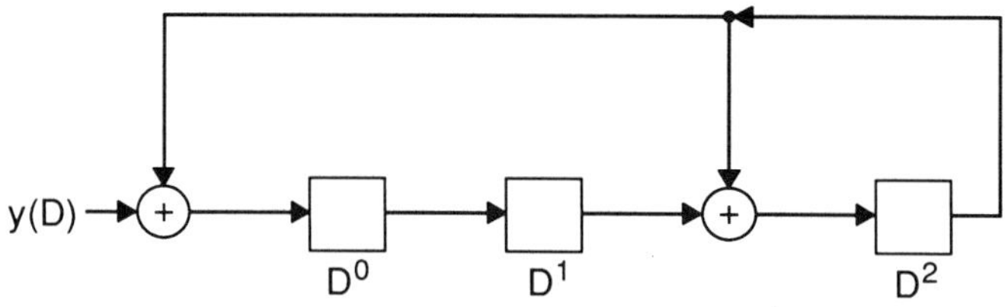


(a)

| Shift | Input | Shift Register Contents $D^2$ | $D^1$ | $D^0$ |
|---|---|---|---|---|
| 0 | | 0 | 0 | 0 |
| 1 | 1 | 0 | 0 | 1 |
| 2 | 0 | 0 | 1 | 0 |
| 3 | 0 | 1 | 0 | 0 |
| 4 | 1 | 1 | 0 | 0 |
| 5 | 1 | 1 | 0 | 0 |
| 6 | 1 | 1 | 0 | 0 |
| 7 | 1 | 1 | 0 | 0 |

(b)

**Figure 3.6-1.** (a) Syndrome computation circuit for the (7,4) binary cyclic Hamming code; (b) contents of the shift register in (a).

## The Meggitt Decoder

A decoder for cyclic codes is easily constructed by observing that if $s(D)$ is the syndrome corresponding to $y(D)$, the syndrome due to cyclically shifting $y(D)$ one position left is $Ds(D) \bmod g(D)$. This is easily shown. Observe that the cyclic shift of $y(D)$ left by one position is given by $y'(D) = Dy(D) \bmod D^N - 1$. Let $s'(D)$ be its syndrome. Then $s'(D)$ is given by

$$\begin{aligned} s'(D) &= y'(D) \bmod g(D) \\ &= \Big(Dy(D) \bmod D^N - 1\Big) \bmod g(D). \end{aligned}$$

Since $g(D)$ divides $D^N - 1$, we get, using the preceding equation and

(3.6-1),

$$\begin{aligned} s'(D) &= Dy(D) \bmod g(D) \\ &= \Big(D\big(q(D)g(D) + s(D)\big)\Big) \bmod g(D) \\ &= Ds(D) \bmod g(D). \end{aligned} \tag{3.6-2}$$

**Example 3.6-2.**

Consider the received sequence of Example 3.6-1. Its cyclic shift one position left is given by $\mathbf{y}'$, where $\mathbf{y}' = (0\ 0\ 1\ 1\ 1\ 1\ 1)$. Using the same circuit of the previous example, we find that the syndrome of $\mathbf{y}'$ is $\mathbf{s}' = (1\ 0\ 1)$ or $s'(D) = D^2 + 1$. Division of $y'(D)$ shows the same thing, that

$$\begin{aligned} y'(D) \bmod g(D) &= D^4 + D^3 + D^2 + D + 1 \bmod g(D) \\ &= D^2 + 1. \end{aligned}$$

Now compute $s'(D)$ using (3.6-2). Computing first $s(D)$ due to $\mathbf{y} = (1\ 0\ 0\ 1\ 1\ 1\ 1)$ yields $s(D) = D^2$, as in Example 3.6-1. Shifting $s(D)$ cyclically once through the circuit in Fig. 3.6-1 yields $D^2+1$ for the polynomial representing the contents of the shift register, which is also $s'(D)$.

We now illustrate how the property of cyclic codes given by (3.6-2) can be used in designing a decoder, called the *Meggitt decoder* [3-19]. Let a single error occur in transmission. Assume further that the error occurs in the position corresponding to the term $D^{N-1}$ of the received polynomial. (Later we will generalize the principle). Let $s(D)$ be computed using a syndrome computing circuit such as the one in Example 3.6-1. By appropriately combining the outputs of the various stages of the syndrome register using a logic gate, the gate's output can be made to be 1 whenever $y_{N-1}$ is in error. By adding the output of the gate to the bit in error, the error can be corrected. Let now an error occur in any other position of the received word, say position $D^l$. By cyclically shifting received word $N-l-1$ times, the erroneously received bit will appear in position $D^{N-1}$; at this time the output of the logic gate will be 1. The received bit in error is corrected by adding to it the logic gate's output. After cyclically shifting the received word $l+1$ more times, the corrected word may be delivered to the user.

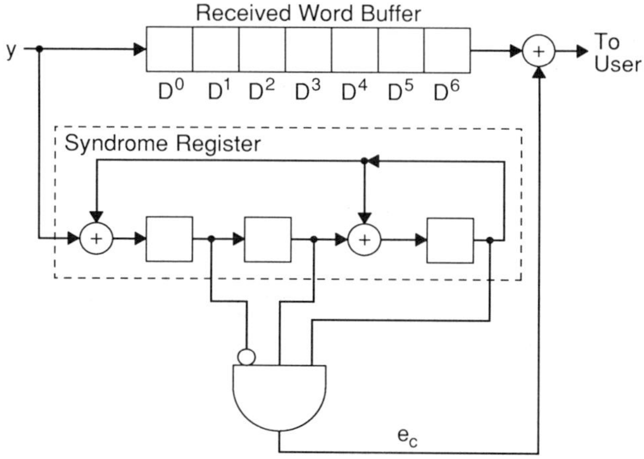


(a)

| Shift | Contents of Received Word Buffer $D^6\ D^5\ D^4\ D^3\ D^2\ D^1\ D^0$ | Contents of Syndrome Register $D^2\ D^1\ D^0$ | Delivered Bit |
|---|---|---|---|
| 7 | 1 0 0 0 0 1 1 | 1 0 1 | |
| 8 | 0 0 0 0 1 1 0 | 1 1 1 | 1 |
| 9 | 0 0 0 1 1 0 0 | 0 1 1 | 0 |
| 10 | 0 0 1 1 0 0 0 | 1 1 0 | 0 |
| 11 | 0 1 1 0 0 0 0 | 0 0 1 | 1 ← Corrected bit delivered |
| 12 | 1 1 0 0 0 0 0 | 0 1 0 | 0 |
| 13 | 1 0 0 0 0 0 0 | 1 0 0 | 1 |
| 14 | 0 0 0 0 0 0 0 | 1 0 1 | 1 |

(b)

**Figure 3.6-2.** (a) A Meggitt decoder for correcting single errors; (b) table illustrating the contents of the received word buffer and the syndrome register in (a) during shifts 7 to 14. The square boxes under the received word buffer column indicate how the erroneous bit moves to position $D^6$.

**Example 3.6-3: A Meggitt Decoder for the (7,4) Cyclic Binary Hamming Code.**

Let the received bit $y_6$ be in error. This implies that for the (7,4) cyclic code with $g(D) = D^3 + D^2 + 1$, the error vector $\mathbf{e}$ equals

(1 0 0 0 0 0 0) or the error polynomial $e(D)$ equals $D^6$. The corresponding syndrome is $s(D) = D^2 + D$. One can test for error $e(D) = D^6$ by an appropriate logic combination of the output of the various stages of the syndrome register. If the output of the logic gate is 1, then an error has occurred in position $D^6$ of the received word, which may then be inverted to yield the corrected bit.

Figure 3.6-2 shows the decoder used to correct a single error. The received word is shifted into the received word buffer and also into the syndrome register. After all the 7 bits of the received word are shifted into the two registers, the syndrome register contains the syndrome corresponding to the received word. Since position $D^6$ of the received word is in error, the syndrome register now contains $s(D) = D^2 + D$. The output of the AND gate is now 1. During the eighth shift, the AND gate's output is added to $y_6$, thus correcting the error in the received word. The corrected received bit $y_6$ is then delivered to the user. In six more shifts, the remaining bits of $\mathbf{y}$ are also delivered.

Let us now assume that $\mathbf{x} = (1\ 0\ 0\ 1\ 0\ 1\ 1)$ is the transmitted word and $y = (1\ 0\ 0\ 0\ 0\ 1\ 1)$ the received word. That is, $y_3$ is the erroneously received bit. Figure 3.6-2(b) shows the contents of the received word buffer and the syndrome register as the received word is shifted into these registers. After the fourteenth shift, the corrected word is completely delivered to the user.

## Meggitt Decoder for Multiple Error Correction

To design a Meggitt decoder to decode a multiple error correcting code, we employ a logic circuit that recognizes the syndromes of all correctable error patterns with an error in position $D^{N-1}$ of the received word. The output of the logic circuit is 1 whenever the syndrome stems from such an error pattern. As in the preceding example, the received bit in position $D^{N-1}$ may be corrected by adding to it the logic circuit's output. The partially corrected received word is then cyclically shifted once within the buffer. The contents of the syndrome register are then corrected so that they now represent the syndrome of the modified and cyclically shifted received word. To see how the contents may be corrected, note that after adding the logic circuit's output $e_c$ to $y_{N-1}$ and rotating the received word once, the contents of the received buffer are given by $y'(D)$, where

$$y'(D) = D\Big(y(D) + e_c D^{N-1}\Big) \bmod D^N - 1.$$

The syndrome corresponding to $y'(D)$ is

$$\begin{aligned} s'(D) &= y'(D) \bmod g(D) \\ &= Dy(D) \bmod g(D) + e_c D^N \bmod g(D) \\ &= Ds(D) \bmod g(D) + e_c, \end{aligned} \tag{3.6-3}$$

where we have made use of (3.6-2) and the fact that $D^N \bmod g(D)$ is 1. Equation (3.6-3) suggests that the syndrome register contents may be corrected by shifting the contents once, performing the mod $g(D)$ operation by using a suitable feedback connection, and adding $e_c$ to stage $D^0$ of the syndrome register. By repeated shifts of the contents of the received word buffer and syndrome register, any correctable error pattern may be corrected. After performing a total of $N$ such steps, the contents of the syndrome register will be zero if a correctable error has occurred in the channel. Otherwise, an uncorrectable error pattern has occurred.

**Example 3.6-4: A Meggitt Decoder for the Two-Error-Correcting (15,7) BCH Code.**

Consider the generator polynomial given by $g(D) = D^8 + D^7 + D^6 + D^4 + 1$. We will show in Section 3.8 that $g(D)$ generates the double error correcting (15,7) BCH code. There are fifteen possible error patterns with a 1 in position $D^{14}$. Of these, one pattern corresponds to a single error and the remaining fourteen to double errors. The error patterns and the corresponding syndromes are shown below.

| Error Pattern | Syndrome |
|---|---|
| 1 0 0 0 0 0 0 0 0 0 0 0 0 0 0 | 1 1 1 0 1 0 0 0 |
| 1 1 0 0 0 0 0 0 0 0 0 0 0 0 0 | 1 0 0 1 1 1 0 0 |
| 1 0 1 0 0 0 0 0 0 0 0 0 0 0 0 | 1 1 0 1 0 0 1 0 |
| 1 0 0 1 0 0 0 0 0 0 0 0 0 0 0 | 1 1 1 1 0 1 0 1 |
| 1 0 0 0 1 0 0 0 0 0 0 0 0 0 0 | 0 0 0 0 1 1 1 0 |
| 1 0 0 0 0 1 0 0 0 0 0 0 0 0 0 | 1 0 0 1 1 0 1 1 |
| 1 0 0 0 0 0 1 0 0 0 0 0 0 0 0 | 0 0 1 1 1 0 0 1 |
| 1 0 0 0 0 0 0 1 0 0 0 0 0 0 0 | 0 1 1 0 1 0 0 0 |
| 1 0 0 0 0 0 0 0 1 0 0 0 0 0 0 | 1 0 1 0 1 0 0 0 |
| 1 0 0 0 0 0 0 0 0 1 0 0 0 0 0 | 1 1 0 0 1 0 0 0 |
| 1 0 0 0 0 0 0 0 0 0 1 0 0 0 0 | 1 1 1 1 1 0 0 0 |
| 1 0 0 0 0 0 0 0 0 0 0 1 0 0 0 | 1 1 1 0 0 0 0 0 |
| 1 0 0 0 0 0 0 0 0 0 0 0 1 0 0 | 1 1 1 0 1 1 0 0 |
| 1 0 0 0 0 0 0 0 0 0 0 0 0 1 0 | 1 1 1 0 1 0 1 0 |
| 1 0 0 0 0 0 0 0 0 0 0 0 0 0 1 | 1 1 1 0 1 0 0 1 |

Let the all-zeros code word be transmitted and let the received word contain two errors in positions $D^{12}$ and $D^5$. Fig. 3.6-3 shows a Meggitt decoder for the (15,7) BCH code. The received polynomial $y(D)$ is shifted into the 15-bit register in fifteen shifts. The table traces the operation of the circuit after these take place. Decoding operations take place during the next set of fifteen shifts; the corrected word shifts out during these fifteen. Observe that the syndrome register contents are zeros since the error pattern is a correctable one.

The steps for a general Meggitt decoder follows. References are made to Fig. 3.6-3 in this algorithm, with the assumption that the received word buffer is an $N$-bit register and the syndrome register and circuit are those for the code being used.

**Meggitt Decoder**

**Step 1.** With gates 1, 2, and 3 closed and all the others open, shift the received word into the buffer and the syndrome register in $N$ shifts.

**Step 2.** With gates 1 and 2 open and 4 and 5 closed, repeat Steps 2a and 2b $N$ times.

> **Step 2a.** The circuit checks if the syndrome is due to one of the correctable error patterns and sets $e_c$ to 1 if it is and to 0 otherwise.
>
> **Step 2b.** Shift once the contents of both the received word buffer and the syndrome register. In the process, $e_c$ is added to the bit of the received word at position $D^{N-1}$ and to that of the syndrome at $D^0$, thus partially correcting the received word and modifying the syndrome.

**Step 3.** With all the gates open except gate 6, the corrected word is shifted out during the next $N$ shifts.

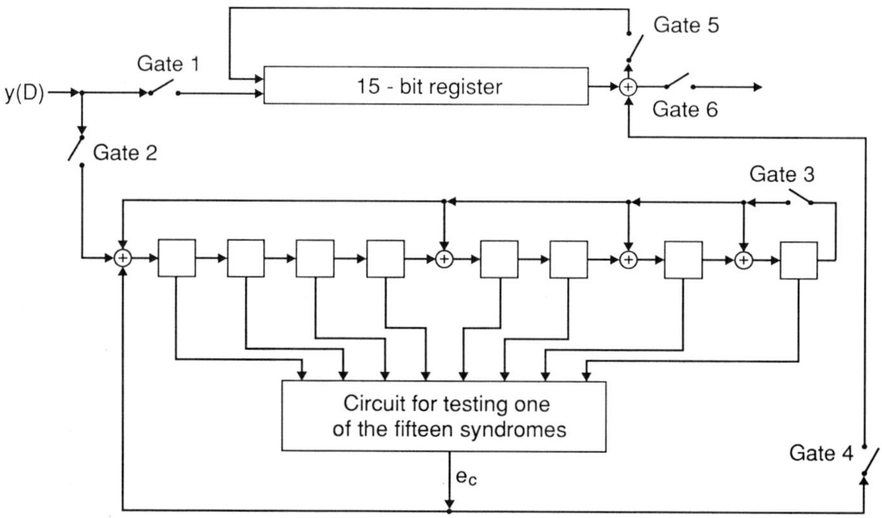


| Shift | Received Word Buffer | Syndrome Register | $e_c$ |
|---|---|---|---|
| 15 | 0 0 1 0 0 0 0 0 0 1 0 0 0 0 0 | 0 0 0 1 1 0 1 0 | 0 |
| 16 | 0 1 0 0 0 0 0 0 1 0 0 0 0 0 0 | 0 0 1 1 0 1 0 0 | 0 |
| 17 | 1 0 0 0 0 0 0 1 0 0 0 0 0 0 0 | 0 1 1 0 1 0 0 0 | 1 |
| 18 | 0 0 0 0 0 0 1 0 0 0 0 0 0 0 0 | 1 1 0 1 0 0 0 1 | 0 |
| 19 | 0 0 0 0 0 1 0 0 0 0 0 0 0 0 0 | 0 1 1 1 0 0 1 1 | 0 |
| 20 | 0 0 0 0 1 0 0 0 0 0 0 0 0 0 0 | 1 1 1 0 0 1 1 0 | 0 |
| 21 | 0 0 0 1 0 0 0 0 0 0 0 0 0 0 0 | 0 0 0 1 1 1 0 1 | 0 |
| 22 | 0 0 1 0 0 0 0 0 0 0 0 0 0 0 0 | 0 0 1 1 1 0 1 0 | 0 |
| 23 | 0 1 0 0 0 0 0 0 0 0 0 0 0 0 0 | 0 1 1 1 0 1 0 0 | 0 |
| 24 | 1 0 0 0 0 0 0 0 0 0 0 0 0 0 0 | 1 1 1 0 1 0 0 0 | 1 |
| 25 | 0 0 0 0 0 0 0 0 0 0 0 0 0 0 0 | 0 0 0 0 0 0 0 0 | 0 |

**Figure 3.6-3** A Meggitt decoder for the (15,7) binary BCH code generated by $g(D) = D^8 + D^7 + D^6 + D^4 + 1$.

## The Error-Trapping Decoder

The Meggitt decoder becomes impractical as the error correction capability of the code increases. However, with certain assumptions, it is possible to simplify its design. Let us assume that errors are confined to only the highest order information bits of the received polynomial, that is, positions $D^{N-1}, D^{N-2}, \ldots, D^K$ of $y(D)$. This implies that the error polynomial $e(D)$ is given by $\sum_{i=K}^{N-1} e_i D^i$. Let $y(D) = x(D) + e(D)$, where $x(D)$ is the transmitted word polynomial. Then

$$s(D) = y(D) \bmod g(D) = e(D) \bmod g(D).$$

Let $y'(D)$ refer to the polynomial $y(D)$ cyclically shifted $(N - K)$ times. The corresponding error polynomial $e'(D)$ is given by

$$e'(D) = e_{N-1}D^{N-K-1} + e_{N-2}D^{N-K-2} + \ldots + e_{K+1}D + e_K.$$

The syndrome corresponding to $y'(D)$ is

$$s'(D) = e'(D) \bmod g(D) = e'(D),$$

since $g(D)$ is a polynomial of degree $N - K$ and $e'(D)$ is of degree at most $N - K - 1$. Now the syndrome register contents are identical to the error pattern. The error pattern is said to be *trapped* in the syndrome register and hence the name *error-trapping decoder* [3-20], [3-21] for this scheme.

Now suppose that the errors are confined to $N - K$ consecutive positions of the received word, but not necessarily the highest order positions. It is possible to shift the contents of the syndrome register sufficiently many times so that the error pattern can be trapped in it.

How can one know when the error pattern gets trapped in the syndrome register? Assume that $t$ or fewer errors occur in $N-K$ consecutive locations. We can show that the the error pattern is trapped in the syndrome register only when the weight of the syndrome $w(\mathrm{s})$ is less than or equal to $t$. Let $e(D) = D^l e_1(D)$ represent the error pattern, where $e_1(D)$ is a polynomial of degree of less than $N - K$, with $t$ or fewer error terms. Since $s(D) = e(D) \bmod g(D)$, we have

$$D^l e_1(D) = e(D) = q(D)g(D) + s(D).$$

We see that $g(D)$ is a factor of $s(D) + D^l e_1(D)$ and hence $s(D) + D^l e_1(D)$ is a code word. For a $t$-error-correcting code, $s(D) + D^l e_1(D)$ must have weight $2t + 1$ or greater if $s(D)$ is different from $D^l e_1(D)$; $w(\mathrm{S})$ is then $t + 1$ or greater. $w(\mathrm{S})$ is $t$ or less only when $s(D)$ equals $D^l e_1(D)$. Thus the test for the error-trapping condition is to check if the weight of the syndrome is $t$ or less.

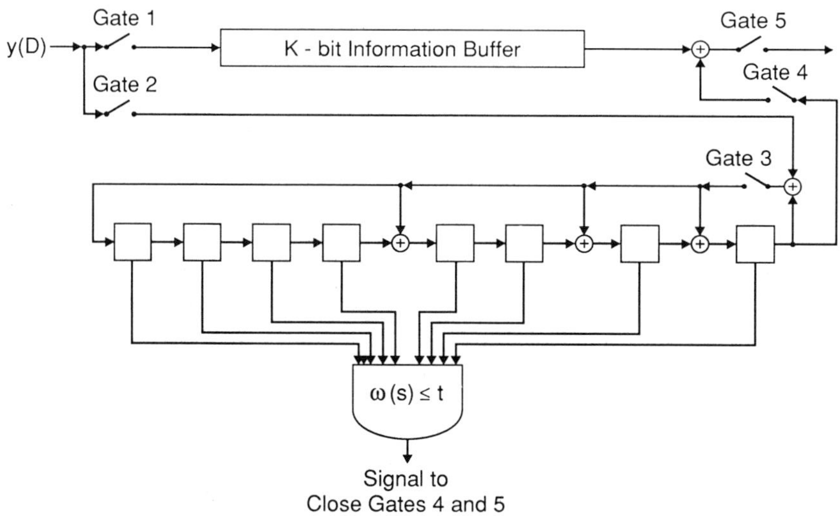


**Figure 3.6-4.** An error-trapping decoder for the (15, 7) BCH code generated by $g(D) = D^8 + D^7 + D^6 + D^4 + 1$.

**Example 3.6-5. The Error-Trapping Decoder for the (15, 7) Double-Error-Correcting BCH Code.**

Consider the (15, 7) double-error-correcting BCH code of the previous example. Figure 3.6-4 shows the error-trapping decoder for this code.

In Fig. 3.6-4, the received polynomial is shifted into the syndrome register from the right end. This is equivalent to premultiplying $y(D)$ by $D^{N-K}$. Initially, gates 1, 2, and 3 are closed and gates 4 and 5 open. After shifting the highest order $K$ bits of the received word into the $K$-bit buffer, gate 1 opened. After all the $N$ bits of the received word are shifted into the syndrome register, gate 2 is opened. The syndrome register now contains the syndrome corresponding to $D^{N-K}y(D)$.

Let the all-zeros code word be transmitted and let the received word contain ones in positions $D^{14}$ and $D^9$, that is y = (1 0 0 0 0 1 0 0 0 0 0 0 0 0 0). Clearly two errors have occurred in the highest $N - K$ (=8) bits. After shifting y into the syndrome register, the error pattern is trapped in it as shown in the following table. Observe that $w(s) \leq 2$. Now with gates 4 and 5 closed and all the others open, the corrected information is put out.

| Shift | Input | Shift Register Contents |
|---|---|---|
| 0 | | 0 0 0 0 0 0 0 0 |
| 1 | 1 | 1 1 0 1 0 0 0 1 |
| 2 | 0 | 0 1 1 1 0 0 1 1 |
| 3 | 0 | 1 1 1 0 0 1 1 0 |
| 4 | 0 | 0 0 0 1 1 1 0 1 |
| 5 | 0 | 0 0 1 1 1 0 1 0 |
| 6 | 1 | 1 0 1 0 0 1 0 1 |
| 7 | 0 | 1 0 0 1 1 0 1 1 |
| 8 | 0 | 1 1 1 0 0 1 1 1 |
| 9 | 0 | 0 0 0 1 1 1 1 1 |
| 10 | 0 | 0 0 1 1 1 1 1 0 |
| 11 | 0 | 0 1 1 1 1 1 0 0 |
| 12 | 0 | 1 1 1 1 1 0 0 0 |
| 13 | 0 | 0 0 1 0 0 0 0 1 |
| 14 | 0 | 0 1 0 0 0 0 1 0 |
| 15 | 0 | 1 0 0 0 0 1 0 0 |
| | | $w(s) \leq 2$ |
| | | error trapped |

Now consider again the all-zeros transmitted word, but with $y(D)$ equal to $D^{12}+D^5$. This is exactly the situation considered in Example 3.6-4. After shifting $y(D)$ into the syndrome register, the register contains the syndrome of $D^{N-K}y(D)$, equal to $D^6+D^4+D^2$. After two more shifts $w(\mathrm{s})$ becomes 2 for the first time, and the error pattern, $\mathbf{e} = (1\ 0\ 0\ 0\ 0\ 0\ 0\ 1)$, is trapped in the syndrome register. After shifting $y(D)$ into the register, the number of shifts required to make $w(\mathrm{s})$ less than or equal to $t$ (in this example 2 shifts) gives the number of highest order bits of $y(D)$ received error free. Now gate 5 is closed and $y_{14}$ and $y_{13}$ are read out. Gate 4 is then closed and the contents of the syndrome register and the remaining contents of the $K$-bit buffer are read out and the corrected information bits are put out. Observe that the contents of stages $D^7$ and $D^0$ (both equal to 1) of the syndrome register correct, respectively, $y_{12}$ and $y_5$. After shifting out the contents of the $K$-bit information buffer, the syndrome register contains the error pattern corresponding to the parity-check bits, which may be discarded.

Now we will consider the last possible situation that may occur, an error pattern that is not confined to $N-K$ consecutive locations. Consider again the all-zeros transmitted word, but with $y(D)$ equal to $D^{13}+D^4$, which corresponds to the error pattern e = (0 1 0 0 0 0 0 0 0 0 0 1 0 0 0 0). This situation can be looked upon as an end-around error pattern. We view positions $D^{14}$ and $D^0$ to be adjacent to each other. The structure of cyclic codes enables us to take this view point. Armed with this insight, we observe that the error pattern is now confined to $N - K$ or fewer consecutive positions.

After shifting the received sequence into the syndrome register in 15 shifts, the register contains $D^6 + D^5 + D^4 + D^3 + D$. After $K$ more shifts, that is after a total of 22 shifts, the syndrome register contains $D^6 + D^5 + D^3$ and so its weight has not gone down to 2 or less. However, after two more shifts, the syndrome register contains the end-around error pattern (0 1 0 0 0 0 1). In general for the end-around case, the weight of the syndrome register contents will become $t$ or less in $K + l$ shifts after the entire received sequence has been shifted in, where $1 \leq l \leq (N - K)$.

The low order $l$ stages of the register contain the error pattern corresponding to the $l$ high order coefficients of the information polynomial. The remaining $N - K - l$ stages of the syndrome register contain the error pattern corresponding to the parity-check bits, which is discarded. In our example, $l$ equals 2. Hence, we discard the contents of the highest 6 stages. Gates 4 and 5 are then closed and all the other gates are opened. The $l$ high order bits of the information bits are corrected using the error pattern trapped in the syndrome register. Observe that the circuit corrects $y_{13}$ by adding $s_0$ (= 1) to it. Now gate 4 is opened, and the remaining $K - l$ error free bits of the information buffer are shifted out.

The reader can work out the details of this example to gain a better understanding of the workings of the error-trapping decoder. We have omitted a formal listing of the algorithm and instead invite the reader to formulate it.

## 3.7 ALGEBRA OF FINITE FIELDS

To study the structure and properties of BCH codes in Section 3.8, we need to introduce some more properties of finite fields. Specifically, we introduce the field GF($2^m$). The field GF($2^m$) is called an *extension field* of GF(2). GF($2^m$) contains GF(2) as a *subfield.*

## Properties of Galois Field GF($2^m$)

In our discussion on cyclic codes in the preceding two sections, we have seen the usefulness of polynomial representation for vectors and of the reduction of these polynomials modulo a polynomial. Similarly, it is useful to view the $2^m$ elements of GF($2^m$) as all the polynomials of degree $m-1$ or less over GF(2). Recall from Appendix B that two operators multiplication and addition are defined in a field and that the field elements must satisfy certain properties with respect to these operators. We see that addition of two polynomials of degree $m-1$ or less still results in another of degree $m-1$ or less and is therefore a valid field element of GF($2^m$). However, multiplication of two such polynomials may result in a polynomial of degree greater than $m-1$. The resulting polynomial is therefore not a field element. If we reduce the product by performing the modulo operation using a degree-$m$ polynomial $p(D)$, the resulting polynomial will have degree $m-1$ or less. If $p(D)$ is not chosen carefully, the requirements of a field, such as the existence of a unique 0 element and a unique multiplicative inverse for each field element, will not be satisfied. If we choose $p(D)$ to be an *irreducible* polynomial, then a field will be formed. A degree-$m$ polynomial is irreducible if it is not divisible by any other polynomial of degree $m-1$ or less. We now illustrate the usefulness of such irreducible polynomials to construct finite fields.

Example 3.7-1: Construction of GF($2^3$) Using $D^3 + D^2 + 1$ over GF(2).

We can verify that $p(D) = D^3 + D^2 + 1$ is irreducible as follows. If a degree-2 polynomial were to be one of the factors of $p(D)$, which is of degree 3, the other factor must be a degree-1 polynomial. The polynomials of degree 1 over GF(2), $D$ and $D+1$, are not factors since $p(0) = p(1) = 1$. Hence $p(D)$ is irreducible.

We begin construction with the two elements 0 and 1, since any field must have these, and another element $\alpha$ that we will define soon. Since multiplication of two field elements results in another field element, we raise $\alpha$ to its successive powers till no new field element can be generated. If $\alpha$ is chosen carefully, we can generate all the non-zero field elements by raising it to its successive powers.

We choose $\alpha$ to be a root of $p(D)$. This implies that $\alpha$ satisfies the equation $\alpha^3 + \alpha^2 + 1 = 0$ or $\alpha^3 = \alpha^2 + 1$. We can now make repeated use of this to construct field elements. For example,

$\alpha^5$ can be reduced as follows:

$$\begin{aligned}\alpha^5 &= \alpha^2\alpha^3 = \alpha^2(\alpha^2+1)\\ &= \alpha^4+\alpha^2 = \alpha\alpha^3+\alpha^2\\ &= \alpha(\alpha^2+1)+\alpha^2 = \alpha^3+\alpha+\alpha^2\\ &= \alpha^2+1+\alpha+\alpha^2 = \alpha+1.\end{aligned}$$

Table 3.7-1 lists the field elements in three different forms; as powers of $\alpha$, as polynomials over GF(2) of degree three or less, and as vectors of 3-tuples. Table 3.7-2 provides addition and multiplication tables in GF($2^3$).

**Table 3.7-1.** THREE DIFFERENT REPRESENTATIONS OF THE FIELD ELEMENTS OF GF($2^3$).

| Power of $\alpha$ | Polynomial | 3-Tuple |
|---|---|---|
| 0 | 0 | 0 0 0 |
| $\alpha^0$ | 1 | 0 0 1 |
| $\alpha^1$ | $\alpha$ | 0 1 0 |
| $\alpha^2$ | $\alpha^2$ | 1 0 0 |
| $\alpha^3$ | $\alpha^2+1$ | 1 0 1 |
| $\alpha^4$ | $\alpha^2+\alpha+1$ | 1 1 1 |
| $\alpha^5$ | $\alpha+1$ | 0 1 1 |
| $\alpha^6$ | $\alpha^2+\alpha$ | 1 1 0 |
| $\alpha^7$ | 1 | 0 0 1 |
| | The cycle repeats | |

We define a *primitive element* as a field element that generates all the nonzero field elements as its successive powers. The field element $\alpha$ in Table 3.7-1 is a primitive element.

A *primitive polynomial* is defined as an irreducible polynomial that has a primitive element as its root. For example, $D^3+D^2+1$ is a primitive polynomial.

The *order* of an element $\beta$ in GF($2^m$) is the smallest positive integer $j$ such that $\beta^j = 1$. The order $\alpha$ in GF($2^3$) of Table 3.7-1 is 7.

We now state several properties of GF($2^m$) necessary for formulating the BCH codes. We state most of these properties without proof but provide examples to illustrate them.

**Table 3.7-2(a).** ADDITION IN GF($2^3$).

| $+$ | $0$ | $1$ | $\alpha$ | $\alpha^2$ | $\alpha^3$ | $\alpha^4$ | $\alpha^5$ | $\alpha^6$ |
|---|---|---|---|---|---|---|---|---|
| $0$ | $0$ | $1$ | $\alpha$ | $\alpha^2$ | $\alpha^3$ | $\alpha^4$ | $\alpha^5$ | $\alpha^6$ |
| $1$ | $1$ | $0$ | $\alpha^5$ | $\alpha^3$ | $\alpha^2$ | $\alpha^6$ | $\alpha$ | $\alpha^4$ |
| $\alpha$ | $\alpha$ | $\alpha^5$ | $0$ | $\alpha^6$ | $\alpha^4$ | $\alpha^3$ | $1$ | $\alpha^2$ |
| $\alpha^2$ | $\alpha^2$ | $\alpha^3$ | $\alpha^6$ | $0$ | $1$ | $\alpha^5$ | $\alpha^4$ | $\alpha$ |
| $\alpha^3$ | $\alpha^3$ | $\alpha^2$ | $\alpha^4$ | $1$ | $0$ | $\alpha$ | $\alpha^6$ | $\alpha^5$ |
| $\alpha^4$ | $\alpha^4$ | $\alpha^6$ | $\alpha^3$ | $\alpha^5$ | $\alpha$ | $0$ | $\alpha^2$ | $1$ |
| $\alpha^5$ | $\alpha^5$ | $\alpha$ | $1$ | $\alpha^4$ | $\alpha^6$ | $\alpha^2$ | $0$ | $\alpha^3$ |
| $\alpha^6$ | $\alpha^6$ | $\alpha^4$ | $\alpha^2$ | $\alpha$ | $\alpha^5$ | $1$ | $\alpha^3$ | $0$ |

**Table 3.7-2(b).** MULTIPLICATION IN GF($2^3$).

| $\cdot$ | $0$ | $1$ | $\alpha$ | $\alpha^2$ | $\alpha^3$ | $\alpha^4$ | $\alpha^5$ | $\alpha^6$ |
|---|---|---|---|---|---|---|---|---|
| $0$ | $0$ | $0$ | $0$ | $0$ | $0$ | $0$ | $0$ | $0$ |
| $1$ | $0$ | $1$ | $\alpha$ | $\alpha^2$ | $\alpha^3$ | $\alpha^4$ | $\alpha^5$ | $\alpha^6$ |
| $\alpha$ | $0$ | $\alpha$ | $\alpha^2$ | $\alpha^3$ | $\alpha^4$ | $\alpha^5$ | $\alpha^6$ | $1$ |
| $\alpha^2$ | $0$ | $\alpha^2$ | $\alpha^3$ | $\alpha^4$ | $\alpha^5$ | $\alpha^6$ | $1$ | $\alpha$ |
| $\alpha^3$ | $0$ | $\alpha^3$ | $\alpha^4$ | $\alpha^5$ | $\alpha^6$ | $1$ | $\alpha$ | $\alpha^2$ |
| $\alpha^4$ | $0$ | $\alpha^4$ | $\alpha^5$ | $\alpha^6$ | $1$ | $\alpha$ | $\alpha^2$ | $\alpha^3$ |
| $\alpha^5$ | $0$ | $\alpha^5$ | $\alpha^6$ | $1$ | $\alpha$ | $\alpha^2$ | $\alpha^3$ | $\alpha^4$ |
| $\alpha^6$ | $0$ | $\alpha^6$ | $1$ | $\alpha$ | $\alpha^2$ | $\alpha^3$ | $\alpha^4$ | $\alpha^5$ |

**Property 1:** Any degree-$m$ irreducible polynomial over GF(2) divides $D^{2^m-1} - 1$.

**Example 3.7-2.**

We have seen in Example 3.7-1 that the polynomial $p(D) = D^3 + D^2 + 1$ is irreducible. It is also primitive since, from Table 3.7-1, one of its roots $\alpha$ generates all the nonzero field elements of GF($2^3$). We have already seen in Example 3.5-1 that $p(D)$ divides $D^{2^3-1} - 1 = D^7 - 1$.

**Property 2:** A primitive element $\alpha$ satisfies the equation $\alpha^{2^m-1} = 1$.

This is easily shown as follows. Any primitive polynomial $p(D)$ with a primitive element $\alpha$ as its root divides $D^{2^m-1} - 1$ (Property 1). Hence

$$D^{2^m-1} - 1 = q(D)p(D), \tag{3.7-1}$$

where $q(D)$ is the quotient polynomial due to division of $D^{2^m-1} - 1$ by $p(D)$. Putting $D = \alpha$ in (3.7-1) yields

$$\alpha^{2^m-1} - 1 = q(\alpha)p(\alpha) = q(\alpha) \cdot 0 = 0$$

or

$$\alpha^{2^m-1} = 1. \tag{3.7-2}$$

Equation (3.7-2) is verified to be true from Table 3.7-1 since $\alpha^7 = 1$.

**Property 3:** All the nonzero elements of GF($2^m$) form all the $2^m - 1$ roots of $D^{2^m-1} - 1$.

Example 3.7-3.

$D^7 - 1$ can be factored into $D^7 - 1 = (D+1)(D^3+D^2+1)(D^3+D+1)$. The roots of $D^3 + D^2 + 1$ are $\alpha, \alpha^2$ and $\alpha^4$ since

$$\begin{aligned}
&(D+\alpha)(D+\alpha^2)(D+\alpha^4) \\
&\quad = D^3 + D^2(\alpha^4 + \alpha^2 + \alpha) + D(\alpha^3 + \alpha^6 + \alpha^5) + \alpha^7 \\
&\quad = D^3 + D^2\big\{(111) + (100) + (010)\big\} \\
&\qquad\qquad + D\big\{(101) + (110) + (011)\big\} + (001) \\
&\quad = D^3 + D^2(001) + D(000) + (001) \\
&\quad = D^3 + D^2 + 1.
\end{aligned}$$

We can similarly verify that the roots of $D^3 + D + 1$ are $\alpha^3, \alpha^5$ and $\alpha^6$, and the root of $D + 1$ is 1. Thus all the seven nonzero elements of GF($2^3$) form all of the roots of $D^7 - 1$.

**Property 4:** All the $2^m$ elements of GF($2^m$) form all the $2^m$ roots of $D^{2^m} + D$.

This follows from Property 3 and the fact that the zero element of GF($2^m$) is the root of $D$.

**Property 5:** If $f(D)$ is any polynomial over GF(2), then

$$f^2(D) = f(D^2).$$

**Example 3.7-4.**

Two examples to verify this property are:

$$\begin{aligned}(1+D)^2 &= 1+D+D+D^2 = 1+D^2;\\ (1+D+D^2)^2 &= [1+(D+D^2)]^2\\ &= 1+(D+D^2)+(D+D^2)+(D+D^2)^2\\ &= 1+(D+D^2)^2\\ &= 1+D^2(1+D)^2\\ &= 1+D^2(1+D^2)\\ &= 1+D^2+D^4.\end{aligned}$$

**Property 6:** Let $f(D)$ be an irreducible polynomial of degree $m$ and $\beta$ be a root of $f(D)$. Then all of the $m$ roots of $f(D)$ are given by $\beta, \beta^2, \beta^{2^2}, \beta^{2^3}, \ldots, \beta^{2^{m-1}}$. The elements $\beta^{2^i}$, $i = 1, 2, \ldots, m-1$, are called conjugates of $\beta$.

This is an important property that will be used in the construction of BCH codes.

**Example 3.7-5.**

Recall that the roots of $D^3+D^2+1$ are $\alpha, \alpha^2$, and $\alpha^4$. We can easily verify that $D^3+D+1$ is also a primitive polynomial and its roots are $\alpha^3$, $(\alpha^3)^2 = \alpha^6$ and $(\alpha^3)^{2^2} = \alpha^{12} = \alpha^7\alpha^5 = \alpha^5$, since $\alpha^7 = 1$.

**Property 7:** Let $\phi(D)$ denote the smallest degree polynomial over GF(2) of which a field element $\beta$ in GF($2^m$) is a root. $\phi(D)$ is called the *minimal polynomial* of $\beta$. $\phi(D)$ is irreducible.

From Property 4, any field element $\beta$ is a root of $D^{2^m}+D$. There may be another polynomial of degree less than $2^m$ of which $\beta$ is a root. $\phi(D)$ is such a polynomial of least degree. $D$ and $D+1$ are, respectively, the minimal polynomials of 0 and 1.

Properties 3, 6, and 7 together provide a way to factor $D^{2^m-1}-1$ into a product of irreducible polynomials (see Problem 3-23).

**Property 8:** If $\phi(D)$ is the minimal polynomial of a field element $\beta$ in GF($2^m$) and $f(D)$ is any polynomial over GF(2) with $\beta$ as a root, then $\phi(D)$ divides $f(D)$.

Example 3.7-6.

Consider the polynomial $f(D) = D^4 + D^2 + D + 1$. One can easily verify that $\alpha$ from GF($2^3$) given in Table 3.7-1 is a root of $f(D)$:

$$\begin{aligned} f(\alpha) &= \alpha^4 + \alpha^2 + \alpha + 1 \\ &= (\alpha^2 + \alpha + 1) + (\alpha^2 + \alpha + 1) \\ &= 0. \end{aligned}$$

The minimal polynomial of $\alpha$ is $\phi(D) = D^3 + D^2 + 1$ and $\phi(D)$ divides $f(D)$ since $f(D) = (D+1)\phi(D)$.

**Property 9:** If $f(D)$ is an irreducible polynomial over GF(2) with $\beta$ in GF($2^m$) as a root, then $f(D)$ is also the minimal polynomial of $\beta$.

From Properties 6 and 9, it follows that if $\phi(D)$ is the minimal polynomial of $\beta$ in GF($2^m$), then $\beta^2, \beta^{2^2}, \ldots, \beta^{2^{\ell-1}}$ are all the distinct conjugates of $\beta$, where $\ell$ is the smallest integer such that $\beta^{2^\ell} = \beta$. From Properties 3 and 4, $\beta^{2^m} = \beta$, and hence $\ell \leq m$. $\beta$ and all its conjugates form the roots of $\phi(D)$. $\phi(D)$ can therefore be formulated as the minimal polynomial defined in Property 10.

**Property 10:** Let $\beta$ be an element in GF($2^m$) and $\beta^2, \beta^{2^2}, \ldots, \beta^{2^{\ell-1}}$ be all its distinct conjugates. Then the minimal polynomial of $\beta$ is given by

$$\phi(D) = \prod_{i=0}^{\ell-1} \left(D + \beta^{2^i}\right). \tag{3.7-3}$$

Example 3.7-7.

All the distinct conjugates of $\beta = \alpha^3$, a field element of GF($2^3$) in Table 3.7-1, are $\beta^2 = \alpha^6$ and $\beta^{2^2} = \alpha^{12} = \alpha^5$. Note that $\alpha^{2^3} = \alpha^{24} = \alpha^3 = \beta$. Thus

$$\begin{aligned} \phi(D) &= (D + \alpha^3)(D + \alpha^6)(D + \alpha^5) \\ &= D^3 + D^2(\alpha^3 + \alpha^6 + \alpha^5) + D(\alpha^9 + \alpha^{11} + \alpha^8) + (\alpha^{14}) \\ &= D^3 + D + 1. \end{aligned}$$

From (3.7-3), the degree of $\phi(D)$, the minimal polynomial of $\beta$ in GF($2^m$), is $\ell$, where $\ell$ is the number of terms of the form $(D + \beta^{2^i})$ on the right side. Note that $\ell$ is the smallest integer such that $\beta^{2^\ell} = \beta$.

Equipped with some knowledge of the intricacies of GF($2^m$), we are now ready to take on the structure and properties of BCH codes.

## 3.8 THE BCH CODES

The BCH codes were discovered by Hocquenghem [3-22] and independently by Bose and Chaudhuri [3-23] in 1959 and 1960, respectively. The important sub-class of BCH codes, the Reed-Solomon Codes, were proposed by Reed and Solomon [3-24] in 1960. Many decoding algorithms for BCH codes are known, the most important of these being the iterative algorithm due to Berlekamp [3-25], a related algorithm due to Massey [3-26], and the search algorithm due to Chien [3-27].

### The Structure of BCH Codes

The BCH codes are defined by their generator polynomial $g(D)$. Let $\alpha$ be a primitive element in GF$(2^m)$. Let $\phi_{\alpha^i}(D)$, $i = 1, 2, \ldots, 2t$, denote the minimal polynomials of $\alpha^i$. The generator polynomial $g(D)$ of a $t$-error-correcting BCH code is defined as

$$g(D) = \text{LCM}\left[\phi_{\alpha^1}(D), \phi_{\alpha^2}(D), \ldots, \phi_{\alpha^{2t}}(D)\right], \tag{3.8-1}$$

where LCM stands for least common multiple. Since $\alpha^i$ and $(\alpha^i)^2 = \alpha^{2i}$ are roots of the same minimal polynomial (Property 10, Section 3.7) we rewrite $g(D)$ as

$$g(D) = \text{LCM}\left[\phi_{\alpha^1}(D), \phi_{\alpha^3}(D), \ldots, \phi_{\alpha^{2t-1}}(D)\right]. \tag{3.8-2}$$

Since the minimal polynomial of a field element in GF$(2^m)$ has degree at most $m$, the generator polynomial of $g(D)$ in (3.8-2) has degree at most $mt$. That is, a $t$-error-correcting BCH code requires no more than $mt$ parity check bits. The block length of this code is the order of the primitive element used in (3.8-2), which is $2^m - 1$. Since a property of algebra states that for any $m$, there is at least one irreducible polynomial that is primitive, we can state the following:

*For any $m$ ($m \geq 3$) and $t < \frac{N}{2}$, there exists a binary $t$-error-correcting BCH code with $N = 2^m - 1$, $N - K \leq mt$ ($K \geq 2^m - 1 - mt$) and $d_{\min} \geq 2t+1$.*

The quantity $2t + 1$ is often called the *designed distance*. The bound above on minimum distance, guaranteeing a minimum distance at least equal to the designed distance is known as the *BCH bound.*

Table 3.8-1 lists the parameters $N, K, t$, and the generator polynomials for some binary BCH codes generated by primitive elements. The BCH codes defined by (3.8-2) are called *primitive* BCH codes. In the examples to follow, we will need the Galois field GF$(2^4)$ constructed in Table 3.8-2.

**Table 3.8-1** PARAMETERS AND GENERATOR POLYNOMIALS FOR SOME BCH CODES GENERATED BY PRIMITIVE POLYNOMIALS; FOR $N \geq 63$, ONLY THOSE CODES WITH RATE ONE-HALF OR GREATER ARE LISTED; POLYNOMIALS ARE LISTED IN OCTAL NOTATION; FOR EXAMPLE, FOR THE (15,11) CODE, $g(D) = (23)_8 = (010011) = D^4 + D + 1$.

| N | K | t | g (D) |
|---|---|---|---|
| 7 | 4 | 1 | 13 |
| 15 | 11 | 1 | 23 |
| | 7 | 2 | 721 |
| | 5 | 3 | 2467 |
| 31 | 26 | 1 | 45 |
| | 21 | 2 | 3551 |
| | 16 | 3 | 107657 |
| | 11 | 5 | 5423325 |
| | 6 | 7 | 313365047 |
| 63 | 57 | 1 | 103 |
| | 51 | 2 | 12471 |
| | 45 | 3 | 1701317 |
| | 39 | 4 | 166623567 |
| | 36 | 5 | 1033500423 |
| 127 | 120 | 1 | 211 |
| | 113 | 2 | 41567 |
| | 106 | 3 | 11554743 |
| | 99 | 4 | 3447023271 |
| | 92 | 5 | 624730022327 |
| | 85 | 6 | 130704476322273 |
| | 78 | 7 | 26230002166130115 |
| | 71 | 9 | 6255010713253127753 |
| | 64 | 10 | 1206534025570773100045 |
| 255 | 247 | 1 | 435 |
| | 239 | 2 | 267543 |
| | 231 | 3 | 156720665 |
| | 223 | 4 | 75626641375 |
| | 215 | 5 | 23157564726421 |
| | 207 | 6 | 16176560567636227 |
| | 199 | 7 | 7633031270420722341 |
| | 191 | 8 | 2663470176115333714567 |
| | 187 | 9 | 52755313540001322236351 |
| | 179 | 10 | 22624710717340432416300455 |
| | 171 | 11 | 15416214212342356077061630637 |
| | 163 | 12 | 7500415510075602551574724514601 |
| | 155 | 13 | 3757513005407665015722506464677633 |
| | 147 | 14 | 1642130173537165525304165305441011711 |
| | 139 | 15 | 461401732060175561570722730247453567445 |
| | 131 | 18 | 215713331471510151261250277442142024165471 |

**Example 3.8-1: The (7,4) Single-Error-Correcting BCH Code.**

From (3.8-2), a single error correcting BCH code is generated by

$$g(D) = \phi_\alpha(D)$$

From Section 3.7, we know that the minimal polynomial of the prim-

Table 3.8-2 THE FIELD GF($2^4$) GENERATED BY $D^4 + D + 1$.

| Power of $\alpha$ | Polynomial of $\alpha$ | 4-tuple |
|---|---|---|
| 0 | 0 | 0000 |
| $\alpha^0$ | 1 | 0001 |
| $\alpha^1$ | $\alpha$ | 0010 |
| $\alpha^2$ | $\alpha^2$ | 0100 |
| $\alpha^3$ | $\alpha^3$ | 1000 |
| $\alpha^4$ | $\alpha+1$ | 0011 |
| $\alpha^5$ | $\alpha^2+\alpha$ | 0110 |
| $\alpha^6$ | $\alpha^3+\alpha^2$ | 1100 |
| $\alpha^7$ | $\alpha^3+\alpha+1$ | 1011 |
| $\alpha^8$ | $\alpha^2+1$ | 0101 |
| $\alpha^9$ | $\alpha^3+\alpha$ | 1010 |
| $\alpha^{10}$ | $\alpha^2+\alpha+1$ | 0111 |
| $\alpha^{11}$ | $\alpha^3+\alpha^2+\alpha$ | 1110 |
| $\alpha^{12}$ | $\alpha^3+\alpha^2+\alpha+1$ | 1111 |
| $\alpha^{13}$ | $\alpha^3+\alpha^2+1$ | 1101 |
| $\alpha^{14}$ | $\alpha^3+1$ | 1001 |

itive element $\alpha$ in GF($2^3$) (Table 3.7-1) is $\phi_\alpha(D)$, where $\phi_\alpha(D) = D^3 + D^2 + 1$. Recall from Section 3.6 that $D^3 + D^2 + 1$ generates the (7, 4) single-error-correcting Hamming code. In fact, if $\alpha$ is a primitive element in GF($2^m$), the generator polynomial for the single error correcting BCH code is $\phi(D)$, which is of degree $m$; that is, $N - K = m$. Since the block length $N$ is the order of the primitive element, $N$ equals $2^m - 1$. The $(N, K)$ single-error-correcting BCH code is actually the $(2^m - 1, 2^m - 1 - m)$ single-error-correcting Hamming code.

**Example 3.8-2: The (15, 7) 2-Error Correcting BCH Code.**

From (3.8-2), $g(D) = \text{LCM}\left[\phi_\alpha(D), \phi_{\alpha^3}(D)\right]$. Note that since GF($2^4$) in Table 3.8-2 is generated by the primitive polynomial $D^4 + D + 1$ with $\alpha$ as a root, $\phi_\alpha(D) = D^4 + D + 1$. To determine $\phi_{\alpha^3}(D)$, we note that all the distinct conjugates of $\alpha^3$ are given by $(\alpha^3)^2 = \alpha^6, (\alpha^3)^{2^2} = \alpha^{12}$ and $(\alpha^3)^{2^3} = \alpha^{24} = \alpha^9$. So $\alpha^3, \alpha^6, \alpha^9$ and $\alpha^{12}$ are the roots of $\phi_{\alpha^3}(D)$. Hence $\phi_{\alpha^3}(D)$ is given by

$$\begin{aligned}\phi_{\alpha^3}(D) &= (D + \alpha^3)(D + \alpha^6)(D + \alpha^9)(D + \alpha^{12}) \\ &= D^4 + D^3 + D^2 + D + 1,\end{aligned}$$

where we make use of Table 3.8-2 to simplify expressions involving sums of powers of $\alpha$. So

$$\begin{aligned} g(D) &= \text{LCM}\left\{\phi_{\alpha}(D), \phi_{\alpha^3}(D)\right\} \\ &= \phi_{\alpha}(D)\phi_{\alpha^3}(D) \\ &= D^8 + D^7 + D^6 + D^4 + 1. \end{aligned}$$

Note that the weight of $g(D)$ is 5 and hence $d_{\min}$ is 5.

Example 3.8-3: The (15, 5) Triple-Error-Correcting BCH Code.

The generator polynomial of the code is

$$g(D) = \text{LCM}\left[\phi_{\alpha}(D), \phi_{\alpha^3}(D), \phi_{\alpha^5}(D)\right].$$

$\phi_{\alpha^5}(D)$ has roots $\alpha^5$ and $\alpha^{10}$ and so

$$\begin{aligned} \phi_{\alpha^5}(D) &= (D+\alpha^5)(D+\alpha^{10}) \\ &= D^2 + D + 1. \end{aligned}$$

To determine the least common multiple, we enumerate the roots of the polynomials $\phi(D), \phi_{\alpha^3}(D)$, and $\phi_{\alpha^5}(D)$ to identify the common roots. The roots of $\phi_{\alpha}(D)$ are $\alpha, \alpha^2, \alpha^4$ and $\alpha^8$. The roots of $\phi_{\alpha^3}(D)$ are $\alpha^3, \alpha^6, \alpha^9$ and $\alpha^{12}$ and those of $\phi_{\alpha^5}(D)$ are $\alpha^5$ and $\alpha^{10}$. Since there are no common roots between these, the generator polynomial is

$$\begin{aligned} g(D) &= \phi_{\alpha}(D)\phi_{\alpha^3}(D)\phi_{\alpha^5}(D) \\ &= (D^4 + D + 1)(D^4 + D^3 + D^2 + D + 1)(D^2 + D + 1) \\ &= D^{10} + D^8 + D^5 + D^4 + D^2 + D + 1. \end{aligned}$$

Note the number of parity bits is given by $N - K$, the degree of $g(D)$; that is, $N - K = 10 < mt = 12$. Since $g(D)$ has consecutive powers of $\alpha$ as roots, $2t$ is 6, and hence $d_{\min}$ is at least $2t + 1 = 7$. $g(D)$ has weight 7 and so $d_{\min}$ is exactly 7. The actual minimum distance of a code may exceed the designed distance.

## Parity-Check Matrix for BCH Codes

We now derive a parity-check matrix for BCH codes. Let $x(D) = x_{N-1}D^{N-1} + x_{N-2}D^{N-2} + \cdots + x_1 D + x_0$ be a code polynomial and $\alpha^i$ a root of the generator polynomial $g(D)$. Since $x(D)$ is a multiple of $g(D), \alpha^i$ is also a root of $x(D)$. Hence

$$\begin{aligned} x(\alpha^i) &= x_{N-1}\alpha^{(N-1)i} + x_{N-2}\alpha^{(N-2)i} + \cdots + x\alpha^i + x_0 \\ &= 0, \qquad 1 \le i \le 2t. \end{aligned} \tag{3.8-3}$$

Expressing the set of $2t$ equations above in matrix form, we have

$$\begin{bmatrix} x_{N-1} & x_{N-2} & \cdots & x_0 \end{bmatrix} \begin{bmatrix} \alpha^{(N-1)} & \alpha^{(N-2)} & \cdots & \alpha & 1 \\ \alpha^{2(N-1)} & \alpha^{2(N-2)} & \cdots & \alpha^2 & 1 \\ \alpha^{3(N-1)} & \alpha^{3(N-2)} & \cdots & \alpha^3 & 1 \\ \vdots & \vdots & \ddots & \vdots & \vdots \\ \alpha^{2t(N-1)} & \alpha^{2t(N-2)} & \cdots & \alpha^{2t} & 1 \end{bmatrix} = \mathbf{0}. \tag{3.8-4}$$

This equation is of the form $\mathbf{x}\mathbf{H}^T = \mathbf{0}$, where $\mathbf{H}$ is the parity-check matrix. Since $\alpha^i$ and $(\alpha^i)^2$ are roots of the same minimal polynomial, they are also roots of $g(D)$ and so are roots of $x(D)$. That is, if $x(\alpha^i) = 0$, then $x[(\alpha^i)^2] = x(\alpha^{2i}) = 0$. We may therefore eliminate all the even rows in (3.8-4), and $\mathbf{H}$ becomes

$$\mathbf{H} = \begin{bmatrix} \alpha^{(N-1)} & \alpha^{(N-2)} & \cdots & \alpha & 1 \\ \alpha^{3(N-1)} & \alpha^{3(N-2)} & \cdots & \alpha^3 & 1 \\ \alpha^{5(N-1)} & \alpha^{5(N-2)} & \cdots & \alpha^5 & 1 \\ \vdots & \vdots & \ddots & \vdots & \vdots \\ \alpha^{(2t-1)(N-1)} & \alpha^{(2t-1)(N-2)} & \cdots & \alpha^{(2t-1)} & 1 \end{bmatrix}.$$

**Example 3.8-4: Parity-Check Matrix for the (15,7) BCH Code.**

Let $\alpha$ be a primitive element of $GF(2^4)$. Then the parity-check matrix for the (15, 7) BCH Code of Example 3.8-2 is

$$\mathbf{H} = \begin{bmatrix} \alpha^{14} & \alpha^{13} & \alpha^{12} & \alpha^{11} & \cdots & \alpha & 1 \\ \alpha^{42} & \alpha^{39} & \alpha^{36} & \alpha^{33} & \cdots & \alpha^3 & 1 \end{bmatrix}$$

$$= \begin{bmatrix} \alpha^{14} & \alpha^{13} & \alpha^{12} & \alpha^{11} & \alpha^{10} & \alpha^9 & \alpha^8 & \alpha^7 & \alpha^6 & \alpha^5 & \alpha^4 & \alpha^3 & \alpha^2 & \alpha & 1 \\ \alpha^{12} & \alpha^9 & \alpha^6 & \alpha^3 & 1 & \alpha^{12} & \alpha^9 & \alpha^6 & \alpha^3 & 1 & \alpha^{12} & \alpha^9 & \alpha^6 & \alpha^3 & 1 \end{bmatrix},$$

where we have made use of the fact that $\alpha^{15} = 1$. Using the vector representation for the powers of $\alpha$ from Table 3.8-2, we get the following $8 \times 15$ parity-check matrix:

$$\mathbf{H} = \begin{bmatrix} 1 & 1 & 1 & 1 & 0 & 1 & 0 & 1 & 1 & 0 & 0 & 1 & 0 & 0 & 0 \\ 0 & 1 & 1 & 1 & 1 & 0 & 1 & 0 & 1 & 1 & 0 & 0 & 1 & 0 & 0 \\ 0 & 0 & 1 & 1 & 1 & 1 & 0 & 1 & 0 & 1 & 1 & 0 & 0 & 1 & 0 \\ 1 & 1 & 1 & 0 & 1 & 0 & 1 & 1 & 0 & 0 & 1 & 0 & 0 & 0 & 1 \\ 1 & 1 & 1 & 1 & 0 & 1 & 1 & 1 & 1 & 0 & 1 & 1 & 1 & 1 & 0 \\ 1 & 0 & 1 & 0 & 0 & 1 & 0 & 1 & 0 & 0 & 1 & 0 & 1 & 0 & 0 \\ 1 & 1 & 0 & 0 & 0 & 1 & 1 & 0 & 0 & 0 & 1 & 1 & 0 & 0 & 0 \\ 1 & 0 & 0 & 0 & 1 & 1 & 0 & 0 & 0 & 1 & 1 & 0 & 0 & 0 & 1 \end{bmatrix}.$$

**Example 3.8-5: Parity-Check Matrix for the (15, 5) BCH Code.**

This is

$$\mathbf{H} = \left[\begin{array}{cccccc} \alpha^{14} & \alpha^{13} & \alpha^{12} & \alpha^{11} & \alpha^{10} & \alpha^{9} \\ \alpha^{12} & \alpha^{9} & \alpha^{6} & \alpha^{3} & 1 & \alpha^{12} \\ \alpha^{10} & \alpha^{5} & 1 & \alpha^{10} & \alpha^{5} & 1 \end{array}\right.$$
$$\left.\begin{array}{ccccccccc} \alpha^{8} & \alpha^{7} & \alpha^{6} & \alpha^{5} & \alpha^{4} & \alpha^{3} & \alpha^{2} & \alpha & 1 \\ \alpha^{9} & \alpha^{6} & \alpha^{3} & 1 & \alpha^{12} & \alpha^{9} & \alpha^{6} & \alpha^{3} & 1 \\ \alpha^{10} & \alpha^{5} & 1 & \alpha^{10} & \alpha^{5} & 1 & \alpha^{10} & \alpha^{5} & 1 \end{array}\right].$$

This H matrix has 12 rows and 15 columns after expressing the $\alpha$ powers as binary 4-tuples. However, one of the rows is an all-zeros row and two others are identical, and so H is actually a $10 \times 15$ matrix.

## 3.9 DECODING BCH CODES

### Syndrome Computation

Let $x(D)$ be the code word polynomial and $y(D)$ the received polynomial, given by

$$y(D) = x(D) + e(D), \tag{3.9-1}$$

where $e(D)$ is the error polynomial introduced by the channel. If $e(D)$ is nonzero, then $y(D)$ evaluated at the roots of the generator polynomial will be nonzero. The syndromes $S_i$, $i = 1, 2, \ldots, 2t$, are defined by

$$\begin{aligned} S_i &= y(\alpha^i) \\ &= x(\alpha^i) + e(\alpha^i) \\ &= e(\alpha^i), \qquad i = 1, 2, \ldots, 2t. \end{aligned} \tag{3.9-2}$$

Note that the code polynomial is zero when evaluated at the roots of the generator polynomial. We can justify (3.9-2) as follows. Recall from Section 3.3 that the syndromes are defined by

$$\mathbf{S} = (S_1, S_2, \ldots, S_{2t}) = \mathbf{y}\mathbf{H}^T.$$

For BCH codes, using $\mathbf{H}$ from (3.8-4), we find that

$$S_i = y(\alpha^i). \tag{3.9-3}$$

We may also compute the syndromes by observing that

$$y(D) = q(D)\phi_{\alpha^i}(D) + r(D), \tag{3.9-4}$$

where $r(D)$ is the remainder obtained when $y(D)$ is divided by $\phi_{\alpha^i}(D)$. From (3.9-3), (3.9-4), and the fact that $\phi_{\alpha^i}(\alpha^i) = 0$, we get

$$S_i = y(\alpha^i) = r(\alpha^i). \tag{3.9-5}$$

**Example 3.9-1: Syndrome Computations for the (15, 5) BCH Code.**

Consider the triple-error-correcting (15, 5) BCH code with $g(D) = D^{10} + D^8 + D^5 + D^4 + D^2 + D + 1$ (see Example 3.8-3). Assume that the all-zeros code word is transmitted and that three errors occur in positions 2, 5, and 14. The received polynomial is therefore

$$y(D) = D^{14} + D^5 + D^2.$$

Divide $y(D)$ by $\phi_\alpha(D)$, where $\phi_\alpha(D) = D^4 + D + 1$, to get $D^3 + D + 1$ as the remainder. From (3.9-5),

$$\begin{aligned} S_1 = y(\alpha) &= r(\alpha) \\ &= \alpha^3 + \alpha + 1 = \alpha^7, \end{aligned}$$

where we have used Table 3.8-2 to equate $\alpha^3 + \alpha + 1$ to $\alpha^7$. We could have computed $y(\alpha)$ directly from the received polynomial; again with Table 3.8-2, we would get

$$\begin{aligned} y(\alpha) &= \alpha^{14} + \alpha^5 + \alpha^2 \\ &= (1001) + (0110) + (0100) \\ &= (1011) = \alpha^7. \end{aligned}$$

The remaining syndromes are:

$$\begin{aligned} S_3 &= y(\alpha^3) \\ &= \alpha^{42} + \alpha^{15} + \alpha^6 = \alpha^{12} + 1 + \alpha^6 = \alpha, \\ S_5 &= y(\alpha^5) \\ &= \alpha^{70} + \alpha^{25} + \alpha^{10} = \alpha^{10} + \alpha^{10} + \alpha^{10} = \alpha^{10}, \\ S_2 &= y(\alpha^2) = [y(\alpha)]^2 = S_1^2 = \alpha^{14}, \\ S_4 &= S_2^2 = \alpha^{28} = \alpha^{13}, \\ S_6 &= S_3^2 = \alpha^2. \end{aligned}$$

## Peterson's Algorithm

An algorithm due to Peterson [3-28] decodes BCH codes. The scheme requires some new notation about error locations. Let $\nu$ errors occur in locations $j_1, j_2, \ldots, j_\nu$, where $0 \leq j_1 < j_2 < \cdots < j_\nu < N$. Then the error polynomial is

$$e(D) = D^{j_\nu} + D^{j_{\nu-1}} + \cdots + D^{j_1}$$

Let us define $\chi_\ell$, called *error locator*, by $\chi_\ell = \alpha^{j_\ell}$, $\ell = 1, 2, \ldots, \nu$. From (3.9-2), the symdrome equations are given by

$$S_i = e(\alpha^i) = \chi_\nu^i + \chi_{\nu-1}^i + \cdots + \chi_1^i, \qquad i = 1, 2, \ldots, 2t. \tag{3.9-6}$$

The error-locator polynomial is

$$\begin{aligned}\sigma(D) &\triangleq \prod_{i=1}^{\nu}\left(1 + \chi_i D\right) \\ &= \sigma_\nu D^\nu + \sigma_{\nu-1} D^{\nu-1} + \cdots + \sigma_1 D + 1,\end{aligned} \tag{3.9-7}$$

where the $\sigma_i$ are given by

$$\begin{aligned}\sigma_1 &= \chi_1 + \chi_2 + \cdots + \chi_\nu \\ \sigma_2 &= \sum_{i<j} \chi_i \chi_j \\ \sigma_3 &= \sum_{i<j<k} \chi_i \chi_j \chi_k \\ &\vdots \\ \sigma_\nu &= \chi_1 \chi_2 \cdots \chi_\nu.\end{aligned} \tag{3.9-8}$$

The $\{\sigma_i\}$ are called the *elementary symmetric functions* of the error locators.

Since the $\{\chi_\ell\}$ are the inverses of the roots of $\sigma(D)$, $\sigma(\chi_\ell^{-1})$ is 0 at each $\ell$. Hence

$$\chi_\ell^\nu \sigma(\chi_\ell^{-1}) = \chi_\ell^\nu + \sigma_1 \chi_\ell^{\nu-1} + \cdots + \sigma_\nu = 0, \qquad \ell = 1, 2, \ldots, \nu.$$

Multiplying the above equation by $\chi_\ell^j$ throughout, we get

$$\chi_\ell^{\nu+j} + \sigma_1 \chi_\ell^{\nu+j-1} + \cdots + \sigma_\nu \chi_\ell^j = 0, \qquad \ell, j = 1, 2, \ldots, \nu.$$

On summing this equation over the index $\ell$ for a fixed $j$, we obtain

$$S_{\nu+j} + \sigma_1 S_{\nu+j-1} + \cdots + \sigma_\nu S_j = 0, \qquad \text{for } j = 1, 2, \ldots, \nu \tag{3.9-9}$$

These $\nu$ equations may be rearranged into the following set of simultaneous equations in the variables $\sigma_1, \ldots, \sigma_\nu$ :

$$
\begin{aligned}
S_1 + \sigma_1 &= 0 \\
S_1 + \sigma_1 + 2\sigma_2 &= 0 \\
S_3 + \sigma_1 S_2 + \sigma_2 S_1 + 3\sigma_3 &= 0 \\
&\vdots \\
S_\nu + \sigma_1 S_{\nu-1} + \cdots + \sigma_{\nu-1} S_1 + \nu\sigma_\nu &= 0 \\
S_{\nu+1} + \sigma_1 S_\nu + \cdots + \sigma_{\nu-1} S_2 + \sigma_\nu S_1 &= 0 \\
&\vdots
\end{aligned}
\tag{3.9-10}
$$

Equations (3.9-10) are known as *Newton's identities.* For binary BCH codes, the rules of arithmetic mean that any odd number here equals 1 and any even number equals 0. Moreover, $S_{2i} = (S_i)^2$ which means that we can eliminate all even rows in (3.9-10). The result is

$$
\begin{bmatrix}
1 & 0 & 0 & \cdots & & & & 0 \\
S_2 & S_1 & 1 & 0 & \cdots & & & 0 \\
S_4 & S_3 & S_2 & S_1 & 1 & 0 & \cdots & 0 \\
\vdots & & & & & & \ddots & \vdots \\
S_{2\nu-4} & S_{2\nu-5} & S_{2\nu-6} & & & & \cdots & S_{\nu-3} \\
S_{2\nu-2} & S_{2\nu-3} & S_{2\nu-5} & & & & \cdots & S_{\nu-1}
\end{bmatrix}
\begin{bmatrix}
\sigma_1 \\ \sigma_2 \\ \sigma_3 \\ \vdots \\ \sigma_{\nu-1} \\ \sigma_\nu
\end{bmatrix}
=
\begin{bmatrix}
S_1 \\ S_3 \\ S_5 \\ \vdots \\ S_{2\nu-3} \\ S_{2\nu-1}
\end{bmatrix}.
\tag{3.9-11}
$$

For a single-error-correcting code, (3.9-11) gives

$$
\sigma_1 = S_1. \tag{3.9-12}
$$

For a double-error-correcting code, the solution to (3.9-11) is

$$
\begin{aligned}
\sigma_1 &= S_1 \\
\sigma_2 &= \frac{S_3 + S_1^3}{S_1}.
\end{aligned}
\tag{3.9-13}
$$

Similarly, for a triple-error-correcting code,

$$
\begin{aligned}
\sigma_1 &= S_1 \\
\sigma_2 &= \frac{S_1^2 S_3 + S_5}{S_1^3 + S_3} \\
\sigma_3 &= S_1^3 + S_3 + S_1\sigma_2,
\end{aligned}
\tag{3.9-14}
$$

and so on.

In the general case of a $t$-error-correcting code, the decoder solves (3.9-11) directly to $t$ or fewer errors. If $t$ or $t-1$ errors have occurred, then the determinant of the $t \times t$ matrix on the left of (3.9-11) will be nonzero. Solution of (3.9-11) yields the error locator polynomial given by $\sigma_1, \sigma_2, \ldots, \sigma_t$; if only $t-1$ errors have occurred $\sigma_t$ will be 0. If less than $t-1$ errors have occurred, the decoder removes two rows and two columns from the $t \times t$ matrix to obtain a $(t-2) \times (t-2)$ matrix. If its determinant is nonzero, then $t-2$ or $t-3$ errors have occurred. If not, two more rows and two more columns are removed from the $(t-2) \times (t-2)$ matrix and the procedure repeats. Having determined the error locator polynomial the decoder then solves for its roots to determine the error locators. This algorithm requires computations proportional to $t^3$.

**Example 3.9-2: The 3-Error-Correcting BCH Code.**

Consider the BCH code in Example 3.9-1, with $g(D) = D^{10} + D^8 + D^5 + D^4 + D^2 + D + 1$. Let the received sequence be $y(D) = D^{14} + D^5 + D^2$. From Example 3.9-1, the syndromes are $S_1 = \alpha^7$, $S_3 = \alpha$, and $S_5 = \alpha^{10}$. Substituting these values for $S_1, S_2$, and $S_3$ into (3.9-14), we may compute $\sigma_1, \sigma_2$, and $\sigma_3$ as

$$\sigma_1 = \alpha^7,$$
$$\sigma_2 = \alpha^9,$$
$$\sigma_3 = \alpha^6.$$

Using these values in (3.9-7), we find the error-locator polynomial to be

$$\sigma(D) = 1 + \alpha^7 D + \alpha^9 D^2 + \alpha^6 D^3.$$

We can now solve for the roots of $\sigma(D)$ to determine the error locators $\chi_1, \chi_2$ and $\chi_3$. An efficient method to do this is due to Chien [3-27]. For now, the reader can verify that $\alpha^{13}, \alpha^{10}$, and $\alpha$ are the roots of $\sigma(D)$. We invert these to obtain $\alpha^2, \alpha^5$ and $\alpha^{14}$ as the error locators. The received word is then decoded into the all-zeros word by correcting the errors in positions 2, 5, and 14.

## The Massey–Berlekamp Algorithm

The direct method of solving the Newton's identities required computations proportional to $t^3, t$ being the error correcting capability of the code. Realizing that the Newton's identities are highly structured, Berlekamp [3-25] found an iterative method of solving them that is more efficient. Subsequently, Massey [3-29] proposed a technique that viewed the BCH decoding problem as a shift register synthesis problem. That is, given the syndromes $S_1, S_2, \ldots,$ find a minimal length shift register that generates the syndromes. Knowing the coefficients of this shift register, construct the error locator polynomial $\sigma(D)$. Then solve for its roots, whose inverses yield the error locators. The algorithm proposed by Massey is known as the *Massey–Berlekamp algorithm.* We now discuss it in more detail.

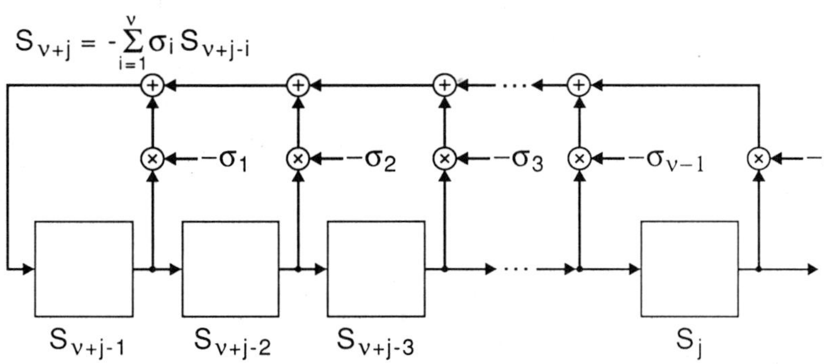


**Figure 3.9-1.** A shift register circuit for generating the syndromes $S_1, S_2, S_3, \ldots$

We begin with (3.9-9), which is

$$S_{\nu+j} + \sigma_1 S_{\nu+j-1} + \cdots + \sigma_\nu S_j = 0, \qquad j = 1, 2, \ldots, \nu.$$

Rewriting this equation as

$$S_{\nu+j} = -\sum_{i=1}^{\nu} \sigma_i S_{\nu+j-i}, \qquad j = 1, 2, \ldots, \nu,$$

we observe that $S_{\nu+1}$ is actually the output of the shift register in Fig. 3.9-1, with its contents initialized with $S_\nu, S_{\nu-1}, \ldots, S_2, S_1$. Similarly, $S_{\nu+2}$

is the output of the register with $S_{\nu+1}, S_\nu, \ldots, S_3, S_2$ as its initial contents. The BCH decoding problem is equivalent to designing a shift register that generates the syndromes in sequence. In general, there exist many such registers. Which one shall we choose? The answer to this becomes obvious when we realize that the many shift registers that generate a given sequence of syndromes correspond to the many error patterns that produce the same syndrome sequence. Out of these error patterns, we choose one with the least weight. Similarly, out of the many shift registers that generate a given syndrome sequence, we choose the one with the least length.

The algorithm begins with $\sigma^{(0)}(D) = 1$ and iteratively constructs a minimum-length shift register generating the given syndrome sequence. The superscript over $\sigma(D)$ denotes the iteration number. Assume that at the end of the $(\gamma - 1)$th iteration, the algorithm has constructed a minimal-length shift register $\sigma^{(\gamma-1)}(D)$ of length $L_{\gamma-1}$ generating the syndromes $S_1, S_2, \ldots, S_{\gamma-1}$. During the next iteration, the algorithm checks to see if $\sigma^{(\gamma-1)}(D)$ generates the $\gamma$th syndrome $S_\gamma$. If it does, the algorithm then checks to see if it also generates the next syndrome, and then the next, and so on.

Assume now that $\sigma^{(\gamma-1)}(D)$ fails to generate $S_\gamma$. Its actual output $S'_\gamma$ is given by

$$S'_\gamma = -\sum_{i=1}^{L_{\gamma-1}} \sigma_i^{(\gamma-1)} S_{\gamma-i}.$$

The discrepancy $\Delta_\gamma$ between the $\gamma$th syndrome and the actual output of the shift register $\sigma^{(\gamma-1)}(D)$ is given by

$$\begin{aligned}\Delta_\gamma &= S_\gamma - S'_\gamma \\ &= S_\gamma + \sum_{i=1}^{L_{\gamma-1}} \sigma_i^{(\gamma-1)} S_{\gamma-i}.\end{aligned}$$

The claim is now that $\sigma^{(\gamma)}(D)$ is given by

$$\sigma^{(\gamma)}(D) = \sigma^{(\gamma-1)}(D) - \Delta_\gamma \Delta_m^{-1} D^{\gamma-m} \sigma^{(m-1)}(D), \tag{3.9-15}$$

where $m$ denotes the most recent iteration, prior to the $\gamma$th, when a nonzero discrepancy was encountered that led to increasing the shift register length. That is,

$$\Delta_j = S_j + \sum_{i=1}^{L_{m-1}} \sigma_j^{(m-1)} S_{j-i},$$

from which

$$\Delta_j = \begin{cases} 0, & \text{for } j = 1, 2, \ldots, m-1 \\ \text{nonzero}, & \text{for } j = m. \end{cases}$$

Observe that, from (3.9-15)

$$S_j + \sum_{i=1}^{L_\gamma} \sigma_i^{(\gamma)} S_{j-i} = S_j + \sum_{i=1}^{L_{\gamma-1}} \sigma_i^{(\gamma-1)} S_{j-i} - \Delta_\gamma \Delta_m^{-1} \left[ S_{j-\gamma+m} + \sum_{i=1}^{L_{m-1}} \sigma_i^{(m-1)} S_{j-\gamma+m-i} \right]. \tag{3.9-16}$$

Within (3.9-16), notice that

$$S_j + \sum_{i=1}^{L_{\gamma-1}} \sigma_i^{(\gamma-1)} S_{j-i} = \begin{cases} 0, & \text{for } j \le \gamma - 1 \\ \Delta_\gamma, & \text{for } j = \gamma; \end{cases}$$

and

$$S_{j-\gamma+m} + \sum_{i=1}^{L_{m-1}} \sigma_i^{(m-1)} S_{j-\gamma+m-i} = \begin{cases} 0, & \text{for } j \le \gamma - 1 \\ \Delta_m, & \text{for } j = \gamma. \end{cases}$$

Thus the right-hand side of (3.9-16) reduces to

$$\begin{cases} 0, & \text{for } j \le \gamma - 1 \\ \Delta_\gamma - \Delta_\gamma \Delta_m^{-1} \Delta_m = 0, & \text{for } j = \gamma. \end{cases}$$

Hence $\sigma^{(\gamma)}(D)$ given by (3.9-15) generates $S_1, S_2, \ldots, S_\gamma$.

We have not yet shown that this is a shift register of least length. To show this we make use of a property of shift registers which we state without proof (Massey [3-29] gives the details).

**Property 3.9-1:** If some shift register of minimal length $L_{\gamma-1}$ generates $S_1, S_2, \ldots, S_{\gamma-1}$ but not $S_1, S_2, \ldots, S_{\gamma-1}, S_\gamma$, the minimal-length shift register that generates $S_1, S_2, \ldots, S_{\gamma-1}, S_\gamma$ has length $L_\gamma$ given by

$$L_\gamma \ge \max[L_{\gamma-1}, \gamma - L_{\gamma-1}]. \tag{3.9-17}$$

With this property, we can show inductively that (3.9-15) guarantees the synthesis of a minimal length shift register producing $S_1, S_2, \ldots, S_\gamma$. Under the inductive hypothesis, assume that we have found some $\sigma^{(1)}(D), \sigma^{(2)}(D), \ldots, \sigma^{(\gamma-1)}(D)$, with length $L_1, L_2, \ldots, L_{\gamma-1}$, that satisfy

(3.9-17) with equality. From (3.9-15), we make the following observation: If $\sigma^{(\gamma-1)}(D)$ has degree greater than or equal to the degree of $D^{\gamma-m}\sigma^{(m-1)}(D)$, then $\sigma^{(\gamma)}(D)$ will have the same degree as $\sigma^{(\gamma-1)}(D)$; otherwise, it will have degree equal to that of $D^{\gamma-m}\sigma^{(m-1)}(D)$. When this observation is expressed in terms of lengths of shift registers, we have

$$L_\gamma = \max\Big[L_{\gamma-1}, \gamma - m + L_{m-1}\Big]. \tag{3.9-18}$$

Since $m$ denotes the most recent iteration prior to the $\gamma$th when $\Delta_m$ was nonzero and $L_m > L_{m-1}$, we have by the inductive hypothesis

$$\begin{aligned} L_{\gamma-1} = L_m &= \max\Big[L_{m-1}, m - L_{m-1}\Big] \\ &= m - L_{m-1}. \end{aligned}$$

Using the above relationship between $L_{\gamma-1}$ and $L_{m-1}$ in (3.9-18), we get

$$L_\gamma = \max\Big[L_{\gamma-1}, \gamma - L_{\gamma-1}\Big],$$

thus guaranteeing that $L_\gamma$ also satisfies (3.9-17) with equality.

The formal listing of the algorithm on page 162 summarizes the discussion thus far. The algorithm is valid for nonbinary codes as well. Later we will point out some simplifications for binary codes.

A few comments are in order about the parameters and polynomials computed by the algorithm. When the algorithm completes, $\sigma(D)$ gives the error locator polynomial. During iterations when the shift register must be lengthened, $\beta(D)$ stores the shift register prior to lengthening, normalized by $\Delta^{-1}$; this is done in Step 2(b) whenever $\Delta$ is nonzero and $2L < \gamma$. Observe that the algorithm accounts for $D^{\gamma-m}$ occurring in (3.9-15) by multiplying $\beta(D)$ by $D$ whenever the shift register computed thus far also generates the next syndrome. This occurs whenever $\Delta$ equals 0 or both $\Delta$ is nonzero and $2L$ is equal to or greater than $\gamma$ (Step 2(b)). Finally, the polynomial $\sigma(D)$ must have degree less than $t$ for it to be a valid error locator polynomial. Otherwise, more than $t$ errors have occurred and the algorithm cannot correct these errors. The reader is encouraged to carry out the various steps of the example below to understand the algorithm's workings.

**The Massey-Berlekamp Algorithm**

**Step 1** (*Initialize*).

$$\gamma = 0, \ \sigma(D) = 1, \ L = 0, \ \beta(D) = 0$$

**Step 2** (*Iterate*). For $\gamma = 1$ to $2t$ in steps of 1, carry out steps 2(a) and 2(b) iteratively.

**Step 2(a)** (*Compute error* $\Delta$).

$$\Delta = \sum_{i=0}^{L} \sigma_i S_{\gamma-i}$$

**Step 2(b)** **if** $\Delta \neq 0$
**then begin** (*update* $\sigma(D)$)
$\sigma'(D) = \sigma(D) - \Delta D\beta(D)$;
**if** $2L < \gamma$
**then begin**
(*lengthen shift register*)
$\beta(D) = \Delta^{-1}\sigma(D)$;
$L = \gamma - L$
**end**
**else** $\beta(D) = D\beta(D)$;
$\sigma(D) = \sigma'(D)$
**end**
**else** $\beta(D) = D\beta(D)$.

**Step 3.** If $\deg \sigma(D) \neq L$, then more than $t$ errors have occurred. Otherwise, $\sigma(D)$ is the inverse error locator polynomial. In this case, solve for the roots of $\sigma(D)$ and then invert them to obtain the error locators.

Example 3.9-3: The 3-Error-Correcting BCH Code.

Consider the (15, 5) code of Example 3.9-1, with $g(D) = D^{10} + D^8 + D^5 + D^4 + D^2 + D + 1$. Just as in the previous examples, assume that the all-zeros code word is transmitted and that three errors occur in

**Table 3.9-1.** THE STEPS OF THE MASSEY-BERLEKAMP ALGORITHM TO DECODE THE 3-ERROR CORRECTING BCH CODE OF EXAMPLE 3.9-3.

| r | Δ | σ'(D) | β(D) | σ(D) | L |
|---|---|---|---|---|---|
| 0 | | | 1 | 1 | 0 |
| 1 | $\alpha^7$ | $1+\alpha^7 D$ | $\alpha^8$ | $1+\alpha^7 D$ | 1 |
| 2 | 0 | $1+\alpha^7 D$ | $\alpha^8 D$ | $1+\alpha^7 D$ | 1 |
| 3 | $\alpha^{11}$ | $1+\alpha^7 D+\alpha^4 D^2$ | $\alpha^4+\alpha^{11} D$ | $1+\alpha^7 D+\alpha^4 D^2$ | 2 |
| 4 | 0 | $1+\alpha^7 D+\alpha^4 D^2$ | $\alpha^4 D+\alpha^{11} D^2$ | $1+\alpha^7 D+\alpha^4 D^2$ | 2 |
| 5 | $\alpha^{10}$ | $1+\alpha^7 D+\alpha^9 D^2+\alpha^6 D^3$ | $\alpha^5+\alpha^{12} D+\alpha^9 D^2$ | $1+\alpha^7 D+\alpha^9 D^2+\alpha^6 D^3$ | 3 |
| 6 | 0 | $1+\alpha^7 D+\alpha^9 D^2+\alpha^6 D^3$ | $\alpha^5 D+\alpha^{12} D^2+\alpha^9 D^3$ | $1+\alpha^7 D+\alpha^9 D^2+\alpha^6 D^3$ | 3 |

| Shift Register σ(D) | Output Sequence | Observation |
|---|---|---|
| $\sigma(D) = 1$ | 0 | Fails to produce $S_1$. |
| 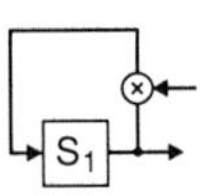  | $\alpha^7 = S_1, \alpha^{14} = S_2, \alpha^{21} = \alpha^6 = S'_3$ | Fails to produce $S_3$. |
| 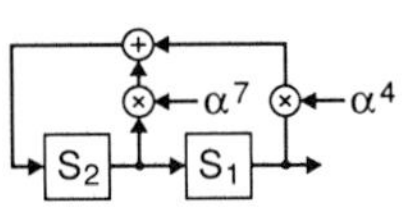  | $S_1, S_2, \alpha = S_3, \alpha^{13} = S_4, 0 = S'_5$ | Fails to produce $S_5$. |
| 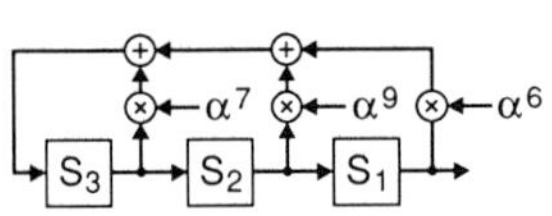  | $S_1, S_2, S_3, S_4, \alpha^{10} = S_5, \alpha^2 = S_6$ | The shift register $\sigma(D) = 1 + \alpha^7 D + \alpha^9 D^2 + \alpha^6 D^3$ generates the syndromes $S_1$, $S_2$, $S_3$, $S_4$, $S_5$, and $S_6$. |

**Figure 3.9-2.** The BCH decoding viewed as shift register synthesis.

positions 2, 5, and 14. In other words, the received polynomial is

$$y(D) = D^{14} + D^5 + D^2.$$

First find the syndromes, which are (see Example 3.9-1):

$$S_1 = \alpha^7, S_2 = \alpha^{14}, S_3 = \alpha,$$
$$S_4 = \alpha^{13}, S_5 = \alpha^{10}, S_6 = \alpha^2.$$

Next, following the steps of the Massey–Berlekamp algorithm, construct Table 3.9-1. Figure 3.9-2 illustrates the shift register synthesis viewpoint of the Massey-Berlekamp algorithm. The reader may verify that $\sigma(D)$ can be factored to

$$\begin{aligned}\sigma(D) &= 1 + \alpha^7 D + \alpha^9 D^2 + \alpha^6 D^3 \\ &= (1 + \alpha^2 D)(1 + \alpha^5 D)(1 + \alpha^{14} D).\end{aligned}$$

Interestingly, Table 3.9-1 reveals that $\Delta$ is zero for even numbered iterations. This is not by accident. We can, in fact, show that for any binary BCH code, $\Delta$ is zero for even numbered iterations and hence they may be skipped. For the binary case, we make the following changes to the Massey-Berlekamp algorithm.

1. In step (2), iterations are carried out for $\gamma = 1$ to $2t - 1$ in steps of 2 instead of 1. We thus skip even numbered iterations.
2. In step (2b), if $\Delta = 0$ or if $2L \geq \gamma$ (the outer and inner else clauses), we let $\beta(D) = D^2\beta(D)$. If $\Delta \neq 0$ and $2L < \gamma$, we let $\beta(D) = \Delta^{-1} D\sigma(D)$ and $L = \gamma - L - 1$.

**Example 3.9-3: (Revisited).**

Consider the $g(D)$ and $y(D)$ of Example 3.9-3. Table 3.9-2 carries out the modified algorithm to decode $y(D)$.

**Table 3.9-2.** THE STEPS OF THE MODIFIED MASSEY-BERLEKAMP ALGORITHM TO DECODE $y(D)$ OF EXAMPLE 3.9-3.

| r | Δ | σ' (D) | β (D) | σ (D) | L |
|---|---|---|---|---|---|
| 0 | | | $1$ | $1$ | 0 |
| 1 | $\alpha^7$ | $1+\alpha^7 D$ | $\alpha^8 D$ | $1+\alpha^7 D$ | 1 |
| 3 | $\alpha^{11}$ | $1+\alpha^{11} D+\alpha^4 D^2$ | $\alpha^4 D+\alpha^{11} D^2$ | $1+\alpha^7 D+\alpha^4 D^2$ | 2 |
| 5 | $\alpha^{10}$ | $1+\alpha^{10} D+\alpha^9 D^2+\alpha^6 D^3$ | $\alpha^5 D+\alpha^{12} D^2+\alpha^9 D^3$ | $1+\alpha^7 D+\alpha^9 D^2+\alpha^6 D^3$ | 3 |

## The Chien Search

We now describe a procedure due to Chien [3-27] for finding the roots of $\sigma(D)$ and decoding the received sequence. Chien's search systematically substitutes $\alpha, \alpha^2, \ldots, \alpha^{N-1}$ for $D$ in $\sigma(D)$ to determine if the result is zero. Suppose $\alpha^l$ is a root of $\sigma(D)$. Then $\alpha^{-l} = \alpha^{N-l}$ is an error location number and the received symbol $y_{N-l}$ is in error. Since $\alpha^l$ is a root, we have

$$1 + \sigma_1\alpha^l + \sigma_2\alpha^{2l} + \cdots + \sigma_\nu\alpha^{\nu l} = 0.$$

Equivalently, $\sigma_1\alpha^l + \sigma_2\alpha^{2l} + \cdots + \sigma_\nu\alpha^{\nu l} = 1$. Hence, if the sum $\sum_{i=1}^{\nu} \sigma_i\alpha^{il}$ is 1, $y_{N-l}$ is in error and it is corrected by adding the sum to it; if the sum is different from 1, then $y_{N-l}$ has been correctly received.

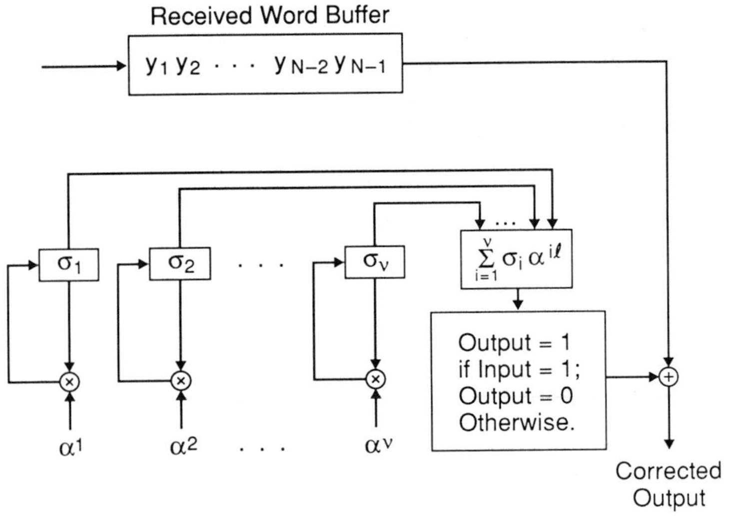


**Figure 3.9-3.** A circuit to perform Chien's Search.

The circuit in Fig. 3.9-3 carries out the above operations systematically. The received sequence is read into a buffer. A set of $\nu$ registers are initialized with $\sigma_1, \sigma_2, \ldots, \sigma_\nu$. A set of multipliers multiply the contents of these registers with $\alpha, \alpha^2, \ldots, \alpha^\nu$ and the outputs of the multipliers are stored back in the registers. To test if $\alpha$ is a root of $\sigma(D)$, the multipliers are clocked once and the outputs $\sigma_1\alpha, \sigma_2\alpha^2, \ldots, \sigma_\nu\alpha^\nu$ are stored back in the registers; the sum $\sum_{i=1}^{\nu} \sigma_i\alpha^i$ is formed and tested to see if it is 1.

A logic circuit puts out 1 if the sum is 1 and 0 otherwise. The output of the logic circuit is added to $y_{N-1}$. Next the multipliers are clocked a second time to form the sum $\sum_{i=1}^{\nu} \sigma_i \alpha^{2i}$, test the sum and correct $y_{N-2}$ if necessary. This process repeats until the entire received sequence shifts out of the buffer.

## 3.10 REED–SOLOMON CODES: ALGEBRA IN NON-BINARY FIELDS

The discussions in the foregoing two sections rely on the binary field GF(2) and its extension field GF($2^m$), which were developed in Section 3.7. Now we extend the discussion to non-binary fields GF($q$) and their extensions GF($q^m$), where $q$ is any prime number. We need this to understand an important subclass of BCH codes, the *Reed–Solomon codes.*

**Example 3.10-1: The Fields GF(3) and GF($3^2$).**

Denoting the elements of GF(3) as 0, 1, and 2, we may construct the following addition and multiplication tables for GF(3). These are just ordinary addition and multiplication reduced modulo 3.

| + | 0 1 2 |
|---|---|
| 0 | 0 1 2 |
| 1 | 1 2 0 |
| 2 | 2 0 1 |

| · | 0 1 2 |
|---|---|
| 0 | 0 0 0 |
| 1 | 0 1 2 |
| 2 | 0 2 1 |

To construct GF($3^2$), first select a primitive polynomial of degree 2 with coefficients from GF(3). $D^2 + D + 2$ is one such polynomial. Using a root $\alpha$ of $D^2+D+2$, we can proceed to construct Table 3.10-1. The construction is similar to that in Example 3.7-1 except that addition and multiplication are now over GF(3) instead of GF(2), and $\alpha$ now satisfies $\alpha^2 + \alpha + 2 = 0$, which is the same as $\alpha^2 = -\alpha - 2 = 2\alpha + 1$ (since $-1 = 2$ and $-2 = 1$ from the addition table).

**Table 3.10-1.** REPRESENTATION OF $GF(3^2)$ USING $D^2+D+2$ OVER $GF(3)$.

| Power of $\alpha$ | Polynomial | 2-tuple Over GF(3) |
|---|---|---|
| 0 | 0 | 00 |
| $\alpha^0$ | 1 | 01 |
| $\alpha^1$ | $\alpha$ | 10 |
| $\alpha^2$ | $2\alpha+1$ | 21 |
| $\alpha^3$ | $2\alpha+2$ | 22 |
| $\alpha^4$ | 2 | 02 |
| $\alpha^5$ | $2\alpha$ | 20 |
| $\alpha^6$ | $\alpha+2$ | 12 |
| $\alpha^7$ | $\alpha+1$ | 11 |

More generally, we can define the *integers modulo* $p$ to be $0, 1, \ldots, p-1$ together with ordinary addition and multiplication performed modulo $p$. This forms a field if and only if $p$ is prime†; we denote this field GF($p$). We construct GF($q$), where $q = p^m$, an integer power of a prime, by using a primitive polynomial of degree $m$ over GF($p$); by raising a root of the primitive polynomial to its successive powers, all the nonzero elements of GF($q$) can be generated. Example 3.10-1 illustrates the procedure. We can extend this procedure to construct the field $GF(q^\gamma)$ using a primitive polynomial of degree $\gamma$ over GF($q$), as illustrated in the following example.

Example 3.10-2: $GF(4^2)$ Using $D^2 + D + 2$ over GF(4).

First, we construct GF(4). Since $4 = 2^2$, construct GF(4) using a degree-2 primitive polynomial over GF(2). Denote the elements of GF(4) as 0, 1, 2, and 3. Generate GF(4) using $D^2 + D + 1$ over GF(2): The elements are $0, \alpha^0 = 1, \alpha^1 = 2$, and $\alpha^2 = \alpha + 1 = 3$. We can now construct the following addition and multiplication tables. These are certainly not ordinary addition and multiplication modulo 4:

| + | 0 | 1 | 2 | 3 |
|---|---|---|---|---|
| 0 | 0 | 1 | 2 | 3 |
| 1 | 1 | 0 | 3 | 2 |
| 2 | 2 | 3 | 0 | 1 |
| 3 | 3 | 2 | 1 | 0 |

| · | 0 | 1 | 2 | 3 |
|---|---|---|---|---|
| 0 | 0 | 0 | 0 | 0 |
| 1 | 0 | 1 | 2 | 3 |
| 2 | 0 | 2 | 3 | 1 |
| 3 | 0 | 3 | 1 | 2 |

We proceed to construct $GF(4^2)$ using the primitive polynomial $D^2 + D + 2$ over GF(4) as in Table 3.10-2, where $\alpha$ is now a root of the

†See Appendix B for its proof

chosen primitive polynomial. We reduce $\alpha^2$ and higher powers of $\alpha$ using $\alpha^2 = -\alpha - 2 = \alpha + 2$. Note that over GF(4), $-1 = 1$ since $1 + 1 = 0$.

**Table 3.10-2.** REPRESENTATION OF GF($4^2$) USING $D^2+D+2$ OVER GF(4).

| Power of $\alpha$ | Polynomial | 2-tuple Over GF(4) |
|---|---|---|
| 0 | 0 | 00 |
| $\alpha^0$ | 1 | 01 |
| $\alpha^1$ | $\alpha$ | 10 |
| $\alpha^2$ | $\alpha+2$ | 12 |
| $\alpha^3$ | $3\alpha+2$ | 32 |
| $\alpha^4$ | $\alpha+1$ | 11 |
| $\alpha^5$ | 2 | 02 |
| $\alpha^6$ | $2\alpha$ | 20 |
| $\alpha^7$ | $2\alpha+3$ | 23 |
| $\alpha^8$ | $\alpha+3$ | 13 |
| $\alpha^9$ | $2\alpha+2$ | 22 |
| $\alpha^{10}$ | 3 | 03 |
| $\alpha^{11}$ | $3\alpha$ | 30 |
| $\alpha^{12}$ | $3\alpha+1$ | 31 |
| $\alpha^{13}$ | $2\alpha+1$ | 21 |
| $\alpha^{14}$ | $3\alpha+3$ | 33 |

We can construct minimal polynomials of field elements in GF($q^\gamma$) by extending Property 10, Section 3.7, to nonbinary fields. Let $\beta$ be an element in GF($q^\gamma$). Let $l$ be the smallest integer such that $\beta^{q^l} = \beta$. That is, $\beta^q, \beta^{q^2}, \ldots, \beta^{q^{l-1}}$ are all the distinct conjugates of $\beta$. Then the minimal polynomial $\phi_\beta(D)$ of $\beta$ is given by

$$\phi_\beta(D) = \prod_{i=0}^{l-1}\left(D + \beta^{q^i}\right).$$

### Example 3.10-3: Minimal Polynomials of GF(16).

Consider GF(16) over GF(4) constructed in Table 3.10-2. We first note that the minimal polynomial of $\alpha$ must have $\alpha$ and $\alpha^4$ as its roots (since $\alpha^{16} = \alpha$). Hence

$$\begin{aligned}\phi_\alpha(D) &= (D - \alpha)(D - \alpha^4)\\ &= D^2 - (\alpha + \alpha^4)D + \alpha^5\\ &= D^2 + D + 2.\end{aligned}$$

Similarly, $\alpha^2$ and $(\alpha^2)^4 = \alpha^8$ must have the same minimal polynomial. Note that $(\alpha^2)^{4^2} = \alpha^{32} = \alpha^2$:

$$\begin{aligned}\phi_{\alpha^2}(D) &= (D-\alpha^2)(D-\alpha^8) \\ &= D^2 - (\alpha^2+\alpha^8)D + \alpha^{10} \\ &= D^2 + D + 3.\end{aligned}$$

Table 3.10-3 lists the minimal polynomials of all the field elements of GF(16).

**Table 3.10-3.** Minimal Polynomials of the Elements of GF(16) Constructed in Table 3.10-2.

| Elements of GF (16) | Minimal Polynomials |
|---|---|
| 0 | D |
| 1 | D+1 |
| $\alpha,\alpha^4$ | $D^2+D+2$ |
| $\alpha^2,\alpha^8$ | $D^2+D+3$ |
| $\alpha^3,\alpha^{12}$ | $D^2+3D+1$ |
| $\alpha^5$ | D+2 |
| $\alpha^6,\alpha^9$ | $D^2+2D+1$ |
| $\alpha^7,\alpha^{13}$ | $D^2+2D+2$ |
| $\alpha^{10}$ | D+3 |
| $\alpha^{11},\alpha^{14}$ | $D^2+3D+3$ |

## Nonbinary BCH Codes

Equipped with the knowledge of algebra in nonbinary fields, we can construct some nonbinary BCH codes. We define the generator polynomial of a $t$-error-correcting BCH code as (see 3.8-1)

$$g(D) = \text{LCM}\left[\phi_\alpha(D)\phi_{\alpha^2}(D),\ldots,\phi_{\alpha^{2t}}(D)\right], \tag{3.10-1}$$

where $\alpha$ is a primitive element of GF($q^m$). The field GF($q$) is usually referred to as the *symbol field* and GF($q^m$) the *locator field.* The code generated by the generator polynomial defined above is called a *primitive* BCH code.

We can generalize the definition of BCH codes by choosing any element $\beta$ from GF$(q^m)$ instead of a primitive element. Define the generator polynomial of a *nonprimitive* $t$-error-correcting BCH code as

$$g(D) = \text{LCM}\left[\phi_{\beta^{n_0}}(D), \phi_{\beta^{n_0+1}}(D), \ldots, \phi_{\beta^{n_0+2t-1}}(D)\right],$$

where $n_0$ is any integer. If we choose $n_0 = 1$, $\beta$ to be a primitive element, and $q = 2$, we get the primitive binary BCH codes defined in (3.8-1). The block length $N$ of a nonprimitive BCH code is the order of $\beta$, the smallest positive integer $N$ such that $\beta^N = 1$. For example, the order of $\alpha^2$ is 15 and that of $\alpha^6$ is 10.

**Example 3.10-4: Nonbinary BCH Codes over GF($4^2$).**

Consider GF($4^2$) in Table 3.10-2. The generator polynomial for a (15,11) single-error-correcting code is

$$\begin{aligned} g(D) &= \text{LCM}\left[\phi_{\alpha}(D)\phi_{\alpha^2}(D)\right] \\ &= (D^2 + D + 2)(D^2 + D + 3) \\ &= D^4 + D + 1. \end{aligned}$$

This code is not a single error correcting Hamming Code. Rather, it corrects a single quaternary symbol error and encodes 11 quaternary symbols (or 22 bits) into 15 quaternary symbols (or 30 bits).

The generator polynomial for the (15, 6) triple-error-correcting BCH code is

$$\begin{aligned} &\text{LCM}\left(\phi_{\alpha}(D), \phi_{\alpha^2}(D), \phi_{\alpha^3}(D), \phi_{\alpha^4}(D), \phi_{\alpha^5}(D), \phi_{\alpha^6}(D)\right) \\ &= (D^2 + D + 2)(D^2 + D + 3)(D^2 + 3D + 1) \\ &\qquad (D + 2)(D^2 + 2D + 1) \\ &= D^9 + 3D^8 + 3D^7 + 2D^6 + D^5 + 2D^4 + D + 2. \end{aligned}$$

This code encodes 6 quaternary information symbols into 15 quaternary channel symbols, with 9 quaternary check symbols. The code is capable of correcting 3 symbol errors in the received symbol stream.

Table 3.10-4, adapted from [3-10] and [3-30], lists some primitive polynomials over several different fields. The elements in a field are represented by $0, 1, 2, \ldots, q-1$ if $q < 10$ and by $0, 1, 2, \ldots, 9, A, B, C, \ldots$ if $q \geq 10$. For some positive integers $q_1, m_1, q_2, m_2$, such that $q_1^{m_1} = q_2^{m_2} = q$, we can construct two different representations for GF($q$), using a degree-$m_1$ primitive polynomial over GF($q_1$) and a degree-$m_2$ primitive polynomial over GF($q_2$). We have already seen two different representations for GF(16) in Tables 3.8-2 and 3.10-2.

**Table 3.10-4.** SOME PRIMITIVE POLYNOMIALS OVER PRIME AND PRIME POWER FIELDS.

| GF (2) | GF (3) | GF (4) | GF (8) | GF (16) |
|---|---|---|---|---|
| D+1 | D+1 | D+2 | D+2 | D+2 |
| $D^2$+D+1 | $D^2$+D+2 | $D^2$+D+2 | $D^2$+2D+2 | $D^2$+D+7 |
| $D^3$+D+1 | $D^3$+2D+1 | $D^3$+$D^2$+D+2 | $D^3$+D+2 | $D^3$+D+7 |
| $D^4$+D+1 | $D^4$+D+2 | $D^4$+$D^2$+2D+3 | $D^4$+D+3 | $D^4$+$D^2$+2D+3 |
| $D^5$+$D^2$+1 | $D^5$+2D+1 | $D^5$+D+2 | $D^5$+D+2 | $D^5$+2D+3 |
| $D^6$+D+1 | $D^6$+D+2 | $D^6$+$D^2$+D+2 | $D^6$+$D^2$+2D+3 | |

## Reed–Solomon Codes

Reed–Solomon codes are BCH codes in which the symbol field GF$(q)$ and the locator field GF$(q^m)$ are the same. The minimal polynomial of some field element $\beta$ in GF$(q)$ is $(D-\beta)$, and we can conclude from (3.10-1) that the generator polynomial for a $t$-error-correcting Reed–Solomon code is given by

$$g(D) = (D-\alpha)(D-\alpha^2)\dots(D-\alpha^{2t}).$$

Since the generator polynomial has degree exactly equal to $2t$, the number of parity-check symbols is $N-K=2t$. The block length of the code is $N=q-1$. Recall the notion of design distance from Section 3.8: For any $t$-error correcting BCH code, we set $2t+1$ to be the design distance. Hence, for Reed–Solomon codes, the design distance equals $N-K+1$. The minimum distance of Reed–Solomon codes is, therefore, greater than or equal to $N-K+1$ (this is shown in [3-31]). In consequence, Reed–Solomon codes have $d_{\min}$ precisely $N-K+1$ and are called *maximum-distance separable* codes.

Example 3.10-5: Generator Polynomials for Some Reed–Solomon Codes.

First consider constructing a (3,2) Reed–Solomon code over GF(4) in Example 3.10-2. The generator polynomial for this code is

$$g(D) = D-\alpha = D-2 = D+2$$

and there are 16 code words in it. The code word corresponding to the input $\mathbf{u} = (2,3)$, or $u(D) = 2D+3$, is given by $x(D) = u(D)g(D) = 2D^2+1$, or $\mathbf{x} = (2\ 0\ 1)$, which in binary becomes

(1 0 0 0 0 1). The reader may generate all the remaining code words and verify that each one of them has even number of 1s. This code is an even-parity-check code.

As another example, consider a (3, 1) Reed–Solomon code with $t = 1$. This code has $g(D) = (D-\alpha)(D-\alpha^2) = (D-2)(D-3) = (D+2)(D+3) = D^2 + D + 1$. The code is a repetition code since the code words generated by it are (000), (111), (222), and (333).

Next we construct the generator polynomial of a (15,9) Reed–Solomon code capable of correcting three symbol errors. We use GF($2^4$) in Table 3.8-2 to construct the generator polynomial given by

$$\begin{aligned}
&(D-\alpha)(D-\alpha^2)(D-\alpha^3)(D-\alpha^4)(D-\alpha^5)(D-\alpha^6)\\
&\quad= (D^2+\alpha^5 D+\alpha^3)(D^2+\alpha^7 D+\alpha^7)(D^2+\alpha^9 D+\alpha^{11})\\
&\quad= (D^4+\alpha^{13}D^3+\alpha^6 D^2+\alpha^3 D+\alpha^{10})(D^2+\alpha^9 D+\alpha^{11})\\
&\quad= (D^6+\alpha^{10}D^5+\alpha^{14}D^4+\alpha^4 D^3+\alpha^6 D^2+\alpha^9 D+\alpha^6).
\end{aligned}$$

Note that $g(D)$ has degree 6 and hence $N - K = 6$. Since the code is over GF(16), $N = 16 - 1 = 15$. The (15,9) Reed–Solomon code encodes 9 16-ary (or hexadecimal) information symbols, or 36 information bits, into a code word with 15 16-ary symbols, or 60 bits. The code can correct 3 symbol errors since $t = (N - K)/2 = 3$.

## Decoding Reed–Solomon Codes

To decode binary BCH codes, we need to find the locations in the received word that are in error. In the case of non-binary BCH codes and Reed–Solomon codes we need to determine the error magnitudes as well. Let $\nu$ errors occur in positions $D^{j_1}, D^{j_2}, \ldots, D^{j_\nu}$ of the received polynomial with error magnitudes $E_1, E_2, \ldots, E_\nu$. Defining $e(D) = e_{N-1}D^{N-1} + e_{N-2}D^{N-2} + \cdots + e_1 D + e_0$, we see that $E_l = e_{j_l}$. We define the error locators $\chi_l = \alpha^{j_l}$, $l = 1, 2, \ldots, \nu$ as in Section 3.9. Now modify (3.9-6) to

$$S_i = e(\alpha^i) = E_\nu \chi_\nu^i + E_{\nu-1}\chi_{\nu-1}^i + \cdots + E_1\chi_1^i, \qquad \text{for } i = 1, 2, \ldots, 2t.$$

Next define the error-locator polynomial $\sigma(D)$ exactly as in (3.9-7):

$$\begin{aligned}
\sigma(D) &= \prod_{i=1}^{\nu}\Big(1+\chi_i D\Big)\\
&= \sigma_\nu D^\nu + \sigma_{\nu-1}D^{\nu-1} + \cdots + \sigma_1 D + 1.
\end{aligned}$$

$\sigma(D)$ may be found by the Massey–Berlekamp algorithm, and the inverses of its roots are the error locations.

To determine the error magnitudes, we now define a new polynomial, the error-evaluator polynomial $\zeta(D)$ to be

$$\zeta(D) = 1 + (S_1 + \sigma_1)D + (S_2 + \sigma_1 S_1 + \sigma_2)D^2 + \cdots + (S_\nu + \sigma_1 S_{\nu-1} + \sigma_2 S_{\nu-2} + \cdots + \sigma_\nu)D^\nu. \qquad (3.10\text{-}2)$$

The error magnitudes then follow from

$$E_l = \frac{\zeta(\chi_l^{-1})}{\prod_{\substack{i=1 \\ i \neq l}}^{\nu} \left(1 + \chi_i \chi_l^{-1}\right)}, \qquad l = 1, 2, \ldots, \nu. \qquad (3.10\text{-}3)$$

We have omitted the steps needed to arrive at (3.10-3); Forney [3-32] gives the details of this.

**Example 3.10-6: Decoding the (15, 9) 3-Symbol-Correcting Reed–Solomon Code.**

From Example 3.10-5, the generator polynomial for the (15, 9) 3-symbol-correcting Reed–Solomon code is

$$g(D) = D^6 + \alpha^{10}D^5 + \alpha^{14}D^4 + \alpha^4 D^3 + \alpha^6 D^2 + \alpha^9 D + \alpha^6.$$

Assume that the all-zero code word is transmitted and that three errors occur in positions $D^{13}$, $D^3$, and $D$ of the received word, with error magnitudes $\alpha^{14}$, $\alpha^5$, and $\alpha^7$, respectively. That is, the received word is

$$y(D) = \alpha^{14}D^{13} + \alpha^5 D^3 + \alpha^7 D.$$

First, we compute the syndromes; computations are done using Table 3.8-2.

$$\begin{aligned} S_1 = y(\alpha) &= \alpha^{12} + \alpha^8 + \alpha^8 = \alpha^{12} \\ S_2 = y(\alpha^2) &= \alpha^{14}\alpha^{26} + \alpha^5\alpha^6 + \alpha^7\alpha^2 \\ &= \alpha^{10} + \alpha^{11} + \alpha^9 \\ &= \alpha^4 \\ S_3 = y(\alpha^3) &= \alpha^{14}\alpha^{39} + \alpha^5\alpha^9 + \alpha^7\alpha^3 \\ &= \alpha^8 + \alpha^{14} + \alpha^{10} \\ &= \alpha^7 \\ S_4 = y(\alpha^4) &= \alpha^{14}\alpha^{52} + \alpha^5\alpha^{12} + \alpha^7\alpha^4 \\ &= \alpha^6 + \alpha^2 + \alpha^{11} \\ &= \alpha^5 \\ S_5 = y(\alpha^5) &= \alpha^{14}\alpha^{65} + \alpha^5\alpha^{15} + \alpha^7\alpha^5 \\ &= \alpha^4 + \alpha^5 + \alpha^{12} \\ &= \alpha^9 \\ S_6 = y(\alpha^6) &= \alpha^{14}\alpha^{78} + \alpha^5\alpha^{18} + \alpha^7\alpha^6 \\ &= \alpha^2 + \alpha^8 + \alpha^{13} \\ &= \alpha^6. \end{aligned}$$

The Massey–Berlekamp algorithm finds $\sigma(D)$, as given in Table 3.10-5.

**Table 3.10-5.** THE MASSEY–BERLEKAMP ALGORITHM TO DETERMINE $\sigma(D)$.

| r | Δ | σ'(D) | β(D) | σ(D) | L |
|---|---|---|---|---|---|
| 0 | | | $1$ | $1$ | 0 |
| 1 | $\alpha^{12}$ | $1+\alpha^{12}D$ | $\alpha^3$ | $1+\alpha^{12}D$ | 1 |
| 2 | $\alpha^{14}$ | $1+\alpha^7 D$ | $\alpha^3 D$ | $1+\alpha^7 D$ | 1 |
| 3 | $\alpha^8$ | $1+\alpha^7 D+\alpha^{11}D^2$ | $\alpha^7+\alpha^{14}D$ | $1+\alpha^7 D+\alpha^{11}D^2$ | 2 |
| 4 | $\alpha^{11}$ | $1+\alpha^4 D+\alpha^{14}D^2$ | $\alpha^7 D+\alpha^{14}D^2$ | $1+\alpha^4 D+\alpha^{14}D^2$ | 2 |
| 5 | $\alpha^6$ | $1+\alpha^4 D+\alpha^2 D^2+\alpha^5 D^3$ | $\alpha^9+\alpha^{13}D+\alpha^8 D^2$ | $1+\alpha^4 D+\alpha^2 D^2+\alpha^5 D^3$ | 3 |
| 6 | $\alpha^8$ | $1+\alpha^{10}D+\alpha^3 D^2+\alpha^2 D^3$ | $\alpha^9 D+\alpha^{13}D^2+\alpha^8 D^3$ | $1+\alpha^{10}D+\alpha^3 D^2+\alpha^2 D^3$ | 3 |

The error locator polynomial $\sigma(D)$ is given by

$$\begin{aligned} \sigma(D) &= 1 + \alpha^{10}D + \alpha^3D^2 + \alpha^2D^3 \\ &= (1+\alpha^{13}D)(1+\alpha^3D)(1+\alpha D) \end{aligned}$$

with roots $\alpha^2, \alpha^{12}$, and $\alpha^{14}$. Their inverses $\alpha^{13}, \alpha^3$ and $\alpha$ give the error locations.

Having identified the locations where errors occurred, we now compute the error magnitudes. First we determine the error evaluation polynomial using (3.10-2):

$$\begin{aligned}\zeta(D) &= 1 + (\alpha^{12} + \alpha^{10})D + (\alpha^4 + \alpha^7 + \alpha^3)D^2 \\ &\quad + (\alpha^7 + \alpha^{14} + 1 + \alpha^2)D^3 \\ &= 1 + \alpha^3 D + \alpha^{10} D^3.\end{aligned}$$

Next we determine the error magnitudes using (3.10-3), with $\chi_1 = \alpha$, $\chi_2 = \alpha^3$ and $\chi_3 = \alpha^{13}$.

$$\begin{aligned}E_1 = e_1 &= \frac{\zeta(\chi_1^{-1})}{(1+\chi_2\chi_1^{-1})(1+\chi_3\chi_1^{-1})} \\ &= \frac{1+\alpha^3\alpha^{-1}+\alpha^{10}\alpha^{-3}}{(1+\alpha^3\alpha^{-1})(1+\alpha^{13}\alpha^{-1})} \\ &= \frac{1+\alpha^2+\alpha^7}{1+\alpha^2+\alpha^{12}+\alpha^{14}} \\ &= \frac{\alpha^{11}}{\alpha^4} = \alpha^7, \\ E_2 = e_3 &= \frac{\zeta(\chi_2^{-1})}{(1+\chi_1\chi_2^{-1})(1+\chi_3\chi_2^{-1})} \\ &= \frac{1+\alpha^3\alpha^{-3}+\alpha^{10}\alpha^{-9}}{(1+\alpha\alpha^{-3})(1+\alpha^{13}\alpha^{-3})} \\ &= \frac{1+1+\alpha}{1+\alpha^{13}+\alpha^{10}+\alpha^8} \\ &= \frac{\alpha}{\alpha^{11}} = \alpha^{-10} = \alpha^5, \\ E_3 = e_{13} &= \frac{\zeta(\chi_3^{-1})}{(1+\chi_1\chi_3^{-1})(1+\chi_2\chi_3^{-1})} \\ &= \frac{1+\alpha^3\alpha^{-13}+\alpha^{10}\alpha^{-39}}{(1+\alpha\alpha^{-13})(1+\alpha^3\alpha^{-13})} \\ &= \frac{1+\alpha^5+\alpha}{1+\alpha^{-12}+\alpha^{-10}+\alpha^{-7}} \\ &= \frac{\alpha^8}{\alpha^9} = \alpha^{-1} = \alpha^{14}.\end{aligned}$$

Hence

$$\begin{aligned}e(D) &= e_{13}D^{13} + e_3 D^3 + e_1 D \\ &= \alpha^{14}D^{13} + \alpha^5 D^3 + \alpha^7 D.\end{aligned}$$

On adding $e(D)$ to $y(D)$, we get the all-zeros code word.

# 3.11 CODES FOR BURST ERROR CORRECTION

Our discussions in this chapter have so far concentrated on codes capable of correcting random errors. Burst errors occur in storage media such as tapes and compact discs because of defects, and in other forms of communication as well. Random error correcting codes are in general not efficient at correcting burst errors, and several special classes of codes have been proposed for burst errors. Abramson [3-33] was the first to study codes for burst error correction. Subsequently, Fire [3-34] proposed a general class of burst-error-correcting codes. A number of researchers have discovered efficient burst error codes by computer search and these have been compiled in, for example, Lin and Costello [3-8] and Peterson and Weldon [3-3].

## Properties of Burst-Error-Correcting Codes

We say that a burst of length $\tau$ occurs whenever its nonzero components are confined to $\tau$ consecutive positions. For example, the vector (000010010101000) is a burst of length 8.

First we observe that a code that can correct burst errors of length $\tau$ cannot have a code word consisting of a burst of length $2\tau$ or less. To see this, suppose that the code has such a burst $\mathbf{x}$ as a word. We can decompose the code word into two bursts $\mathbf{b}_1$ and $\mathbf{b}_2$ each of length $\tau$ or less. If the code is to correct both $\mathbf{b}_1$ and $\mathbf{b}_2$, they must both be coset leaders in a standard array (recall the discussion in Section 3.3). This cannot be the case since $\mathbf{x}$ equals $\mathbf{b}_1 + \mathbf{b}_2$, implying that both $\mathbf{b}_1$ and $\mathbf{b}_2$ occur in the same row of the standard array and that only one of the two can be a coset leader. This means that only one of them can be a correctable burst error and that the code cannot correct all bursts of length $\tau$ or less. Hence, a code capable of correcting all bursts of length $\tau$ or less must have code words that are bursts of length greater than $2\tau$.

Next we obtain a bound on the burst-error-correcting capability of an $(N, K)$ code over GF($q$). Consider all bursts of length $2\tau$ or less occurring in the first $2\tau$ positions. There are $q^{2\tau}$ such bursts. Observe that the difference (or sum if $q$ equals 2) between any two such bursts must be another burst of length $2\tau$ or less. Two such bursts must occur in different rows of a standard array; otherwise their difference, (sum for codes over GF(2)) will be a code word of burst $2\tau$ or less. Hence, there are at least $q^{2\tau}$ rows in the standard array. We know by generalizing the discussion in Section 3.3 to codes over GF($q$) that a standard array has $q^{N-K}$ rows. Thus we have the bound

$$2\tau \leq N - K$$

or

$$\tau \leq \frac{N-K}{2}. \tag{3.11-1}$$

The above bound, due to Reiger [3-35] is known as the *Reiger bound.*

**Table 3.11-1.** SOME COMPUTER-SEARCHED BINARY CYCLIC CODES FOR BURST ERROR CORRECTION.

| Burst Error Correcting Capability | (N, K) | g (D) in Octal |
|---|---|---|
| 2 | (7, 3) | 35 |
| 3 | (15, 9) | 171 |
| 4 | (15, 7) | 721 |
| | (19, 11) | 1151 |
| 5 | (15, 5) | 2467 |
| | (27, 17) | 2671 |
| 6 | (21, 9) | 14515 |
| | (34, 22) | 15173 |
| 7 | (21, 7) | 47343 |
| | (38, 24) | 114361 |
| 8 | (21, 5) | 214537 |
| | (50, 34) | 224531 |
| 9 | (21, 3) | 1647235 |
| | (56, 38) | 1505773 |
| 10 | (59, 39) | 40003351 |

## Codes for Burst Error Correction

Table 3.11-1 gives computer-searched cyclic codes satisfying the Reiger bound with equality. The generator sequences are in octal notation: For example, the generator for the (7,3) code in the table is 35, which in binary becomes (011101), or in polynomial notation $D^4 + D^3 + D^2 + 1$.

### *Fire Codes*

A Fire code over GF($q$), capable of correcting bursts of length $\tau$, is constructed using the generator polynomial

$$g(D) = (D^{2\tau-1} - 1)p(D),$$

where $p(D)$ is a degree-$m$ polynomial such that $m$ is greater than or equal to $\tau$ and $p(D)$ and $D^{2\tau-1} - 1$ are relatively prime; that is $p(D)$ does not divide $D^{2\tau-1} - 1$. The block length $N$ of this code is

$$N = l(2\tau - 1),$$

where $l$ is the smallest integer such that $p(D)$ divides $D^l - 1$.

**Example 3.11-1: The (105, 94) Fire Code Using $D^4 + D + 1$ over GF(2).**

> Consider $p(D) = D^4 + D + 1$ over GF(2). $p(D)$ is a primitive polynomial (see Table 3.10-4) and consequently the least degree polynomial it divides is $D^{15} - 1$ (Property 1, Section 3.7). Let $\tau = 4$. $p(D)$ does not divide $D^{2\tau-1} - 1 = D^7 - 1$. The generator polynomial of the code is
>
> $$\begin{aligned} g(D) &= (D^7 + 1)(D^4 + D + 1) \\ &= D^{11} + D^8 + D^7 + D^4 + D + 1. \end{aligned}$$
>
> The block length is $N = 15(8 - 1) = 105$. This Fire code is a (105, 94) cyclic code.

We know from Section 3.7 that a primitive polynomial $p(D)$ over GF(2) divides $D^{2^m-1} - 1$. In general, if $p(D)$ is over GF($q$), it divides $D^{q^m-1} - 1$. Furthermore, the smallest $l$ for which $p(D)$ divides $D^l - 1$ is $l = q^m - 1$. So the block length of the resulting code is $N = (q^m - 1)(2\tau - 1)$. Since the generator polynomial of a Fire code is of degree $m + 2\tau - 1, N - K = m + 2\tau - 1$. Consequently, using a primitive polynomial for $p(D)$ results in a $((q^m-1)(2\tau-1), (q^m-1)(2\tau-1)-m-2\tau+1)$ code. Note that since $m$ must be greater than or equal to $\tau$, the redundancy $N - K$ is the least for $m = \tau$. In the above example, we used $m = \tau = 4$.

*Reed–Solomon Codes*

Consider a Reed–Solomon code over GF($q$), capable of correcting $t$ symbol errors. Taking $q$ to be $2^m$, we see that each symbol can be represented in binary notation using $m$ bits. We can look upon the set of all the code words of the original code transformed into their binary equivalent as constituting a binary code with

$$N = m(2^m - 1)$$

and

$$N - K = 2mt.$$

This binary code can be viewed as a multiple burst-error-correcting code, capable of correcting $t$ bursts, each of length $m$. Equivalently, it can be viewed as a single burst-error-correcting code, capable of correcting a burst of length $(t-1)m+1$, since such a burst amounts to $t$ symbols in the original code.

*Interleaved, Concatenated, and Product Codes*

These codes are long block length codes derived from short block codes. An *interleaved code* is obtained by taking $A$ code words of a given code and forming a new code word by interleaving the $A$ code words; that is, if the $j$th code word of the original code is $\mathbf{x}_j = (x_{j1}, x_{j2}, \ldots, x_{jN})$, $j = 1, 2, \ldots, A$, the new code word is $\mathbf{x}_I = (x_{11}, x_{21}, \ldots, x_{A1}, x_{12}, x_{22}, \ldots, x_{A2}, \ldots, x_{1N}, x_{2N}, \ldots, x_{AN})$. The paramater $A$ is called the *degree* of interleaving. The code is said to be interleaved to a depth of $A$. It is possible to show that interleaved codes possess the following properties.

1. If the original code is an $(N, K)$ cyclic code, the interleaved code is a $(AN, AK)$ cyclic code.
2. If $g(D)$ is the generator polynomial of the original cyclic code, then $g(D^A)$ is the generator polynomial of the interleaved code.
3. If the original code can correct a single burst of length $\tau$ or less, the interleaved code can correct a single burst of length $A\tau$ or less.
4. If the original code has maximum burst-error-correcting capability, that is, the code satisfies the Reiger bound, the interleaved code also has maximum burst-error-correcting capability.

We can visualize interleaving a code as the process of forming an $A \times N$ matrix of $A$ code words row by row and transmitting the matrix column by column. Transmission cannot begin until all the $A$ code words arrive. In practice, it is desirable to be able to transmit a column whenever a new code word becomes available. This is accomplished by *delayed interleaving.*

In delayed interleaving, instead of arranging code words along rows, we arrange them down the diagonals as shown below. Here the $i$th code word $\mathbf{x}_i$ is denoted $(x_{i,1}, x_{i,2}, \ldots, x_{i,N})$.

$$\begin{array}{cccccccc}
x_{i-(N-1),1} & & \cdots & x_{i-2,1} & x_{i-1,1} & & x_{i,1} & \\
& x_{i-(N-1),2} & \cdots & & x_{i-2,2} & & x_{i-1,2} & x_{i,2} \\
& & \ddots & & \vdots & & \vdots & \vdots \\
& & & & x_{i-(N-1),(N-1)} & x_{i-(N-2),(N-1)} & & \\
& & & & & x_{i-(N-1),N} & x_{i-(N-2),N} &
\end{array}$$

When arranged in this manner, each new code word completes a column of this array. For example, the code word $\mathbf{x}_{i+1}$ completes the column containing $x_{i+1,1}, x_{i,2}, \ldots, x_{i-(N-2),N}$. Often the term *frame* is used to denote a column of the interleaved array. The interleaving technique described here is known as *1-frame delayed interleaving.* A generalization of this technique is the *$\lambda$-frame delayed interleaving.* In this technique, successive elements of a code word, such as $x_{i,j}$ and $x_{i,j+1}$, appear $\lambda$ frames apart in the interleaved array; that is $x_{i,j+1}$ is placed $\lambda$ columns right and one row down with respect to $x_{i,j}$.

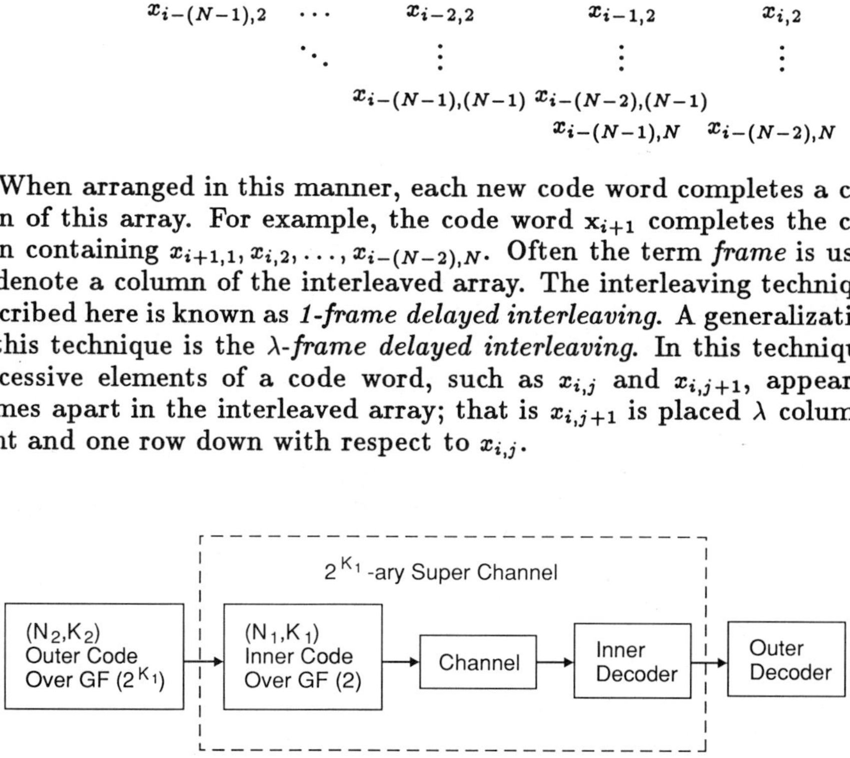


**Figure 3.11-1.** The concatenated coding idea.

The principle behind *concatenated codes* is easily explained. Consider a code word generated by an $(N_1, K_1)$ binary code $\mathcal{C}_1$. After transmission through a channel, the code word arrives at the decoder, possibly with errors. The decoder decodes the erroneously received word and often decodes it correctly. But occasionally the received word is incorrectly decoded. We may, therefore, view the encoder, the channel, and the decoder as a super channel whose input and output alphabet consist of elements from GF$(2^{K_1})$. Now consider transmitting $K_2K_1$ information digits. We can split the $K_2K_1$ information digits into $K_2$ blocks of $K_1$ information digits, that is, $K_2$ symbols from GF$(2^{K_1})$. We can encode the

$K_2$ symbols using another code $C_2$, an $(N_2, K_2)$ code over GF$(2^{K_1})$, and the $N_2$ output symbols can then be transmitted over the super channel. The resulting scheme, called a concatenated code (initially by Forney [3-36]), has an $(N_2, K_2)$ *outer code* and an $(N_1, K_1)$ *inner code* producing an overall $(N_1N_2, K_1K_2)$ code. The scheme is illustrated in Fig. 3.11-1. It can be generalized to $q$-ary channels. The inner code will then be over GF$(q)$ and the outer code over GF$(q^{K_1})$. One can show that the minimum distance of the concatenated code is the product of the minimum distances of the inner and outer codes.

**Example 3.11-2: Compact Disc Interleaving.**

> As an example of a system incorporating interleaving and concatenation, consider the compact disc digital audio recording system [3-37]. The audio signal is sampled at the rate of 44.1kHz and each sample is quantized to 16 bits. At each sampling instant, actually two samples are taken, one corresponding to the left and the other to the right channel. The samples are grouped into frames, with each frame consisting of twelve samples, six from the left and six from the right channel; a frame thus contains 24 8-bit symbols (or bytes). These 24 are encoded by a (28,24) Reed–Solomon outer code $C_2$ over GF$(2^8)$. The code words of $C_2$ are then interleaved to a depth of 28 using the 4-frame delayed interleaving technique. Each column of the interleaved array, consisting of 28 symbols, is encoded by a (32,38) Reed–Solomon inner coder $C_1$ over GF$(2^8)$. Code words of $C_1$ are scrambled such that odd-numbered symbols of each code word are grouped with even-numbered symbols of the one that immediately follows it. The overall scheme of interleaving and concatenating creates a *cross-interleaved* Reed–Solomon code.

A *product code* is an $(N_1N_2, K_1K_2)$ code in which each code word forms an $N_1 \times N_2$ rectangular array such that each row is a code word from an $(N_1, K_1)$ code $C_1$ and each column is a code word from an $(N_2, K_2)$ code $C_2$. This is illustrated in Fig. 3.11-2. Parity-check digits on rows are formed using rules of $C_1$ and those on columns using rules of $C_2$. The bottom right-hand rectangle in Fig. 3.11-2, giving parity-check digits on parity-check digits, may be formed using either rules given by $C_1$ on column check digits or those given by $C_2$ on row check digits. Either method will lead to the same set of $(N_1 - K_1) \times (N_2 - K_2)$ check digits.

We can easily show how a product code extends the burst-error-correcting capability of $C_1$ and $C_2$ to longer bursts. Let $C_1$ and $C_2$ be capable of correcting bursts of length $\tau_1$ and $\tau_2$, respectively. Let us assume that we transmit a code word from the product code formed by $C_1$ and $C_2$ on a row-by-row basis and decode the received word on

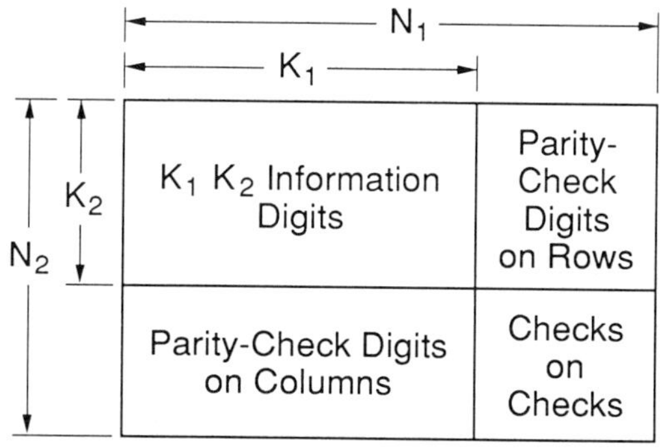


**Figure 3.11-2.** The structure of a code word of a product code.

a column-by-column basis. Suppose a burst error of length $N_1\tau_2$ or less occurs during transmission of a code word. Regardless of when the burst error occurs during transmission, each column is affected by a burst of length $\tau_2$ or less and is therefore correctable by $\mathcal{C}_2$ when decoding column by column. The product code can therefore correct burst errors of length $N_1\tau_2$ or less. Similarly, if we transmit each code word on a column-by-column basis and decode the received word on a row-by-row basis, we can argue that the product code can correct burst errors of length $N_2\tau_1$ or less. The product code is therefore capable of correcting burst errors of length $\tau'$ or less, where $\tau' = \max(N_1\tau_2, N_2\tau_1)$.

It is possible to show that the minimum distance of a product code is the product of the minimum distances of the two codes forming the product code. Product codes were proposed by Elias [3-37] and they can be extended to higher dimensions. For example, in three dimensions an $(N_1N_2N_3, K_1K_2K_3)$ code may be visualized as a box with $N_3$ arrays like the one in Fig. 3.11-2.

## 3.12 AUTOMATIC REPEAT REQUEST PROTOCOLS

Our attention in this chapter has so far focused on *forward error correction* schemes. These employ error correcting codes but do not deal further with incorrectly decoded information. They simply deliver decoded information to the user, whether or not it is correctly decoded. In systems where a reverse channel is available, as in computer networks, a different strategy is employed. A good error-detecting code is used to encode information. The code word, usually called a frame, is then transmitted. On detecting errors in the received frame, the receiver requests the transmitter to retransmit that frame. It does so as many times as necessary until the receiver acknowledges correct reception. Many different protocols† can be used by the transmitter-receiver pair in transmitting frames and requesting retransmission of frames in error. These are referred to as *automatic repeat request*, or ARQ, protocols.

In an error-detection system, the receiver computes the syndrome of the received word and declares error detection if the syndrome is nonzero. The receiver assumes error-free transmission whenever the syndrome is zero. Hence, the receiver commits a decoding error whenever the channel error changes the transmitted code word to another code word. Based on random coding arguments, it is possible to show for an $(N, K)$ block code that the probability of undetected error $P_{\text{ud}}$ is bounded by

$$P_{\text{ud}} \leq 2^{-(N-K)} \left[1 - (1-p)^N\right], \tag{3.12-1}$$

where $p$ is the crossover probability of the BSC. A proof appears in ref. [3-8]; a stronger result known as the *expurgated* random coding bound appears in [3-3].

In addition to $P_{\text{ud}}$, we will find it useful to define the following:

$$\begin{aligned} P_{\text{cd}} &= P\{\text{correctly decoding the received word}\} \\ &= P\{\text{No channel error occurs}\} \\ &= (1-p)^N; \end{aligned} \tag{3.12-2}$$

$$\begin{aligned} P_{\text{ed}} &= P\{\text{Error detection in the received word}\} \\ &= 1 - (P_{\text{ud}} + P_{\text{cd}}). \end{aligned} \tag{3.12-3}$$

The codes most widely used in error detection systems are cyclic codes and the parity check bits they generate are called *cyclic redundancy check* (CRC) bits. In addition to their ability to detect random errors, cyclic codes are effective in detecting burst errors. In fact, an $(N, K)$ cyclic code is capable of detecting bursts of length $N - K$ or less (Problem 3-37). In the computer networking field, two standard cyclic codes are

†The term protocol refers to the rules and conventions used in the interaction between two entities such as a transmitter and a receiver or two computer programs executing on two different machines in a computer network.

used that are generated by a polynomial of degree 16. They are called CRC-16 and CRC-CCITT‡ and the polynomials are

$$g(D) = D^{16} + D^{15} + D^2 + 1, \qquad \text{[for CRC-16];}$$
$$g(D) = D^{16} + D^{12} + D^5 + 1, \qquad \text{[for CRC-CCITT].}$$

Another code recommended by CCITT for use in broadband networks has $g(D) = D^8 + D^2 + D + 1$.

These cyclic codes are actually the product of a primitive polynomial and the term $(D + 1)$. $(D + 1)$ is a factor of the generator polynomials given above since g(1) is 0 for each one. By making $(D + 1)$ a factor of $g(D)$, the design ensures that all odd numbers of errors are detected (see Problem 3-20).

There exist three basic types of ARQ schemes, *the stop-and-wait, go-back-n,* and *selective-repeat*, which we now describe.

## Stop-and-Wait ARQ

This is the simplest of the three ARQ schemes. The transmitter sends a frame and waits till the receiver acknowledges error-free reception of that frame before it transmits the next frame. On receiving a frame without errors, the receiver acknowledges error-free reception by sending an ACK message; otherwise it acknowledges negatively by sending a NAK. If the transmitter received an ACK it transmits the next frame. On receiving a NAK, it retransmits the previous frame. Figure 3.11-1 illustrates the stop-and-wait ARQ scheme. This scheme, though simple, is inefficient on channels, such as the satellite channel, that have a long round-trip delay.

Let $T_f$ be the time required to transmit one frame and $T_I$ the time between successive frame transmissions. $T_I$ is the round-trip delay during which the transmitter remains idle.

Ideally, if the channel is noiseless, one frame can be transmitted in $T_f + T_I$ seconds, but with channel noise and the resulting retransmissions, the average time required to successfully transmit a frame error-free will be greater than $T_f + T_I$. A frame will be retransmitted whenever the receiver detects errors. This occurs with probability $P_{\text{ed}}$ given by (3.12-3). For most codes used in practice, $P_{\text{ud}} << P_{\text{ed}}$ and hence

‡The Comité Consultatif International Télégraphique et Téléphonique (CCITT) is a world standards body that sets out recommendations for communications systems.

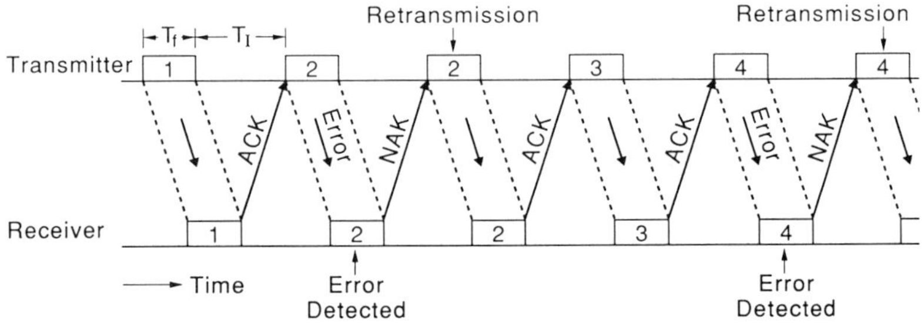


**Figure 3.12-1.** **Example illustrating the stop-and-wait ARQ scheme.**

$$\begin{aligned} P_{\mathrm{ed}} &\approx 1 - P_{\mathrm{cd}} \\ &= 1 - (1-p)^N. \end{aligned}$$

We now compute the average time required for the successful sending of a frame by the transmitter. The original transmission of a frame is successfully received with probability $(1 - P_{\mathrm{ed}})$ and, from Fig. 3.12-1, the time for transmission is $T_t = T_I + T_f$. If the original transmission is erroneously received and the first retransmission is successful, the time taken is $2T_t$ and the probability that this is the case is $(1 - P_{\mathrm{ed}})P_{\mathrm{ed}}$. Similarly, if successful reception of a frame occurs only on the second retransmission, the total time taken is $3T_t$ and the probability is $(1 - P_{\mathrm{ed}})P_{\mathrm{ed}}^2$, and so on. Hence, the average time required for the successful transmission of a frame, or equivalently, successful reception of a frame is given by

$$\begin{aligned} T_{\mathrm{av}} &= T_t(1 - P_{\mathrm{ed}}) + 2T_t(1 - P_{\mathrm{ed}})P_{\mathrm{ed}} + 3T_t(1 - P_{\mathrm{ed}})P_{\mathrm{ed}}^2 \\ &= T_t(1 - P_{\mathrm{ed}}) \sum_{\ell=1}^{\infty} \ell P_{\mathrm{ed}}^{\ell-1} \\ &= \frac{T_t}{1 - P_{\mathrm{ed}}}. \end{aligned}$$

How efficiently does the stop-and-wait ARQ scheme utilize the channel? $T_{\mathrm{av}}/T_f$ frames would pass through the channel in $T_{\mathrm{av}}$ seconds if the ARQ scheme were not used. Thus the *throughput efficiency* of the scheme is

$$\rho_{SW} = \frac{1}{T_{av}/T_f}\left(\frac{K}{N}\right),$$

where the factor $K/N$ accounts for the fact that each $N$-bit frame carries only $K$ information bits. Substituting for $T_{av}$ into $\rho_{SW}$ yields

$$\begin{aligned}\rho_{SW} &= \left(\frac{1 - P_{ed}}{1 + \dfrac{T_I}{T_f}}\right)\left(\frac{K}{N}\right) \\ &= \frac{(1-p)^N}{1+F_I}R, \qquad (3.12\text{-}4)\end{aligned}$$

where $R = \frac{K}{N}$ is the code rate and $F_I = \frac{T_I}{T_f}$ is the number of frames that could have been transmitted during time $T_I$ had the transmitter not been idle.

## Go-Back-n ARQ

Unlike the stop-and-wait scheme, the go-back-$n$ scheme does not wait for ACKs or NAKs from the receiver to transmit the next frame. The frames are instead transmitted continuously, one after another, until a NAK is received. The transmitter then begins to send the frame that is negatively acknowledged by the receiver and all the frames following it, whether or not the following frames were correctly received.

Figure 3.12-2 provides an example of the go-back-$n$ scheme. The parameter $n$ is the number of frames that can be transmitted in one round-trip delay. In this example, $n$ equals 4. The figure shows frame 4 as being received in error, after which the receiver sends a NAK to the transmitter. The NAK arrives when the transmitter has just completed sending frame 7. Instead of transmitting the next frame, frame 8, the transmitter goes back by 4 frames (in general, by $n$ frames) and transmits frame 4 and all the successive frames.

Let us now compute the throughput efficiency of the go-back-$n$ scheme. As in the stop-and-wait scheme, the original transmission of a frame succeeds with probability $(1 - P_{ed})$. But now the time required is just $T_f$ instead of $T_t$. If the original transmission is detected to be erroneous, then the $(n-1)$ frames that follow the frame in error will be discarded, whether in error or not. If the first retransmission is successful, the total time for successful transmission is $(n+1)T_f$, and the

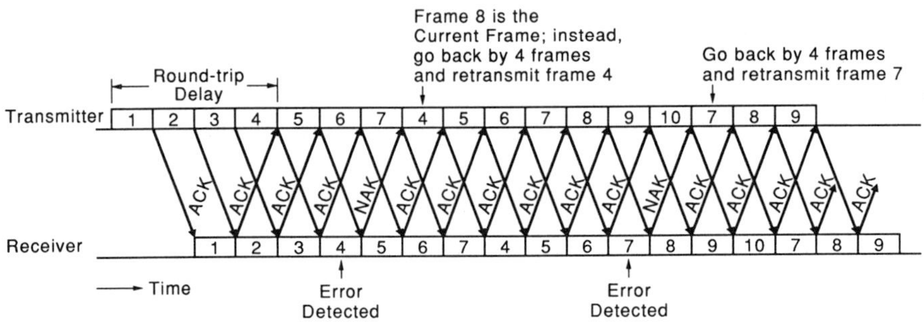


**Figure 3.12-2.** Example to illustrate the working of the go-back-$n$ protocol; $n$ equals four frames.

probability of this case is $P_{\text{ed}}(1-P_{\text{ed}})$. Similarly, if successful reception of the frame takes place only on the second retransmission, then the time required is $(2n+1)T_f$ with probability $P_{\text{ed}}^2(1-P_{\text{ed}})$, and so on.

The average time required for successful transmission of a frame is

$$\begin{aligned} T_{\text{av}} &= T_f(1-P_{\text{ed}}) + (n+1)T_f(1-P_{\text{ed}})P_{\text{ed}} \\ &\quad + (2n+1)T_f(1-P_{\text{ed}})P_{\text{ed}}^2 + \cdots \\ &= T_f\left[1+\frac{nP_{\text{ed}}}{1-P_{\text{ed}}}\right]. \end{aligned}$$

The throughput efficiency of the go-back-$n$ ARQ scheme is

$$\begin{aligned} \rho_{\text{GB}} &= \frac{1}{T_{\text{av}}/T_f}R \\ &= \frac{1-P_{\text{ed}}}{1+(n-1)P_{\text{ed}}}R. \end{aligned} \tag{3.12-5}$$

Observe that $\rho_{\text{SW}}$ in (3.12-4) and $\rho_{\text{GB}}$ in (3.12-5) are similar in form. In fact, $F_I$, the number of frames that can be transmitted during the idle time, equals $n-1$. Hence $\rho_{\text{SW}}$ can be rewritten as

$$\rho_{\text{SW}} = \frac{1-P_{\text{ed}}}{1+(n-1)}R. \tag{3.12-6}$$

For channels with small $p$, $P_{\text{ed}}$ is small and, hence, the go-back-$n$ scheme is highly efficient compared to the stop-and-wait scheme. For channels with large error rate, the performance of the go-back-$n$ scheme deteriorates. Observe that for $n = 1$ the stop-and-wait and go-back-$n$ schemes have the same throughput efficiency.

## Selective-Repeat ARQ

Here the transmitter sends frames continuously but, unlike the go-back-$n$ scheme, retransmits only the negatively acknowledged frames. Fig. 3.12-3 illustrates the selective-repeat ARQ scheme. Usually, the receiver must deliver the frames in the order in which they were transmitted and the receiver must, therefore, keep a buffer in which to store frames that follow an erroneous frame. The buffer space required for an ideal scheme is infinite.

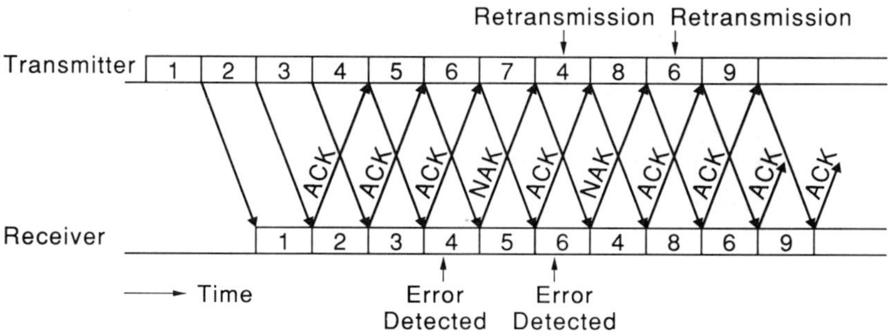


**Figure 3.12-3.** **Example illustrating the selective-repeat ARQ scheme.**

We now compute the throughput efficiency of this scheme. Since the scheme retransmits only the frame in error, the probability that $\ell$ retransmissions are required is $P_{\mathrm{ed}}^{\ell}(1-P_{\mathrm{ed}})$ and the channel time required is $(\ell+1)T_f$. The average time required for a successful frame transmission is

$$T_{\mathrm{av}} = T_f \sum_{\ell=0}^{\infty}(\ell+1)P_{\mathrm{ed}}^{\ell}(1-P_{\mathrm{ed}})$$
$$= \frac{T_f}{1-P_{\mathrm{ed}}}.$$

Hence, the throughput efficiency of the selective-repeat ARQ scheme is

$$\rho_{SR} = \frac{1}{T_{av}/T_f} R$$
$$= (1 - P_{ed}) R. \tag{3.12-7}$$

With a finite receiver buffer, it is possible that the earliest frame in the buffer is received in error and no error-free copy of the earliest frame arrives until after the buffer fills up with correct frames. Once the buffer is full, the receiver must reject any new frame other than the earliest frame in the buffer. Consequently, the throughput for the finite buffer selective-repeat scheme is upper-bounded by (3.12-7). Refer to [3-8] for an analysis of this scheme.

We have not treated *hybrid ARQ* schemes that employ a basic ARQ scheme together with forward error correction. Lin and Costello [3-8] analyze hybrid ARQ schemes. Our treatment of ARQ schemes has assumed ideal conditions such as ACKs and NAKs that always arrive at the transmitter, processing time that is negligible, and round-trip delay that does not vary. Bertsekas and Gallager [3-39] give a rigorous treatment of non-ideal conditions. Refs. [3-40] and [3-41] also treat the role of ARQ in data networking.

# PROBLEMS

3-1 a) Derive (3.1-3) from (3.1-1).

b) Show the following equalities concerning mutual information:

$$I(X;Y) = I(Y;X),$$
$$I(X;Y) = H(Y) - H(Y|X),$$
$$I(X;Y) = H(X) + H(Y) - H(X,Y),$$

where $H(X,Y)$, called the *joint entropy* of $X$ and $Y$, is given by

$$H(X,Y) = \sum_x \sum_y P(X,Y) \log \frac{1}{P(x,y)}.$$

3-2 a) Compute $I(X;Y)$ for each of the two channels given below. The input alphabet has the distribution $p_x(0) = \frac{1}{4}, p_x(1) = \frac{1}{4}$, and

$p_x(2) = \frac{1}{2}$.

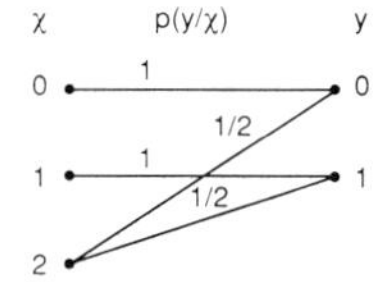


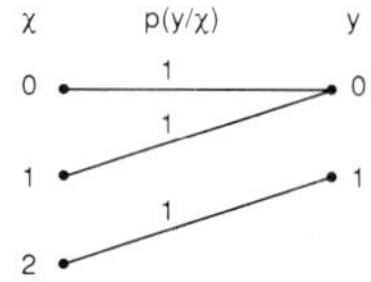


b) Consider the binary erasure channel (BEC) mentioned in Section 3.1. Calculate $I(X;Y)$ and then determine the input distribution that maximizes $I(X;Y)$; this is the channel capacity of the BEC. Let the input distribution be given by $p_x(0) = p_o$ and $p_x(1) = p_1 = 1 - p_o$; the outputs are 0, 1, and e; the channel is given by $p_{Y|X}(0|0) = p_{Y|X}(1|1) = 1 - \epsilon$ and $p_{Y|X}(e|0) = p_{Y|X}(e|1) = \epsilon$.

3-3 Let $\mathcal{V}$ be a vector space over a field $\mathcal{F}$ and $\mathcal{S}$ a subset of $\mathcal{V}$. $\mathcal{S}$ is called a subspace of $\mathcal{V}$ if $\mathcal{S}$ forms a vector space over $\mathcal{F}$. We showed in Section 3.2 that the set of all $N$-tuples forms a vector space over GF(2). Find a three-dimensional subspace of the space formed by the set of all 4-tuples over GF(2).

3-4 Consider the following set of 5-tuples:

$$\mathcal{S} = \Big\{(00000), (01111), 01001), (00110), (10011), \\ (11100), (11010), (10101)\Big\}.$$

a) Show that $\mathcal{S}$ is a vector space over GF(2).
b) What is the dimension of $\mathcal{S}$?
c) Find a set of basis vectors.

3-5 Consider again the set $\mathcal{S}$ from Problem 3-4 and view this as a binary code over GF(2).
a) Express each code word digit as a linear combination of the information digits.
b) Find the generator and parity-check matrices for the code.
c) Is this a systematic code? If not, obtain the systematic form for the generator matrix.
d) Determine the minimum distance of this code.

3-6 For the ($N$,1) repetition code, the (6,3) code, and the single-parity-check code of Section 3.2, verify that the relation $\mathbf{GH}^T = \mathbf{0}$ holds.

3-7 Let a given binary $(N, K)$ code have odd minimum distance. We may *extend* this code by adding an overall check bit equal to the sum of all the other digits in the code word.
a) Show that, if $\mathbf{H}$ is the parity-check matrix of the original code, then the parity-check matrix of the extended code is

$$\mathbf{H}' = \begin{bmatrix} 1 & 1 \ldots & 1 \\ 0 & & \\ 0 & \mathbf{H} & \\ \vdots & & \\ 0 & & \end{bmatrix}.$$

b) Show that the minimum distance of the extended code is $d_{\min} + 1$, where $d_{\min}$ is the minimum distance of the original code.
c) Show that the extended code is an $(N+1, K)$ linear code.

3-8 a) Obtain an (8,4) code by extending the (7,4) Hamming code.
b) Determine its minimum distance.
c) Show that the (8,4) code can correct all single errors and detect all double errors.
d) Show that the (8,4) code and its dual are identical. Such codes are called *self-dual* codes.

3-9 a) Let $\mathbf{G} = [\mathbf{I}_K; \mathbf{P}]$ be the generator matrix of an $(N, K)$ code. Show that the code is a self-dual code if and only if $\mathbf{P}$ is square and $\mathbf{P}\mathbf{P}^T = \mathbf{I}$.
b) Construct self-dual codes with $N = 4, 6$, and 8.
c) Show that $N = 2K$ for self-dual codes.

3-10 Section 3.3 lists eight properties of the standard array and shows that Property 6 holds. Show that all the other properties hold.

3-11 a) For the (8,4) code of Problem 3-8, construct a standard array.
b) Construct a syndrome decoding table for this code.
c) Use the standard array to decode the two received words

$$\mathbf{y}_1 = (0\ 1\ 1\ 0\ 1\ 1\ 0\ 0),$$
$$\mathbf{y}_2 = (1\ 1\ 0\ 0\ 1\ 0\ 0\ 1).$$

d) Decode $\mathbf{y}_1$ and $\mathbf{y}_2$ using the syndrome decoding table.
e) Compute the error probability for the code when used to correct error patterns caused by the BSC with crossover probability $p = 10^{-3}$.

3-12 An $\ell$th order Reed-Muller† code has the generator matrix

$$\mathbf{G} = \begin{bmatrix} \mathbf{G}_0 \\ \mathbf{G}_1 \\ \vdots \\ \mathbf{G}_\ell \end{bmatrix}.$$

†Reed-Muller codes were used by Mariner 9 in 1972 to transmit photographs and also by Voyager 1 and Voyager 2 spacecrafts to transmit pictures of Jupiter and Saturn during 1979-1981.

Here $\mathbf{G}_0$ is an all 1s vector of length $N = 2^m$, $m$ any positive nonzero integer and $\ell < m$; $\mathbf{G}_1$ is an $m \times 2^m$ matrix whose columns are the $2^m$ binary $m$-tuples; and any $\mathbf{G}_n, 1 \leq n \leq \ell$, has for its rows all possible products of rows of $\mathbf{G}_1$ when taken $n$ at a time. The product of two vectors is the result of component-wise (bit-wise) multiplication.

a) Show that the 0th order Reed-Muller code is the ($N$, 1) repetition code.

b) Construct the first order Reed-Muller code for $m = 3$. What is its relationship to the (8, 4) code of Problem 3-8? Is this code self-dual?

c) Construct the second-order Reed-Muller code for $m = 4$. How is it related to the (15, 11) Hamming code?

d) For any Reed-Muller code, show that

$$N = 1 + \binom{m}{1} + \ldots + \binom{m}{\ell},$$

$$N - K = 1 + \binom{m}{1} + \ldots + \binom{m}{m - \ell - 1}.$$

e) Show that the minimum distance for the first-order Reed-Muller code is $2^{m-1}$ and that it is a $(2^m, m + 1)$ code.

3-13 A code can be *shortened* by deleting information symbols. This can be done by deleting columns of $\mathbf{P}^T$ in $\mathbf{H} = [\mathbf{P}^T; \mathbf{I}_{N-K}]$. Consider deleting all columns of $\mathbf{P}^T$ of even weight from the parity-check matrix of the $(2^m - 1, 2^m - m - 1)$ Hamming code.

a) Derive the new code parameters $N'$ and $K'$ in terms of those of the original code.

b) Show that the minimum distance of the new code is 4.

c) Obtain a decoding algorithm to correct all single errors and simultaneously detect all double errors.

3-14 a) Show that a linear code can correct all error patterns containing $t_c$ or fewer errors and simultaneously detect all error patterns containing $t_d$ ($t_d \geq t_c$) or fewer errors provided $d_{\min} \geq t_c + t_d + 1$, where $d_{\min}$ is the minimum distance of the code.

b) Assume that the binary erasure channel is used to transmit code words of a linear code and that $t_c$ errors and $t_e$ erasures occur. Note that the positions of the erasures are known but not those of the errors. Show that the code can correct the $t_c$ errors and fill in the erasures provided $d_{\min} \geq 2t_c + t_e + 1$.

3-15 a) Show that the (10,6) code whose generator matrix is given below is a cyclic code.

$$\mathbf{G} = \begin{bmatrix} 1&0&0&0&0&0&1&1&1&1 \\ 0&1&0&0&0&0&1&0&0&0 \\ 0&0&1&0&0&0&0&1&0&0 \\ 0&0&0&1&0&0&0&0&1&0 \\ 0&0&0&0&1&0&0&0&0&1 \\ 0&0&0&0&0&1&1&1&1&1 \end{bmatrix}.$$

b) Determine the generator polynomial of the code.
c) Verify that $g(D)$ is a factor of $D^{10} - 1$.

3-16 a) Obtain a parity-check matrix for the (10,6) code in Problem 3-15 in a form suitable for implementation by a shift register circuit.
b) Draw two encoders for the (10,6) code, one based on $g(D)$ obtained in Problem 3-15(c) and the other based on the parity-check matrix obtained in (a).

3-17 a) Show that $g(D) = D^4 + D^3 + 1$ generates the (15,11) binary cyclic Hamming code.
b) Determine the parity-check polynomial $h(D)$.
c) Express this code in terms of its generator and parity-check matrices in systematic form.
d) Design a systematic encoder for the code.

3-18 Design a single-error-correcting decoder for the (15,11) code in Problem 3-17.

3-19 a) Determine the code word corresponding to $u(D) = D^3+1$ generated by $g(D) = D^4 + D^3 + 1$.
b) Compute the syndrome corresponding to $y(D) = D^5 + D^4 + D^3 + D^2$.
c) Using the decoder designed in Problem 3-18, decode the received sequence $y(D) = D^5 + D^4 + D^3 + D^2$.

3-20 Consider a binary $(N, K)$ cyclic code with $g(D)$ as the generator polynomial.
a) Show that if $g(D)$ does not contain $(D + 1)$ as a factor, the code has only code words of even weight. Hence, any odd number of errors can be detected.
b) If $N$ is odd, and $g(D)$ does not have $D + 1$ as a factor, then the code generated by $g(D)$ has the all-1s code word.

3-21 Consider the (15,7) two-error-correcting BCH code of Example 3.6-4, whose generator polynomial is $D^8 + D^7 + D^6 + D^4 + 1$. Using the Meggitt decoder decode each of the following received word polynomials:

i) $y(D) = D^{12} + D^4 + D^3 + D^2 + 1$,
ii) $y(D) = D^{11} + D^2 + 1$,
iii) $y(D) = D^{14} + D^{11} + D^4 + 1$.

3-22 Decode the received polynomials of Problem 3-21 using the error-trapping decoder.

3-23 a) Factor $D^{15} - 1$ as a product of irreducible polynomials. (*Hint*: Use Properties 3 and 6 (Section 3.7). Find the minimal polynomials of $1, \alpha, \alpha^3, \alpha^5$, and $\alpha^7$.)
b) How many binary cyclic codes of length 15 exist?
c) Determine their generator polynomials.

3-24 Decode the received polynomials of Problem 3-21 using Peterson's algorithm.

3-25 Rewrite the Massey-Berlekamp algorithm that incorporates modifications for the binary case.

3-26 Use the Massey-Berlekamp algorithm to decode the received polynomials of Problem 3-21.

3-27 Consider the 3-error-correecting (15,5) BCH code with $g(D) = D^{10} + D^8 + D^5 + D^4 + D^2 + D + 1$. Decode the following received word polynomials using the Massey-Berlekamp algorithm.
i) $y(D) = D^7 + D^3 + D + 1$,
ii) $y(D) = D^{13} + D^{11} + D^2 + D + 1$,
iii) $y(D) = D^{13} + D^{10} + D^7 + D^6 + D^5 + D^2 + 1$.

3-28 Consider decoding the (15,5) BCH code of Problem 3-27 using the error-trapping decoder.
a) Show that all error patterns of weight 3 or less can be confined to $15 - 5 = 10$ consecutive positions (end-around included) except $\mathbf{e} = (1\ 0\ 0\ 0\ 0\ 1\ 0\ 0\ 0\ 0\ 1\ 0\ 0\ 0\ 0)$ and its cyclic shifts.
b) Decode $y(D) = D^9 + D^7 + D^6 + D^5 + 1$ using the error-trapping decoder.

3-29 Consider again the (15,5) BCH code of Problem 3-27. We may modify the error-trapping decoder to decode all the error patterns of weight 3 or less. To see how, we observe that the syndrome of $\mathbf{e} = (1\ 0\ 0\ 0\ 0\ 1\ 0\ 0\ 0\ 0\ 1\ 0\ 0\ 0\ 0)$ is $s(D) = D^{14} \bmod g(D) + D^9 + D^4$, from which $w(s(D) - D^{14} \bmod g(D)) = 2$. The strategy is therefore to check if $w(\mathrm{s}) \leq 3$ or $w(s(D) - D^{14} \bmod g(D)) \leq 2$. The received word is shifted as many times as necessary, and the tests are performed each time. If the second of the two tests is satisfied the error pattern is $D^{15-\ell}(D^{14} + s_\ell(D) - D^{14} \bmod g(D)) \bmod D^{15} - 1$, where $\ell$ is the number of cyclic shifts and $s_\ell(D)$ is the syndrome register contents after $\ell$ shifts.
a) Show that the modifications do indeed work.
b) Use the modified error-trapping decoder to decode $y(D) = D^{14} + D^{10} + D^9 + D^7 + D^5 + D^4 + D^3 + D$.

3-30 a) Construct GF($2^5$) using the primitive polynomial $p(D) = D^5 + D^2 + 1$.

b) Determine the generator polynomials for the (31,26), (31,21), and (31,16) BCH codes.

c) Construct circuits for these codes that perform Chien's search.

3-31 Consider the 3-symbol-correcting Reed–Solomon code in Example 3.10-5. Decode the received polynomial $y(D) = \alpha D^{10} + \alpha^{10} D^5 + \alpha D$ by first determining the error-location polynomial and then the error-evaluator polynomial.

3-33 Consider the (15,9) binary cyclic code with a burst error correcting capability equal to 3 and given by $g(D) = D^6 + D^5 + D^4 + D^3 + 1$ (see Table 3.11-1).

a) Show that there are 60 bursts of length 3 or less.

b) Show that the code does correct all bursts of length 3 or less. (*Hint*: The syndromes of these bursts must all be distinct.)

3-34 Consider the primitive polynomial $p(D) = D^6 + D + 1$ over GF(2). This generates GF($2^6$). Use $p(D)$ to determine the Fire code that can correct bursts of lengh 7 or less.

3-35 Consider the (7,4) Hamming code with

$$\mathbf{G} = \begin{bmatrix} 1\,0\,0\,0\,1\,1\,0 \\ 0\,1\,0\,0\,0\,1\,1 \\ 0\,0\,1\,0\,1\,1\,1 \\ 0\,0\,0\,1\,1\,0\,1 \end{bmatrix}.$$

The code is interleaved to a depth of 5. Determine the transmitted sequence if the following information words are encoded:

(0 0 1 0), (1 0 0 1), (1 1 1 1), (0 1 0 0), (1 1 1 0).

3-36 Construct a concatenated code using the (7,4) binary Hamming code as the inner code and the (15,9) 3-symbol-correcting Reed–Solomon code as the outer code.

a) What is the overall length of the code?

b) How many binary information symbols does each code word contain?

c) Discuss its error correcting capability.

3-37 Show that any $(N, K)$ cyclic code is capable of detecting bursts of length $N - K$ or less. (*Hint*: Any burst of length $N - K$ or less can be expressed in the form $D^{\ell}\gamma(D)$, where $\gamma(D)$ is of degree $N - K - 1$ or less. Show that $g(D)$ is not a factor of $D^{\ell}\gamma(D)$ and the syndrome is nonzero.)

3-38 Consider the use of the BEC of Problem 3-2 to transmit an independent and identically distributed binary source. Whenever an erasure occurs, assume that the receiver informs the transmitter over a reverse noise-free channel that an erasure has occurred. The transmitter then repeats the digit that was erased. Determine the expected number of digits transmitted per channel use.

3-39 Consider using the 3-error-correcting (255,239) binary BCH code (see Table 3.8-1) for use in ARQ systems. The channel used for transmission of frames can be modeled as a BSC with $p = 10^{-2}$. The channel has a round trip delay equal to 7 frames. Compute the throughput of i) stop-and-wait, ii) go-back-$n$, with $n = 7$, and iii) selective-repeat ARQ schemes. Compare the schemes in this context.

# REFERENCES


[3-1] W. W. Peterson, *Error-Correcting Codes*, M.I.T. Press, Cambridge, Mass., 1961.

[3-2] E. R. Berlekamp, *Algebraic Coding Theory*, McGraw-Hill, New York, 1968; also revised 1984 edition, Aegean Park Press.

[3-3] W. W. Peterson and E. J. Weldon, *Error-Correcting Codes*, M.I.T. Press, Cambridge, Mass., 1972.

[3-4] I. F. Blake and R. C. Mullin, *An Introduction to Algebraic and Combinatorial Coding Theory*, Academic Press, New York, 1976.

[3-5] F. J. MacWilliams and N. J. A. Sloan, *The Theory of Error-Correcting Codes*, North-Holland, New York, 1977.

[3-6] G. C. Clark and J. Bibb Cain, *Error-Correction Coding for Digital Communications*, Plenum, New York, 1981.

[3-7] V. Pless, *Introduction to the Theory of Error-Correcting Codes*, Wiley-Interscience, New York, 1982.

[3-8] S. Lin and D. J. Costello, Jr., *Error Control Coding: Fundamentals and Applications*, Prentice-Hall, Englewood Cliffs, N.J., 1983.

[3-9] R. E. Blahut, *Theory and Practice of Error Control Codes*, Addison-Wesley, Reading, Mass., 1984.

[3-10] A. M. Michelson and A. H. Levesque, *Error-Control Techniques for Digital Communications*, John Wiley & Sons, New York, 1985.

[3-11] S. A. Vanstone and P. C. van Oorschot, *An Introduction to Error Correcting Codes with Applications*, Kluwer Academic, Boston, Mass., 1989.

[3-12] R. G. Gallager, *Information Theory and Reliable Communication*, McGraw-Hill, New York, 1968.

[3-13] R. W. Lucky, J. Salz, and E. J. Weldon, Jr., *Principles of Data Communication*, McGraw-Hill, New York, 1968.

[3-14] R. J. McEliece, *The Theory of Information and Coding*, Addison-Wesley, Reading, Mass., 1977.

[3-15] C. E. Shannon, "A mathematical theory of communication", *Bell Syst. Tech. J.*, vol. 27, pp. 379–429, (Part I), pp. 623–656 (Part II), July 1948.

[3-16] R. W. Hamming, "Error detecting and error correcting codes", *Bell Syst. Tech. J.*, vol. 29, pp. 147–160, April 1950.

[3-17] M.J.E. Golay, "Notes on digital coding", *Proc. IRE*, vol. 37, p. 657, June 1949.

[3-18] E. Prange, "Cyclic error-correcting codes in two symbols", Report AFCRC-TN-57, 103, Air Force Cambridge Research Center, Cambridge, Mass., Sept. 1957.

[3-19] J. E. Meggitt, "Error correcting codes and their implementation", *IRE Trans. Information Theory*, IT-7, pp. 232–244, October 1961.

[3-20] T. Kasami, "A decoding procedure for multiple-error-correcting cyclic codes", *IEEE Trans. Information Theory*, IT-10, pp. 134–139, April 1964.

[3-21] L. Rudolph and M. E. Mitchell, "Implementation of decoders for cyclic codes, *IEEE Trans. Information Theory*, IT-10, pp. 259–250, July 1964.

[3-22] A. Hocquenghem, "Codes correcteurs d'erreurs", *Chiffres 2*, pp. 147–156, 1959.

[3-23] R. C. Bose and D. K. Ray-Chaudhuri, "On a class of error-correcting binary group codes", *Information and Control*, vol. 3, pp. 68–79, March 1960.

[3-24] I. S. Reed and G. Solomon, "Polynomial codes over certain finite fields", *J. Soc. Ind. Appl. Math.*. vol. 8, pp. 300–304, June 1960.

[3-25] E. R. Berlekamp, "On decoding binary Bose-Chaudhuri-Hocquenghem codes", *IEEE Trans. Information Theory*, IT-11, pp. 577–580, October 1965.

[3-26] J. L. Massey, "Shift-register synthesis and BCH decoding", *IEEE Trans. Information Theory*, IT-15, pp. 122–127, January 1969.

[3-27] R. T. Chien, "Cyclic decoding procedure for the Bose-Chaudhuri-Hocquenghem codes", *IEEE Trans. Information Theory*, IT-10, pp. 357–363, October 1964.

[3-28] W. W. Peterson, "Encoding and error-correction procedures for the Bose-Chaudhuri codes", *IRE Trans. Information Theory*, IT-6, pp. 459–470, 1960.

[3-29] J. L. Massey, "Shift register synthesis and BCH decoding", *IEEE Trans. Information Theory*, IT-15, pp. 122–127, January 1969.

[3-30] D. H. Green and I. S. Taylor, "Irreducible polynomials over composite Galois fields and their applications in coding techniques", *Proc. IEE*, vol. 121, pp. 935–939, 1974.

[3-31] R. C. Singleton, "Maximum distance $q$-ary codes", *IEEE Trans. Information Theory*, IT-10, pp. 116–118, January 1964.

[3-32] G. D. Forney, Jr., "On decoding BCH codes", *IEEE Trans. Information Theory*, IT-11, pp. 549–557, 1965.

[3-33] N. Abramson, "A class of systematic codes for non-independent errors", *IRE Trans. Information Theory*, IT-4, pp. 150–157, December 1959.

[3-34] P. Fire, "A class of multiple-error-correcting codes for non-independent errors", Sylvania Report No. RSL-E-2, Reconnaissance Systems Division, Mountain View, Calif., 1959.

[3-35] S. H. Reiger, "Codes for the correction of 'clustered' errors", *IRE Trans. Information Theory*, IT-6, pp. 16–21, March 1960.

[3-36] G. D. Forney, Jr., *Concatenated codes*, MIT Press, Cambridge, Mass., 1966.

[3-37] J. B. H. Peek, "Communication aspects of the compact disc digital audio system", *IEEE Commun. Mag.*, vol. 23, no. 2, pp. 7–15, February 1985.

[3-38] P. Elias, "Error-free coding", *IRE Trans. Information Theory*, PGIT-4, pp. 29–37, September 1954.

[3-39] D. Bertsekas and R. G. Gallager, *Data Networks*, Prentice-Hall, Englewood Cliffs, N.J., 1987.

[3-40] A. S. Tanenbaum, *Computer Networks*, 2nd edition, Prentice-Hall, Englewood Cliffs, N.J., 1988.

[3-41] M. Schwartz, *Telecommunications Networks: Protocols, Modeling, and Analysis*, Addison-Wesley Publishing Co., Reading, Mass., 1987.

# CHAPTER 4

# Trellis Codes

A trellis is a portrayal of state transitions of a finite-state machine over time. Any practical code generator is a finite-state computing machine that traverses from state to state over time, and a code trellis is a graph of all possible traversals. The trellis idea was suggested around 1970 by G.D. Forney, who was inspired by the ordinary garden trellis, an interlaced lattice of pathways for plants. Some of our code trellises have indeed been spotted in the garden.

All realizable codes have a trellis, including those of Chapter 3, but the term trellis code refers to a code with a highly regular trellis like those we will begin to see in Section 4.2. The regularity eases implementation and it is the key to the popularity of trellis codes.

We will begin with the oldest class of trellis codes, the convolutional codes. These are parity-check codes like those in Chapter 3 but the trellis idea is a newer way to view them. Sections 4.3–4.5 discuss *bandwidth-efficient* trellis codes, a more recent type of coding that is not based on parity checking. This concept is introduced for a few pages at the beginning of Section 4.3. Ideally, one needs a grounding in signal space theory and data modulation in order to read these sections easily, subjects that lie outside the scope this book. A brief tutorial on these subjects forms Appendix A.

It is interesting to observe that all the codes in this chapter somehow relate to convolution. Some wag has said that communications engineers don't know how to do much else. He may be right.

We need to point out a slight notational variance in this chapter from the rest of the book. A tradition has grown up in trellis coding to number the data symbols beginning with 0, i.e., $u_0, u_1, \ldots$, rather than to call the first symbol $u_1$. This simplifies delay operator and algebra notations, and we will continue the tradition when working with trellis channel codes.

## 4.1 CONVOLUTIONAL CODES

Convolutional codes were proposed by Elias [4-1] as a possible alternative to block codes. Unlike the block codes of Chapter 3, convolutional codes need not be restricted to transmit code words in blocks; these codes can accept a semi-infinite stream of information symbols, operate upon them in a *sliding-block* manner, and transmit a continuous stream of encoded symbols. Each information symbol remains within the sliding-block for a finite amount of time, affecting the transmitted symbols during that time.

In contrast to the codes of the preceding chapter, which depended strongly on the algebraic properties of finite fields, convolutional codes have been traditionally studied from a non-algebraic point of view. The interested reader may refer to Forney [4-2] for a rigorous treatment of the algebraic structure of convolutional codes. We will adopt a non-algebraic approach here in studying encoding and decoding algorithms for these codes.

In Section 2.1 we introduced the idea of a tree code. Convolutional codes are tree codes, although we will hold off discussing their tree and trellis nature until Section 4.2. For now, we need only the parameters of the tree branching structure, the number of code word bits on a branch, $a$, and the number of branches out of a node, $b$. In codes used in practice, $b$ is a power of 2; it will be convenient to reserve the notation $B$ for $\log_2 b$, the number of bits that select a branch.

## Encoding with Convolutional Codes

A convolutional encoder accepts data bits in groups of $B = \log_2 b$ and produces outputs in groups of $a$, and so it generates a code with rate $R = (\log b)/a$ input bits per channel use. For example, the encoder in Fig. 4.1-1 has one input and three outputs and generates a rate $R = 1/3$ code. The information sequence, denoted by $\mathbf{u} = (u_0, u_1, u_2, \ldots)$, enters the encoder one bit at a time to produce three output bits denoted by $x_{1,i}, x_{2,i}$, and $x_{3,i}$ at time instant $i$. The adders carry out modulo-2 addition and may be implemented using Exclusive-OR gates. Each square block denotes a delay unit.

The encoding relations expressing the output bits $x_{j,i}, j = 1, 2, \ldots, a$, as linear combinations of input bits are

$$\begin{aligned} x_{1,i} &= u_i, \\ x_{2,i} &= u_i + u_{i-1}, \\ x_{3,i} &= u_i + u_{i-1} + u_{i-2}. \end{aligned} \tag{4.1-1}$$

Each sequence $\mathbf{x}_j = (x_{j,0}, x_{j,1}, x_{j,2}, \ldots), j = 1, 2, \ldots, a$, may also be expressed as the convolution of the information sequence and the impulse response of the encoder generating that output sequence. On applying $\mathbf{u} = (1\ 0\ 0\ \ldots)$ to the encoder in Fig. 4.1-1, we see that the shift register generating $\mathbf{x}_1$ has impulse response $\mathbf{g}_1 = (1\ 0\ 0)$. Similarly, the impulse responses of the encoders generating $\mathbf{x}_2$ and $\mathbf{x}_3$ are $\mathbf{g}_2 = (1\ 1\ 0)$ and $\mathbf{g}_3 = (1\ 1\ 1)$, respectively. The impulse responses $\mathbf{g}_j, j = 1, 2, 3$, are called *generator sequences*. In general, the generator sequences of the $a$ encoders are denoted by

$$\mathbf{g}_j = (g_{j,0}, g_{j,1}, \ldots, g_{j,\nu-1}), \qquad j = 1, 2, \ldots, a. \tag{4.1-2}$$

The parameter $\nu$, representing the length of the generator sequences, is called the *constraint length* of the code. The code generated by the encoder in Fig. 4.1-1 has $\nu = 3$. The $a$ output sequences $\mathbf{x}_j, j = 1, 2, \ldots, a$ are

$$\mathbf{x}_j = \mathbf{u} * \mathbf{g}_j, \tag{4.1-3}$$

where $*$ denotes discrete convolution and the $i$th bit of $x_j$ is

$$x_{j,i} = \sum_{k=0}^{\nu-1} g_{j,k} u_{i-k}, \qquad j = 1, 2, \ldots, a. \tag{4.1-4}$$

It is possible to express (4.1-4) in terms of matrix operations as

$$\mathbf{x} = \mathbf{u}\mathbf{G}, \tag{4.1-5}$$

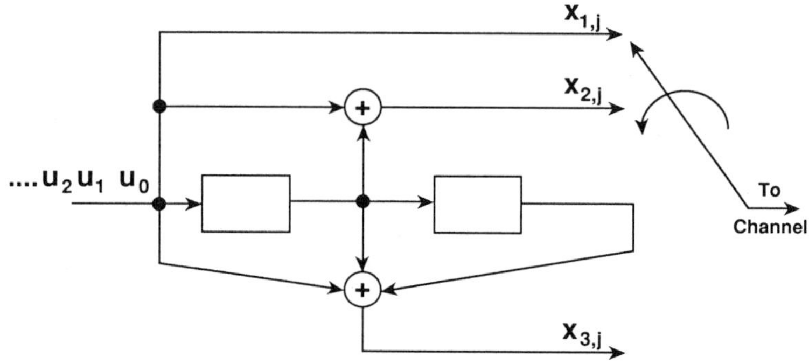


**Figure 4.1-1.** Encoder for a rate-1/3 convolutional code with constraint length $\nu = 3$.

where $\mathbf{u} = (u_0, u_1, u_2, \ldots)$ denotes the semi-infinite input sequence, $\mathbf{x} = (x_{1,0}, x_{2,0}, \ldots, x_{a,0}, x_{1,1}, x_{2,1}, \ldots, x_{a,1}, \ldots)$ denotes the code word, and

$$\mathbf{G} = \begin{bmatrix} \mathbf{G}_0 & \mathbf{G}_1 & \cdots & \mathbf{G}_{\nu-1} & & & \\ & \mathbf{G}_0 & \mathbf{G}_1 & \cdots & \mathbf{G}_{\nu-1} & & \\ & & \mathbf{G}_0 & \cdots & \mathbf{G}_{\nu-2} & \mathbf{G}_{\nu-1} & \\ & & & \ddots & & & \end{bmatrix}, \tag{4.1-6}$$

where $\mathbf{G}_l = (g_{1,l}, g_{2,l}, \ldots, g_{a,l}), l = 0, 1, 2, \ldots, \nu - 1$ and the blank spaces are zero entries. We call $\mathbf{G}$ the generator matrix. Observe that each row of $\mathbf{G}$ in (4.1-6) is identical to the one above it, but shifted right by $a$ positions.

We now illustrate these concepts with some examples.

**Example 4.1-1.**

Consider the rate-1/3 convolutional code of Fig. 4.1-1. The generator sequences are given by

$$\mathbf{g}_1 = (1\ 0\ 0), \quad \mathbf{g}_2 = (1\ 1\ 0), \text{ and } \quad \mathbf{g}_3 = (1\ 1\ 1).$$

Let the input to the encoder be $\mathbf{u} = (1\ 0\ 0\ 1\ 1)$. Using (4.1-4), the code word generated is

$$\mathbf{x} = (1\ 1\ 1\ 0\ 1\ 1\ 0\ 0\ 1\ 1\ 1\ 1\ 1\ 0\ 0\ 0\ 1\ 0\ 0\ 0\ 1).$$

In producing this code word we assume that zeros are shifted into the register after the last bit of $\mathbf{u}$ has entered the encoder until all the delay units contain zeros. This is equivalent to adding $\nu-1$ zeros to the input sequence. We may also obtain the output sequence $\mathbf{x}$ by applying the convolution operation given by (4.1-3) which gives

$$\begin{aligned}
\mathbf{x}_1 &= (1\ 0\ 0\ 1\ 1) * (1\ 0\ 0) \\
&= (1\ 0\ 0\ 1\ 1\ 0\ 0), \\
\mathbf{x}_2 &= (1\ 0\ 0\ 1\ 1) * (1\ 1\ 0) \\
&= (1\ 1\ 0\ 1\ 0\ 1\ 0), \\
\mathbf{x}_3 &= (1\ 0\ 0\ 1\ 1) * (1\ 1\ 1) \\
&= (1\ 1\ 1\ 1\ 0\ 0\ 1).
\end{aligned}$$

The output sequence obtained by multiplexing $\mathbf{x}_1$, $\mathbf{x}_2$ and $\mathbf{x}_3$ is

$$\mathbf{x} = (1\ 1\ 1\ 0\ 1\ 1\ 0\ 0\ 1\ 1\ 1\ 1\ 1\ 0\ 0\ 0\ 1\ 0\ 0\ 0\ 1),$$

the same as before.

Alternatively, we may also use (4.1-5) to obtain the code word. First we form the $\mathbf{G}$ matrix for the encoder in Fig. 4.1-1 by using (4.1-6), which gives

$$\mathbf{G} = \begin{bmatrix}
111 & 011 & 001 & & & & \\
 & 111 & 011 & 001 & & & \\
 & & 111 & 011 & 001 & & \\
 & & & 111 & 011 & 001 & \\
 & & & & 111 & 011 & 001
\end{bmatrix}.$$

Applying (4.1-5), we get

$$\begin{aligned}
\mathbf{x} &= (1\ 0\ 0\ 1\ 1)\mathbf{G} \\
&= (1\ 1\ 1\ 0\ 1\ 1\ 0\ 0\ 1\ 1\ 1\ 1\ 1\ 0\ 0\ 0\ 1\ 0\ 0\ 0\ 1),
\end{aligned}$$

which is again the same as the result obtained earlier.

In the previous example one information bit enters the encoder during each cycle. In general, for a rate $B/a$ encoder, $B$ consecutive information bits enter the encoder in parallel during each cycle. The encoder puts out bits which are multiplexed into a single sequence called the code word. We again denote the information sequence by u, but now u is given by

$$\mathbf{u} = (u_{1,0}, u_{2,0}, \ldots, u_{B,0}, u_{1,1}, u_{2,1}, \ldots, u_{B,1}, \ldots).$$

We let $\mathbf{u}_j = (u_{j,0}, u_{j,1}, u_{j,2}, \ldots)$, $j = 1, 2, \ldots, B$, denote the input sequence fed to the $j$th shift register. Each input sequence $\mathbf{u}_j, j = 1, 2, \ldots, B$, contributes differently to each output sequence $\mathbf{x}_j$, $j = 1, 2, \ldots, a$. Each such sequence $\mathbf{x}_j$ is given by the sum of $B$ convolutions as

$$\mathbf{x}_j = \sum_{i=1}^{B} \mathbf{u}_i * \mathbf{g}_{i,j}, \tag{4.1-7}$$

where $\mathbf{g}_{i,j}$ represents the generator sequence $(g_{i,j,0}, g_{i,j,1}, g_{i,j,2}, \ldots, g_{i,j,\nu-1})$ and $\nu$ is the constraint length of the code.

Alternatively, we may formulate the generator matrix and use (4.1-5) to determine the code word corresponding to any input sequence u. The generator matrix for a rate $B/a$ convolutional encoder is

$$\mathbf{G} = \begin{bmatrix} \mathbf{G}_0 & \mathbf{G}_1 & \mathbf{G}_2 & \cdots & \mathbf{G}_{\nu-1} & & \\ & \mathbf{G}_0 & \mathbf{G}_1 & \cdots & \mathbf{G}_{\nu-2}, & \mathbf{G}_{\nu-1} & \\ & & \mathbf{G}_0 & \cdots & \mathbf{G}_{\nu-3} & \mathbf{G}_{\nu-2}, & \mathbf{G}_{\nu-1} \\ & & & \ddots & & & \end{bmatrix}, \tag{4.1-8}$$

where

$$\mathbf{G}_l = \begin{bmatrix} g_{1,0,l} & g_{1,1,l} & \cdots & g_{1,a,l} \\ g_{2,0,l} & g_{2,1,l} & \cdots & g_{2,a,l} \\ \vdots & & & \\ g_{B,0,l} & g_{B,1,l}, & \cdots & g_{B,a,l} \end{bmatrix}.$$

Observe that each row of G in (4.1-8) is identical to the one above but shifted right by $a$ bits. The G matrix exhibits a block structure where each block $\mathbf{G}_k$ is a $B \times a$ matrix as defined above. We now illustrate this case.

**Example 4.1-2.**

Consider the rate-2/3 convolutional code of Fig. 4.1-2. The generator sequences are: $\mathbf{g}_{1,1} = (1\ 0\ 0)$, $\mathbf{g}_{1,2} = (0\ 1\ 0)$, $\mathbf{g}_{1,3} = (1\ 1\ 0)$, $\mathbf{g}_{2,1} = (0\ 1\ 1)$, $\mathbf{g}_{2,2} = (0\ 0\ 0)$, $\mathbf{g}_{2,3} = (1\ 0\ 0)$.

Let the input sequence be given by $\mathbf{u} = (1\ 0\ 0\ 1\ 1\ 1)$. Then $\mathbf{u}_1 = (1\ 0\ 1)$ and $\mathbf{u}_2 = (0\ 1\ 1)$. Applying (4.1-7), we obtain following output sequences:

$$\begin{aligned}
\mathbf{x}_1 &= (1\ 0\ 1) * (1\ 0\ 0) + (0\ 1\ 1) * (0\ 1\ 1) \\
&= (1\ 0\ 0\ 0\ 1) + (0\ 0\ 1\ 0\ 1) \\
&= (1\ 0\ 0\ 0\ 1), \\
\mathbf{x}_2 &= (1\ 0\ 1) * (0\ 1\ 0) + (0\ 1\ 1) * (0\ 0\ 0) \\
&= (0\ 1\ 0\ 1\ 0) + (0\ 0\ 0\ 0\ 0) \\
&= (0\ 1\ 0\ 1\ 0), \\
\mathbf{x}_3 &= (1\ 0\ 1) * (1\ 1\ 0) + (0\ 1\ 1) * (1\ 0\ 0) \\
&= (1\ 0\ 0\ 1\ 0).
\end{aligned}$$

The code word obtained by multiplexing these three sequences is

$$\mathbf{x} = (1\ 0\ 1\ 0\ 1\ 0\ 0\ 0\ 0\ 0\ 1\ 1\ 1\ 0\ 0).$$

Alternatively, formulating the $\mathbf{G}$ matrix as

$$\mathbf{G} = \begin{bmatrix}
101 & 011 & 000 & & \\
001 & 100 & 100 & & \\
 & 101 & 011 & 000 & \\
 & 001 & 100 & 100 & \\
 & & 101 & 011 & 000 \\
 & & 001 & 100 & 100
\end{bmatrix},$$

and using $\mathbf{x} = \mathbf{u}\mathbf{G}$ yields $\mathbf{x} = (1\ 0\ 1\ 0\ 1\ 0\ 0\ 0\ 0\ 0\ 1\ 1\ 1\ 0\ 0)$, the same as that obtained using convolution.

In both these examples, we have assumed that the input sequence is terminated with two zeros. In general, any finite input sequence is terminated with $\nu - 1$ zeros. This reduces the rate to less than $B/a$. If the input sequence is of length $BL$, the code word is of length $a(L+\nu-1)$. Viewing the convolutional code as an $(N, K)$ block code with $N = a(L+\nu-1)$ and $K = BL$, we get a rate

$$R' = \frac{BL}{a(L+\nu-1)}. \tag{4.1-9}$$

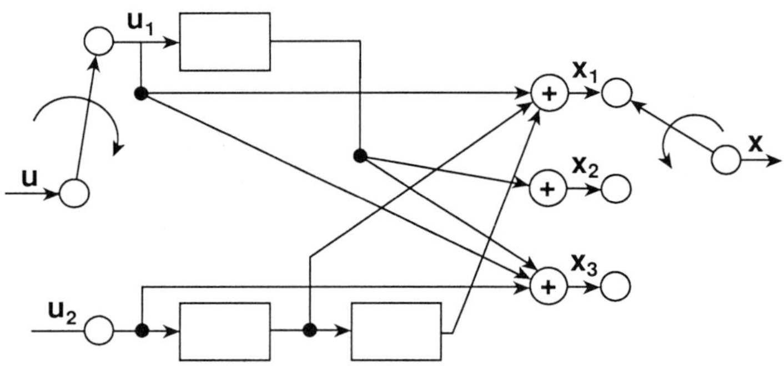


**Figure 4.1-2.** A rate-2/3 convolutional encoder with $\nu = 3$.

Note that if $L$ is much larger than $\nu$ then $R'$ tends to $B/a$. With finite length input sequences a convolutional code is said to suffer a *fractional rate loss* defined by

$$\frac{R - R'}{R} = \frac{\nu - 1}{L + \nu - 1}. \tag{4.1-10}$$

In Example 4.1-1, with $L = 5$ and $\nu = 3$, the fractional rate loss is $2/7 = 28.5\%$ and in Example 4.1-2, with $L = 3$ and $\nu = 3$, it is $2/5 = 40\%$. In practice, however, $L$ will be much larger than in these two examples and the fractional rate loss will be much smaller.

As in the block codes of Chapter 3, we may use generator polynomials to represent a convolutional encoder. We know from Chapter 3 that vectors can be represented by polynomials. We denote an arbitrary input sequence $\mathbf{u}_i = (u_{i,0}, u_{i,1}, u_{i,2}, \ldots)$ by the polynomial

$$u_i(D) = u_{i,0} + u_{i,1}D + u_{i,2}D^2 + \ldots, \qquad i = 1, 2, \ldots, B,$$

the $aB$ generator polynomials by

$$g_{i,j}(D) = g_{i,j,0} + g_{i,j,1}D + g_{i,j,2}D^2 + \ldots, \qquad \text{for } i = 1, 2, \ldots, B, \text{ and } j = 1, 2, \ldots, a,$$

and the output sequences $\mathbf{x}_j$ by

$$x_j(D) = \sum_{i=1}^{B} u_i(D) g_{i,j}(D), \qquad j = 1, 2, \ldots, a. \tag{4.1-11}$$

Observe that (4.1-11) represents the same set of encoding relations as does (4.1-7). The set of $a$ equations given by (4.1-11) may be expressed by the following matrix equation:

$$[x_1(D)\ x_2(D)\ \ldots\ x_a(D)] = [u_1(D)\ u_2(D)\ \ldots\ u_B(D)]\ \mathbf{G}(D), \qquad (4.1\text{-}12)$$

where $\mathbf{G}$ is the generator of polynomials

$$\mathbf{G}(D) = [g_{i,j}(D)]. \qquad (4.1\text{-}13)$$

The code word polynomial $x(D)$, obtained by multiplexing the individual sequences $x_j$, $j = 1, 2, \ldots, a$, is given by

$$x(D) = x_1(D^a) + Dx_2(D^a) + \ldots + D^{a-1}x_a(D^a). \qquad (4.1\text{-}14)$$

**Example 4.1-3.**

First consider the rate-1/3 encoder of Fig. 4.1-1. The generator polynomials for this code are $g_1(D) = 1$, $g_2(D) = 1+D$, and $g_3(D) = 1 + D + D^2$. The information polynomial $u(D)$ is given by

$$u(D) = 1 + D^3 + D^4.$$

For the special case of $B = 1$, (4.1-13) and (4.1-14) simplify to

$$\begin{aligned}
&[\,x_1(D) \quad x_2(D) \quad x_3(D)\,] \\
&\quad = u(D)G(D) \\
&\quad = u(D)\,[\,g_1(D) \quad g_2(D) \quad g_3(D)\,] \\
&\quad = [1 + D^3 + D^4]\,[1 \quad 1 + D \quad 1 + D + D^2\,] \\
&\quad = [\,1 + D^3 + D^4 \quad 1 + D + D^3 + D^5 \quad 1 + D + D^2 + D^3 + D^6\,].
\end{aligned}$$

From (4.1-14) the code polynomial is

$$\begin{aligned}
x(D) &= 1 + D^9 + D^{12} + D(1 + D^3 + D^9 + D^{15}) \\
&\qquad + D^2(1 + D^3 + D^6 + D^9 + D^{18}) \\
&= 1 + D + D^2 + D^4 + D^5 + D^8 + D^9 + D^{10} \\
&\qquad + D^{11} + D^{12} + D^{16} + D^{20},
\end{aligned}$$

which is the polynomial corresponding to the code word of Example 4.1-1.

Next we consider the rate-2/3 encoder of Fig. 4.1-2. The matrix of generator polynomials $\mathbf{G}(D)$ is

$$\mathbf{G}(D) = \begin{bmatrix} 1 & D & 1+D \\ D+D^2 & 0 & 1 \end{bmatrix}.$$

From (4.1-13) and (4.1-14), we get

$$\begin{aligned}
&[\,x_1(D) \quad x_2(D) \quad x_3(D)\,] \\
&\quad = [\,u_1(D) \quad u_2(D)\,]\,\mathbf{G}(D) \\
&\quad = [\,1+D^2 \quad D+D^2\,]\,\mathbf{G}(D) \\
&\quad = [\,1+D^2+D^2+D^4 \quad D+D^3 \quad 1+D^2+D+D^3+D+D^2\,] \\
&\quad = [\,1+D^4 \quad D+D^3 \quad 1+D^3\,].
\end{aligned}$$

Using (4.1-14), we obtain

$$\begin{aligned}
x(D) &= 1+D^{12}+D(D^3+D^9)+D^2(1+D^9) \\
&= 1+D^2+D^4+D^{10}+D^{11}+D^{12},
\end{aligned}$$

which represents the same code word as in Example 4.1-2.

*Systematic codes* form an important subclass of convolutional codes. In a rate-$B/a$ systematic code, the $B$ information bits entering the encoder are also the first $B$ of the $a$ bits put out by the encoder. The code generated by the encoder in Fig. 4.1-1 is an example of a systematic code. The generator matrix for a general rate-$B/a$ systematic code is given by

$$\mathbf{G} = \begin{bmatrix} \mathbf{I} & \mathbf{P}_0 & \mathbf{0} & \mathbf{P}_1 & \mathbf{0} & \mathbf{P}_2 & \cdots & \mathbf{0} & \mathbf{P}_{\nu-1} & & & & \\ & & \mathbf{I} & \mathbf{P}_0 & \mathbf{0} & \mathbf{P}_1 & \cdots & \mathbf{0} & \mathbf{P}_{\nu-2} & \mathbf{0} & \mathbf{P}_{\nu-1} & & \\ & & & & \mathbf{I} & \mathbf{P}_0 & \cdots & \mathbf{0} & \mathbf{P}_{\nu-3} & \mathbf{0} & \mathbf{P}_{\nu-2} & \mathbf{0} & \mathbf{P}_{\nu-1} \\ & & & & & & \ddots & & & & & & \end{bmatrix},$$

where $\mathbf{I}$ and $\mathbf{0}$ represent the $B \times B$ identity and all zeros matrices, respectively, and $\mathbf{P}_l$ is the $B \times (a-B)$ matrix given by

$$\mathbf{P}_l = \begin{bmatrix} g_{1,l,B+1} & g_{1,l,B+2} & \cdots & g_{1,l,\nu-1} \\ g_{2,l,B+1} & g_{2,l,B+2} & \cdots & g_{2,l,\nu-1} \\ \vdots & \vdots & \ddots & \vdots \\ g_{B,l,B+1} & g_{B,l,B+2} & \cdots & g_{B,l,\nu-1} \end{bmatrix}.$$

## 4.2 TREES, TRELLISES AND THE VITERBI ALGORITHM

A convenient way to represent the encoding relations is by means of the code tree. The idea of such a tree was introduced in Section 2.1. Convolutional codes are tree codes and Fig. 4.2-1 shows an example, for the code in Example 4.1-1.

The code tree has two branches out of each node. In general, for a rate-log $b/a$ convolutional code, the code tree will have $b$ branches out of each node and each branch will have $a$ binary symbols on it, representing the output of the encoder. An upward transition in the code tree of Fig 4.2-1 corresponds to input 0 and a downward transition to input 1. Assume the encoder memory stages are initialized so that they contain zeros. If a 0 is applied to the encoder, the output is 000. If instead a 1 enters the encoder, the output is 111. The two branches at level 1 of the code tree are assigned these outputs. If the first input symbol is 0, the two possible encoder outputs at level 2 are 000 and 111, respectively, corresponding to 0 or 1 for the second input symbol. Similarly, if the first input symbol is 1, the the two possible outputs at level 1 are 011 and 100. The four possible outputs are assigned to the four branches of the code tree at level 2. The code tree proceeds to develop in this manner indefinitely. One can verify that the code tree generates the same code word corresponding to the input 10011 as in Example 4.1-1, which is 111 011 001 111 100 010 001. We have assumed that the finite input sequence is terminated with $\nu - 1$ zeros. In this example $\nu$ is 3.

How does one use this code tree to decode a received sequence, which is a noisy version of a transmitted code word? Each group of three received bits is compared against all the branches of the tree at that level. Each branch is then assigned a metric which is the Hamming distance between the three bits of the code word on that branch and the group of received bits at that level. For example, assume the received sequence to be 001, 110, .... Comparing the first group of three received bits with the two branches at level 1, we assign a metric of 1 to the top branch and 2 to the bottom branch. Similarly, by comparing 110 with the four branches at level 2, we assign the numbers 2, 1, 2 and 1 as metrics to these branches. We now accumulate the metrics of the individual branches on a path in the tree and call it the cumulative metric of that path. For example, the cumulative metric of the path corresponding to the topmost two branches at levels 1 and 2 of the tree is 3. Proceeding in this manner, we can compute the metrics of all the branches of the tree. If the received sequence is of length $3L$, we must compute $2[2^L - 1]$ branch metrics. There are now $2^L$ paths whose cumulative metrics have been computed with respect to the received sequence. Out of these $2^L$ paths, the decoder chooses the most likely path, that is, the one with the least cumulative metric. Such a decoder is called a *maximum likelihood decoder.*

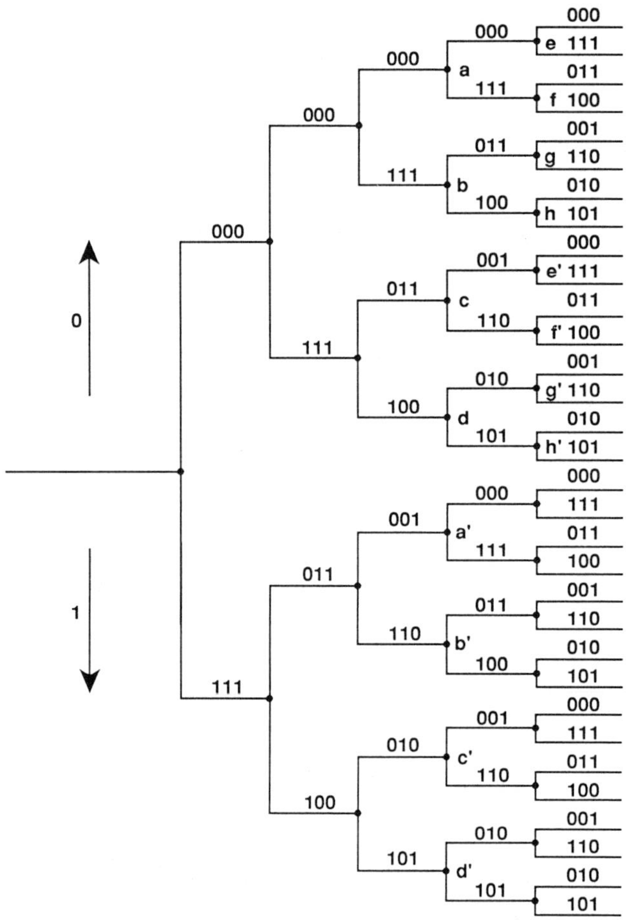

Figure 4.2-1. Code tree generated by the convolutional coder of Fig. 4.1-1.

The computations required of the exhaustive decoding procedure described above grow exponentially with $L$, the number of levels in the tree. Fortunately, an efficient algorithm, known as the Viterbi algorithm [4-3], exists that exploits repetitions in the code tree to reduce the computations. Observe that at level 4 in Fig. 4.2-1 the two subtrees with root nodes at $a$ and $a'$ are identical. Therefore the cumulative metrics of paths in the top subtree with root node at $a$ will be same as those of the corresponding ones in the bottom subtree with root node at $a'$. Assuming that the cumulative metric of the tree branch path from the root node of the tree to node $a$ is smaller than that of the path to node $a'$, we conclude that any path through node $a'$ will have a larger metric compared to the corresponding path through $a$. In other words, the subtree at node $a'$ can be completely discarded from further consideration. We observe that these arguments hold good for the pairs of nodes $(b, b')$, $(c, c')$ and $(d, d')$, and that only one node from each of these pairs needs to be retained at level 4. Extending the four paths at level 4 produces eight paths and again half of these will be discarded.

An equivalent way to describe the Viterbi algorithm is by means of a state transition or trellis diagram. First we introduce the state transition diagram for the encoder of Fig. 4.1-1. This is shown in Fig. 4.2-2. We define the state of the encoder at time instant $i$ as the pair $(u_{i-1}, u_{i-2})$, representing the outputs of the two delay units in the figure. There are four possible states $(0\ 0), (0\ 1), (1\ 0)$ and $(1\ 1)$ or equivalently, 0,1,2 and 3 in decimal notation.

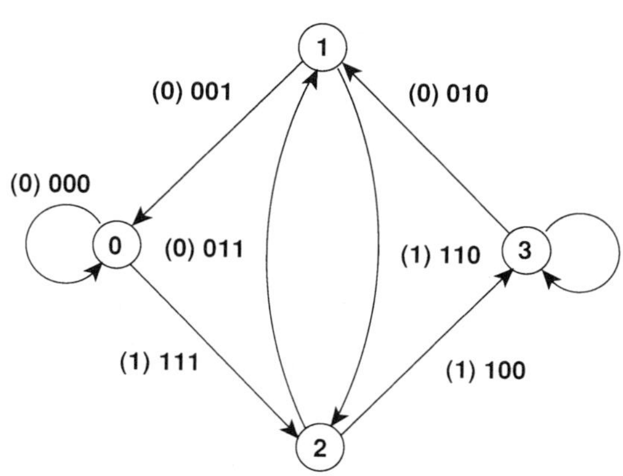


Figure 4.2-2. The state transition diagram for the encoder of Fig. 4.1-1. Circles represent encoder states.

In Fig. 4.2-2, the information bit that causes state transition is shown in parentheses. This bit is the current information bit that enters the encoder at the left end in Fig. 4.1-1. If the current state is $(u_{i-1}\ u_{i-2})$, the next state is $(u_i\ u_{i-1})$. Also associated with each branch is a 3-tuple representing the output of the encoder.

Starting with the encoder in state 0, and depicting all possible state transitions in time as the data bits arrive, we get the state trellis diagram shown in Figure 4.2-3. The state trellis diagram is really the state transition diagram with a time element incorporated. Each state in the trellis diagram has two branches leaving and two branches entering it. Starting from state 0, the first data bit $u_0$ can drive the encoder to only two states, state 0 or 2, depending upon whether $u_0$ is 0 or 1. From either of these states, the second data bit $u_1$ drives the encoder to the states indicated in the trellis diagram. By selecting the two bits $u_0\ u_1$ appropriately, the encoder can reach any of the four possible states starting from state 0. In general, for a rate-$1/a$ convolutional code with constraint length $\nu$, there are $2^{\nu-1}$ possible states in the state trellis diagram. Starting from state 0, the encoder can be driven to any one of these states by selecting the appropriate input sequence $u_0\ u_1\ \ldots\ u_{\nu-2}$ of length $\nu-1$. In fact applying the sequence $u_0\ u_1\ \ldots\ u_{\nu-2}$ drives the encoder to the state whose binary expansion is given by the length $\nu-1$ input sequence. When the $\nu$th bit arrives (in our example $u_2$), the trellis is complete and from then onwards the trellis stages are identical, continuing forever.

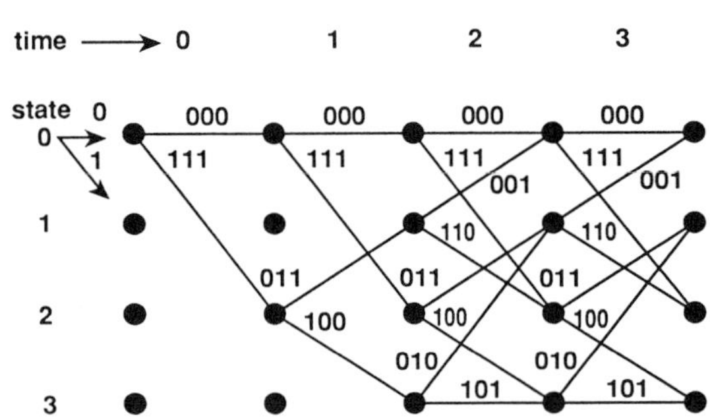


Figure 4.2-3. State trellis diagram for the encoder of Figure 4.1-1.

The Viterbi algorithm explained earlier using the code tree of Fig. 4.2-1 can now be illustrated with the trellis diagram in Fig. 4.2-3. Figure 4.2-4 illustrates the progression of the algorithm in time. At time 0, the

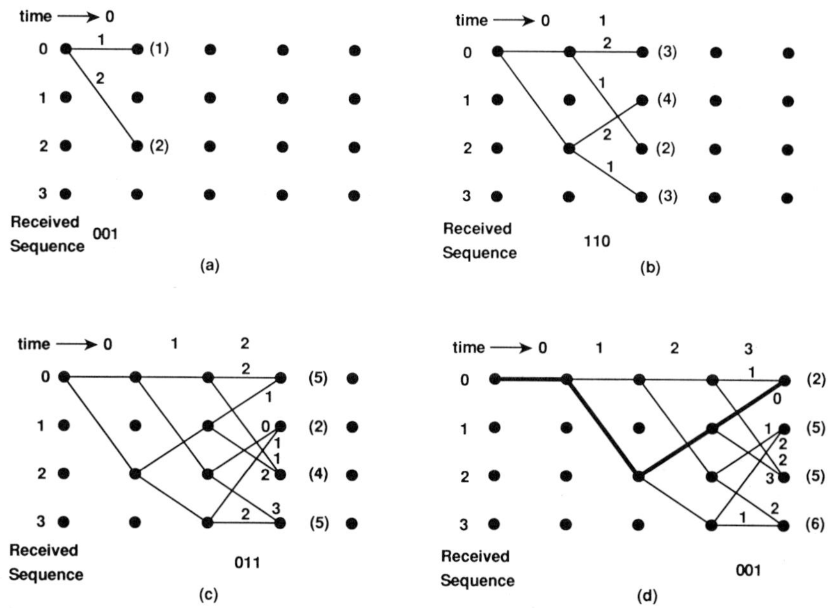


Figure 4.2-4. Example illustrating the Viterbi algorithm. Numbers on branches indicate branch metrics; cumulative metrics of surviving paths are given in parentheses.

sequence 001 arrives at the receiver. The decoder computes the Hamming distance of the received sequence to the code words assigned to the two branches out of state 0. This step is shown in Fig. 4.2-4(a). The numbers in parenthesis denote the cumulative metric of the two paths. The cumulative metric of a path is defined as the sum of metrics of all branches on that path. Next, the sequence 110 arrives. The decoder now computes four branch metrics as shown in Fig. 4.2-4(b); the cumulative metrics of the four paths are again shown in parentheses. Next the decoder computes eight branch metrics corresponding to the received sequence at time 2 (see Fig. 4.2-4(c)). Two paths enter each state in this figure, of which the Viterbi algorithm retains the one with the smaller cumulative metric. The other path is discarded from further consideration and is shown crossed out in Fig. 4.2-4(c). Observe that discarding the bottom path entering state 0 in this figure is equivalent to discarding the subtree with root node $a'$ in Fig. 4.1-3. Of the two paths entering each state, the process of discarding the one with a larger cumulative metric is in fact the *principle of optimality* employed in dynamic programming. Hence the Viterbi algorithm is a dynamic programming algorithm. Readers familiar with graph theory will realize that this is indeed an algorithm to find a shortest path in a graph. Fig. 4.2-4(d) shows the set of operations done by the decoder at time 3. The most likely transmitted code word up to time 3 corresponds to the path shown by thick lines in this figure. The cumulative metric of the path indicates the number of errors (two in this case) corrected by the Viterbi algorithm. The decoded code word is 000 111 011 001 up to time 3.

## Distance Measures for Convolutional Codes

In the case of block codes, we defined the minimum distance as the smallest Hamming distance between any two code words. One can similarly define a distance measure for convolutional codes. But what is the length of the code words? A distance measure called the *minimum free distance*, $d_{\text{free}}$, arises when the code words are infinitely long.

Let $\mathbf{x}'$ and $\mathbf{x}''$ be two code words generated by a code corresponding to inputs $\mathbf{u}'$ and $\mathbf{u}''$, respectively. Then $d_{\text{free}}$ is defined as

$$\begin{aligned} d_{\text{free}} &= \min\{w_H(\mathbf{x}' + \mathbf{x}'') : \mathbf{u}' \neq \mathbf{u}''\} \\ &= \min\{w_H(\mathbf{x}) : \mathbf{u} \neq 0\}. \end{aligned}$$

We see that $d_{\text{free}}$ may also be defined the Hamming weight of the minimum weight code word generated by a nonzero information word of infinite length. References [4-4] and [4-5] give efficient algorithms for computing $d_{\text{free}}$. Table 4.2-1 lists codes of several rates and constraint lengths with maximal $d_{\text{free}}$. These codes were obtained by computer search and can be found in [4-6] – [4-11].

Readers desiring more advanced material on convolutional codes are referred to the book by Viterbi and Omura [4-12]. In addition, several text books on coding listed in Chapter 3 devote a chapter or more to the study of convolutional codes.

Suppose now that the decoder views $L$ code word branches. Another distance measure, called the *decoding minimum distance*, $d_L$, is defined as the minimum distance between any two nonzero code words of length $La$. $L$ is called the *decision depth*, or the *decoding constraint length*, or sometimes the *decoding window*. Let

$$\mathbf{x} = (x_{1,0}, x_{2,0}, \ldots, x_{a,0}, x_{1,1}, \ldots, x_{a,1}, \ldots, x_{1,L-1}, \ldots, x_{a,L-1})$$

**Table 4.2-1.** SOME GOOD CONVOLUTIONAL CODES; GENERATORS ARE GIVEN IN OCTAL NOTATION; FOR EXAMPLE, THE RATE-1/2 CODE WITH GENERATORS $\mathbf{g}_1 = 23$ AND $\mathbf{g}_2 = 35$ SHOULD BE EXPANDED AS $\mathbf{g}_1 = 10011$, $\mathbf{g}_2 = 11101$ AFTER OMITTING ANY LEADING ZEROS.

Generator Sequences for Rate - 1/2 Codes

| $g_1$ | $g_2$ | $d_{free}$ | $G_h$ (dB) eqn. (4.2-17) |
|---|---|---|---|
| 5 | 7 | 5 | 0.97 |
| 15 | 17 | 6 | 1.76 |
| 23 | 35 | 7 | 2.43 |
| 65 | 57 | 8 | 3.01 |
| 133 | 171 | 10 | 3.98 |
| 345 | 237 | 10 | 3.98 |
| 561 | 753 | 12 | 4.77 |
| 1167 | 1545 | 12 | 4.77 |
| 2335 | 3661 | 14 | 5.44 |
| 4355 | 5723 | 15 | 5.74 |
| 10533 | 17661 | 16 | 6.02 |
| 21675 | 27123 | 16 | 6.02 |
| 56721 | 61713 | 18 | 6.53 |
| 111653 | 145665 | 19 | 6.77 |
| 347241 | 246277 | 20 | 6.99 |

Generator Sequences for Rate - 1/3 Codes

| $g_1$ | $g_2$ | $g_3$ | $d_{free}$ | $G_h$ (dB) eqn. (4.2-17) |
|---|---|---|---|---|
| 5 | 7 | 7 | 8 | 1.25 |
| 13 | 15 | 17 | 10 | 2.21 |
| 25 | 33 | 37 | 12 | 3.01 |
| 47 | 53 | 75 | 13 | 3.35 |
| 133 | 145 | 175 | 15 | 3.98 |
| 113 | 155 | 173 | 16 | 4.26 |
| 557 | 663 | 711 | 18 | 4.77 |
| 1117 | 1365 | 1633 | 20 | 5.22 |
| 2353 | 2671 | 3175 | 22 | 5.64 |
| 4767 | 5723 | 6265 | 24 | 6.02 |
| 10533 | 10675 | 17661 | 24 | 6.02 |
| 21645 | 35661 | 37133 | 26 | 6.36 |

Generator sequences for rate - 1/4 codes

| $g_1$ | $g_2$ | $g_3$ | $g_4$ | $d_{free}$ | $G_h$(dB) eqn.(4.2-17) |
|---|---|---|---|---|---|
| 5 | 7 | 7 | 7 | 10 | 0.97 |
| 13 | 15 | 15 | 17 | 13 | 2.11 |
| 25 | 27 | 33 | 37 | 16 | 3.01 |
| 53 | 67 | 71 | 75 | 18 | 3.52 |
| 135 | 135 | 147 | 163 | 20 | 3.98 |
| 235 | 275 | 313 | 357 | 22 | 4.39 |
| 463 | 535 | 733 | 745 | 24 | 4.77 |
| 1117 | 1257 | 1633 | 1653 | 27 | 5.28 |
| 2327 | 2313 | 2671 | 3175 | 29 | 5.59 |

denote a code word of length $La$,

$$\mathbf{u} = (u_{1,0}, u_{2,0}, \ldots, u_{B,0}, u_{1,1}, \ldots, u_{B,1}, \ldots, u_{1,L-1}, \ldots, u_{B,L-1})$$

the corresponding information sequence, and $\mathbf{u}_i = (u_{1,i}, \ldots, u_{B,i})$ the $i$th information segment of length $B$ bits. We now define

$$\begin{aligned} d_L &= \min\{d_H(\mathbf{x}', \mathbf{x}'') : \mathbf{u}_0' \neq \mathbf{u}_0''\} \\ &= \min\{w_H(\mathbf{x}' + \mathbf{x}'') : \mathbf{u}_0' + \mathbf{u}_0'' \neq \mathbf{0}\} \\ &= \min\{w_H(\mathbf{x}) : \mathbf{u}_0 \neq 0\}, \end{aligned}$$

where $\mathbf{x}'$ and $\mathbf{x}''$ are code words of length $La$ bits, and $\mathbf{u}'$ and $\mathbf{u}''$ the corresponding information words of length $LB$ bits. Hence $d_L$ is defined as the weight of the minimum-weight code word of length $La$ bits produced by an information word with nonzero initial segment. The distance measure $d_L$, represented as a function of $L$, is referred to as the *column distance function.* Clearly, by letting $L$ tend to $\infty$, we get

$$\lim_{L \to \infty} = d_\infty = d_{\text{free}}.$$

Alternately, we can define the inverse function to $d_L$, the decision depth function $L_D(d)$, which is the code depth at which minimum distance $d$ is achieved for the first time. Convolutional codes closely follow the relation

$$L_D(d) \approx \frac{d}{a}\, h_B^{-1}(1 - R) \quad \text{levels,} \tag{4.2-1}$$

as will be discussed in Section 6.5. At rate 1/2, for example, this function is $4.54d$, and at rate 1/3 it is $1.92d$. Suppose a code at either of these rates has a certain $d_{\text{free}}$; then the decoding window $L$ needs to be about $4.54d_{\text{free}}$ trellis levels at rate 1/2 and $1.92d_{\text{free}}$ levels at rate 1/3.

The decoding window (or decision depth) is an important aspect of sequential decoding in Chapter 6, since it defines the depth of path needed. The precise shape of its inverse, the column distance function, can affect decoder computation (see Fig. 6.5-3). A fast initial growth in $d_L$ reduces computation. A code is said to possess an *optimum distance profile* (ODP) if $d_l$, $l = 1, 2, \ldots, \nu - 1$, are the largest among all codes of the same rate and constraint length $\nu$. Table 4.2-2 lists codes of various rates and constraint lengths possessing an optimum distance profile.

**Table 4.2-2.** SOME SYSTEMATIC CONVOLUTIONAL CODES WITH CONSTRAINT LENGTH GREATER THAN 10 POSSESSING AN OPTIMUM DISTANCE PROFILE.

Generator Sequences for rate-1/2 codes

| $g_2$ | $d_{free}$ |
|---|---|
| 3465 | 8 |
| 7153 | 9 |
| 67115 | 10 |
| 334473 | 12 |
| 34623073 | 14 |
| 163114353 | 15 |
| 671151433 | 16 |
| 7144760535 | 18 |
| 71447605247 | ≥ 18 |
| 714461626555 | ≥ 19 |

Generator Sequences for rate–1/3 codes

| $g_2$ | $g_3$ | $d_{free}$ |
|---|---|---|
| 2545 | 3373 | 16 |
| 15261 | 16257 | 17 |
| 32543 | 34537 | 18 |
| 65305 | 71273 | 19 |
| 152775 | 162543 | 20 |
| 653761 | 712611 | 22 |
| 1526137 | 1625661 | 24 |
| 6530547 | 7127375 | 26 |
| 65305477 | 71273753 | 28 |

## Punctured Convolutional Codes

Decoding a rate-$B/a$ code is similar to decoding a rate-$1/a$ code, with the difference there are now $b = 2^B$ branches entering each trellis state. The Viterbi algorithm must therefore select one of these $b$ branches, which makes its implementation somewhat difficult, especially at high data rates. Fortunately, this difficulty can be overcome with the use of *punctured* convolutional codes.

The principle behind punctured codes is easily explained using the rate-1/2 code and its trellis shown in Fig. 4.2-5(a) and 4.2-5(b). If we delete, or *puncture*, every fourth bit put out by the coder, the resulting code produces three output bits for every two input bits and hence has rate 2/3. The trellis for this code is shown in Fig. 4.2-5(c), where an $\times$ indicates a punctured output bit. The trellis for the rate-2/3 code shown in Fig. 4.2-6 is clearly equivalent to the trellis in Fig. 4.2-5(c), although one stage of the former corresponds to two stages of the latter.

Moreover, the trellis in Fig. 4.2-5(c) has only two branches entering each state instead of four as in Fig. 4.2-6. This enables the Viterbi algorithm to be implemented using binary, instead of quaternary, comparators. Recently, punctured convolutional codes have been applied to speech transmission in the mobile environment [4-14] and for automatic repeat request (ARQ) protocols [4-15]. Table 4.2-3 gives some good punctured convolutional codes.

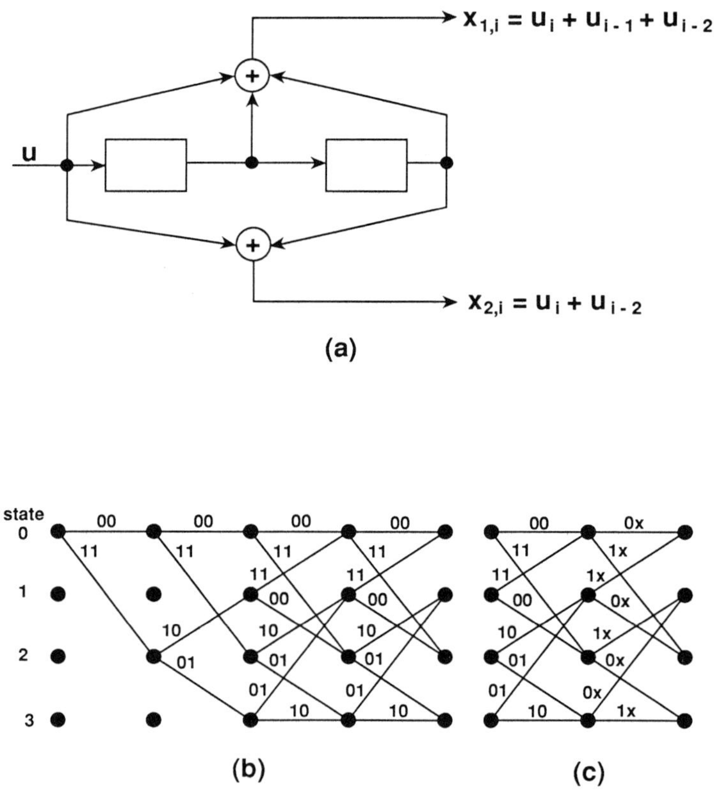


Figure 4.2-5. (a) A rate-1/2 code; (b) The trellis; (c) Trellis section for the punctured code.

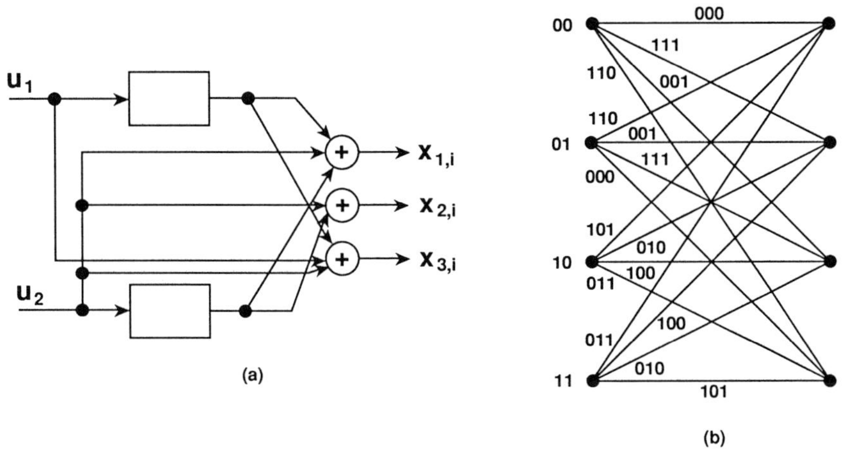


Figure 4.2-6. (a) A rate-2/3 code; (b) A trellis section.

**Table 4.2-3.** SOME GOOD PUNCTURED CONVOLUTIONAL CODES.

| Rate - 2/3 Codes | |
|---|---|
| Generators | $d_{free}$ |
| 7, 5, 7, x | 3 |
| 15, 13, 15, x | 4 |
| 31, 33, 31, x | 5 |
| 73, 41, 73, x | 6 |
| 163, 135, 163, x | 6 |
| 337, 251, 337, x | 8 |
| 661, 473, 661, x | 8 |

| Rate - 3/4 Codes | |
|---|---|
| Generators | $d_{free}$ |
| 5, 7, 5, x, x, 7 | 3 |
| 15, 17, 15, x, x, 17 | 4 |
| 25, 37, x, 37, x, 37 | 4 |
| 61, 53, x, 53, x, 53 | 5 |
| 135, 163, x, 163, x, 163 | 6 |
| 121, 165, 121, x, x, 165 | 5 |
| 205, 307, x, 307, x, 307 | 6 |
| 515, 737, x, 737, x, 737 | 6 |

## Weight Structure and Its Transfer Function

The weight distribution of a convolutional code enables us to determine several properties of the code such as its free distance and error rate. We use the rate-1/3 convolutional code in Fig. 4.1-1 and its trellis in Fig. 4.2-2 to illustrate the ideas.

The state diagram is redrawn in Fig. 4.2-7. State 0 is split into two states, the initial state $0_s$ and the final state $0_f$. Each branch is labeled as $D^n$, $n = 0, 1, 2, \text{or } 3$, where $n$ is the number of 1s on the corresponding branch in Fig. 4.2-2. The reader may observe several properties from the modified state diagram. For example, the minimum weight code word has weight 6. One can indeed identify the two code words of weight 6 from this diagram. One path departing from state 0, traverses states 2 and 1 and then reenters state 0. The other path, again departing from state 0, traverses states 2, 3, and 1 before reentering state 0. The minimum free distance $d_{\text{free}}$ is therefore 6. Similarly we can identify and enumerate paths of different weights. Enumerating by inspection soon becomes difficult with increasing constraint length. Instead, we compute the transfer function of the code; using this we can enumerate in a simple manner paths of different weights.

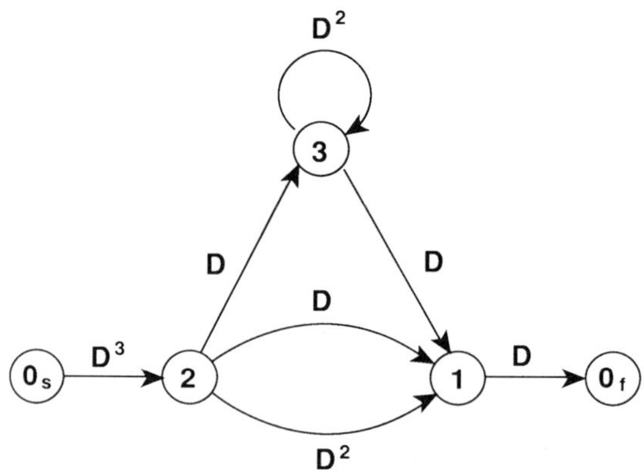


Figure 4.2-7. The modified state diagram of the rate-1/3 encoder in Fig. 4.1-1.

With the help of the modified state diagram, we write down the following four simultaneous equations in the five node gains $X_i, i = 0_s, 1, 2, 3,$ and $0_f$:

$$\begin{aligned} X_1 &= D^2 X_2 + D X_3, \\ X_2 &= D^3 X_{o_s} + D^2 X_1, \\ X_3 &= D X_2 + D^2 X_3, \\ X_{o_f} &= D X_1. \end{aligned} \tag{4.2-2}$$

Defining the transfer function $T(D)$ as

$$T(D) \triangleq \frac{X_{0_f}}{X_{0_s}},$$

and solving (4.2-2) simultaneously, we get

$$T(D) = \frac{D^6(2 - D^2)}{(1 - D^2)(1 - 2D^2)}.$$

Expanding $T(D)$ using long division, we obtain

$$T(D) = 2D^6 + D^8 + 5D^{10} + 5D^{12} + 14D^{14} + \cdots. \tag{4.2-3}$$

From (4.2-3) we see that there are two paths of weight 6, as pointed out in the preceding paragraph, one path of weight 8 given by $0_s$, 2, 3, 3, 1, $0_f$, and five paths of weight 10 given by

$0_s$, 2, 1, 2, 1, $0_f$,
$0_s$, 2, 1, 2, 3, 1, $0_f$,
$0_s$, 2, 3, 1, 2, 1, $0_f$,
$0_s$, 2, 3, 1, 2, 3, 1, $0_f$,
$0_s$, 2, 3, 3, 3, 1, $0_f$.

For large constraint length convolutional codes the method of solving $2^{\nu-1}$ equations simultaneously becomes unwieldy. In this case, the method of Mason [4-13] is used to compute the transfer function.

By labeling each branch with $D^i I^j \beta$, we obtain the generalized transfer function $T(D, I, \beta)$, where $i$ is the weight of the code word on that branch, $j$ the weight of the input sequence that produced the code word and $\beta$ a variable indicating that the code word is one branch long. The generalized transfer function for the code in Fig. 4.1-1 is (Problem 4-5)

$$\begin{aligned} T(D, I, \beta) &= \frac{D^6 I\beta^3 + D^6 I^2\beta^4 - D^8 I^2 \beta^4}{1 - D^2 I\beta - D^4 I\beta^2 - D^4 I^2\beta^3 + D^6 I^2 \beta^3} \\ &= D^6 I\beta^3 + D^6 I^2\beta^4 + D^8 I^3\beta^5 + D^{10} I^2 \beta^5 \\ &\quad + 2D^{10} I^3 \beta^6 + D^{10} I^4 \beta^6 + D^{10} I^4 \beta^7 \\ &\quad + 2D^{12} I^4 \beta^7 + D^{12} I^5 \beta^7 + 2D^{12} I^5 \beta^8 + \cdots. \end{aligned} \tag{4.2-4}$$

The above equation reveals that not only are there two code words of weight 6 but also one is three branches long and produced by a data sequence of weight 1 and the other four branches long and produced by a data sequence of weight 2. One can similarly find out information about code words of other weights.

## Error Probability Bounds

Without loss of generality, assume that the all-zero code word ($\mathbf{x} = \mathbf{0}$) is transmitted, corresponding to the path through all-zero states in the trellis. An *error event* is said to occur if a nonzero code word $\mathbf{x}'$ (an incorrect path) survives at state 0, thereby eliminating the all-zero code word (the correct path). Fig. 4.2-8 shows some possible error events.

Consider the first error event bcd that occurs in Fig. 4.2-8. If the weight of this path is $d$, then for odd $d$, the received sequence must have had $(d+1)/2$ or more errors. Otherwise the all-zero sequence would have survived and the error event bcd would not have occurred. For even values of $d$, if exactly $d/2$ errors occur then paths abd and abcd will have equal metrics and one of them will be selected with probability 1/2. If more than $d/2$ errors occur, then the path bcd will be selected in preference to the all-zero path.

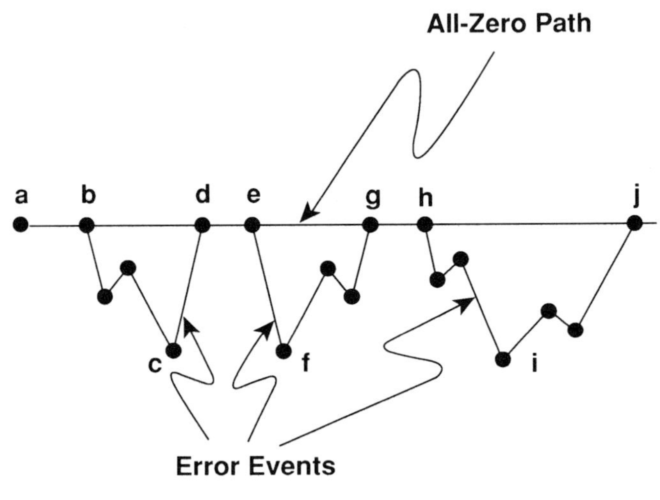


Figure 4.2-8 The all-zero and decoded paths and error events.

Let $P(E, d)$ denote the probability that the first error event of weight $d$ occurs. Assuming the BSC transition probability is $p$, the first error event probability is

$$P(E,d) = \begin{cases} \displaystyle\sum_{\frac{d+1}{2}}^{d} \binom{d}{i} p^i (1-p)^{d-i}, & \text{if } d \text{ is odd,} \\ \displaystyle\frac{1}{2}\binom{d}{d/2} p^{d/2}(1-p)^{d/2} & \\ \displaystyle\qquad + \sum_{\frac{d}{2}+1}^{d} \binom{d}{i} p^i (1-p)^{d-i}, & \text{if } d \text{ is even.} \end{cases} \tag{4.2-5}$$

Since the first error event may be $d_{\text{free}}$ or more in weight, we can over-bound the event error probability $P_E$ by the following union bound:

$$P_E < \sum_{d=d_{\text{free}}}^{\infty} \alpha_d P(E,d). \tag{4.2-6}$$

Here $\alpha_d$ is the number of code words of weight $d$ and it can be obtained from the transfer function $T(D)$. It is possible to show that a bound on $P(E,d)$ is (see Problem 4-5)

$$P(E,d) < 2^d p^{d/2}(1-p)^{d/2}. \tag{4.2-7}$$

Combining (4.2-6) and (4.2-7) yields,

$$\begin{aligned} P_E &< \alpha_d \big[4p(1-p)\big]^{d/2} \\ &= T(D)\,\big|_{D=\sqrt{4p(1-p)}}\,. \end{aligned} \tag{4.2-8}$$

For small $p$, (4.2-6) may be approximated by its first term as

$$P_E \approx \alpha_{d_{\text{free}}} (4p)^{d_{\text{free}}/2}. \tag{4.2-9}$$

Refer to [4-12] for details.

To determine the bit error probability $P_b$, we observe that each error event path carries with it a certain number of nonzero data bits and exactly these many data bits are in error. Let $\beta_d$ refer to the number of nonzero data bits carried by all the $\alpha_d$ paths of weight $d$. Multiplying $P(E,d)$ by $\beta_d$ and summing over all $d$ yields a bound on the expected number of data bits in error during one time unit (or a trellis branch). We may divide this by $B$, the number of data bits carried by the code each time unit, to get

$$P_b < \frac{1}{B} \sum_{d=d_{\text{free}}}^{\infty} \beta_d P(E,d). \tag{4.2-10}$$

Define $T(D,I)$ as

$$T(D,I) \triangleq T(D,I,\beta)\;\Big|_{\beta=1} = \sum_{d=d_{\text{free}}}^{\infty} \sum_{i=0}^{\infty} \alpha_{d,i} D^d I^i.$$

Now differentiate $T(D,I)$ with respect to $I$, and equate $I$ to 1 to obtain

$$\left.\frac{\partial T(D,I)}{\partial I}\right|_{I=1} = \sum_{d=d_{\text{free}}}^{\infty} \sum_{i=1}^{\infty} i\alpha_{d,i} D^d = \sum_{d=d_{\text{free}}}^{\infty} \beta_d D^d, \tag{4.2-11}$$

where $\beta_d = \sum_{i=1}^{\infty} i\alpha_{d,i}$. From (4.2-7), (4.2-10), and (4.2-11), we obtain

$$P_b < \frac{1}{B} \left.\frac{\partial T(D,I)}{\partial I}\right|_{D=[4p(1-p)]^{\frac{1}{2}},\; I=1}. \tag{4.2-12}$$

For small $p$, $P_b$ may be approximated by

$$P_b \approx \frac{1}{B}\beta_{d_{\text{free}}} \left[4p(1-p)\right]^{d_{\text{free}}/2} \approx \frac{1}{B}\beta_{d_{\text{free}}} [4p]^{d_{\text{free}}/2}. \tag{4.2-13}$$

If the BSC is implemented by an AWGN channel with BPSK or QPSK modulation, coherent detection, and hard-decision decoding, the transition probability $p$ may be expressed as (see (3.4-9) and (3.4-15))

$$p = Q\left(\sqrt{\frac{2RE_b}{N_0}}\right) \approx \frac{1}{2} e^{-RE_b/N_0}, \tag{4.2-14}$$

where $R$ is the rate of the code, $E_b$ the energy per data bit, and $N_0/2$ the two-sided noise power spectral density.

Substitute for $p$ from (4.2-14) into (4.2-13) to get

$$P_b \approx \frac{1}{B}\beta_{d_{\text{free}}} 2^{d_{\text{free}}/2} e^{-Rd_{\text{free}}E_b/2N_0}. \tag{4.2-15}$$

From (4.2-14), the expression for bit error probability without coding, i.e., $R = 1$, is

$$P_b \approx \frac{1}{2} e^{-E_b/N_0}. \tag{4.2-16}$$

Comparing (4.2-15) and (4.2-16), the expressions for bit error probability with and without coding, we see that the exponent in (4.2-15) is $Rd_{\text{free}}/2$ times that in (4.2-16). Since the exponential term dominates both these expressions, the ratio $E_b/N_0$ for the uncoded case must be increased by the factor $Rd_{\text{free}}/2$ to achieve the same asymptotic error probability behavior as in the coded case. The factor $Rd_{\text{free}}/2$ is called the *asymptotic coding gain* $G_h$. Expressed in decibels, the coding gain is

$$G_h = 10\log_{10}\left(\frac{Rd_{\text{free}}}{2}\right). \tag{4.2-17}$$

In the above expression we have assumed the use of hard decision at the decoder. With soft decision decoding that uses unquantized channel outputs, one can show that the asymptotic coding gain $G_s$ is given by

$$G_s = 10\log_{10}(Rd_{\text{free}}). \tag{4.2-18}$$

Thus $G_s$ is 3 dB more than $G_h$, pointing to the advantage of soft decision decoding. However, the decoder complexity increases since the decoder must now consider unquantized samples put out by the channel. In practice, an 8-level quantizer is found to yield most of the 3 dB gain. We have already noticed in Chapter 3 (see (3.4-17) and (3.4-18)) that soft decision decoding of block codes provided the same 3 dB gain over hard decision decoding.

## 4.3 CONTINUOUS PHASE MODULATION CODES

Continuous phase modulation, or CPM, codes carry data in the coded phase of a sine wave. Since a modulated sine has a distinct bandwidth and energy, the code words have these quantities explicitly. In the section just finished, we related the error probability of parity-check codes to energy by transmitting the code symbols by BPSK or QPSK modulation. The end result was a coding gain, a reduction in energy per data bit for a given probability, compared to uncoded modulation. In what follows the energy–error probability relation will be explicit because the code words themselves have energy.

They also have bandwidth. This means something deeper, that the new coding schemes gain and consume in *two dimensions*, energy and bandwidth. We can use them to save or spend energy or bandwidth, or both, or to trade one for the other. Schemes that perform explicitly in both these dimensions are called *bandwidth-efficient* codes.

The plan of this section is to explore more carefully the ideas of energy and bandwidth and then turn to CPM codes. The next two sections study two other kinds of bandwidth-efficient coding, codes based on choosing subsets from a larger signal set and codes derived from intersymbol interference patterns. In all three kinds of coding the trellis plays a central role.

## Energy and Bandwidth in Coding

We have already introduced the ideas of energy, bandwidth and coding in Sections 1.1 and 1.2. Channel coding was defined as the application of extra symbols to correct errors in data symbols, and modulation was the conversion of all these symbols into a transmittable waveform; these ideas underlie Chapter 3 and Sections 4.1 and 4.2. We must drop them now and turn to a more general definition of channel coding: *The imposition of set patterns onto the transmitted signal.* The decoder knows which patterns are possible and it chooses the closest one to the noisy signal that it receives. There is no separate encoding and modulating, only the choice of a patterned signal. Similarly, demodulating and decoding are replaced by one decision about what pattern was sent.

In this more general concept of coding there is no idea of redundant symbols *per se*. A BSC contains only two symbols, and if there are to be more patterns to the coding than the data sequences themselves, there must be additional non-data symbols. A channel with a wider alphabet can sustain coding without extra symbols. *Parity-check symbols are an artifact of the binary symmetric channel.*

Figure 4.3-1 is a plot of energy and bandwidth as we shall use them in the rest of this chapter. The energy axis is the ratio $E_b/N_0$ of the signal energy per data bit and the noise density of a Gaussian channel (these are defined in Section 1.3 and Appendix A). $E_b$ is the energy needed to produce a data symbol error rate of $10^{-5}$, a benchmark chosen for the figure. The bandwidth axis is the bandwidth of the signal as an RF transmission times the period devoted to a data bit, $B_{RF}T_b$. Figure 1.3-1 plots the upper sideband of this normalized spectrum for a kind of QPSK; our present measure, the "RF bandwidth", includes both upper and lower sidebands. In Fig. 4.3-1, we will assume that a channel filter limits the signal to bandwidth $B_{RF}$. The derivation of the normalizing constant $T_b$ is similar to that of the bit energy $E_b$ in Section 1.3: If $m$ data bits are carried per interval, and $T$ is the interval length, then $T_b = T/m$.

The reason for these normalizations is that transmission schemes carry differing numbers of data bits per signal interval and because of coding

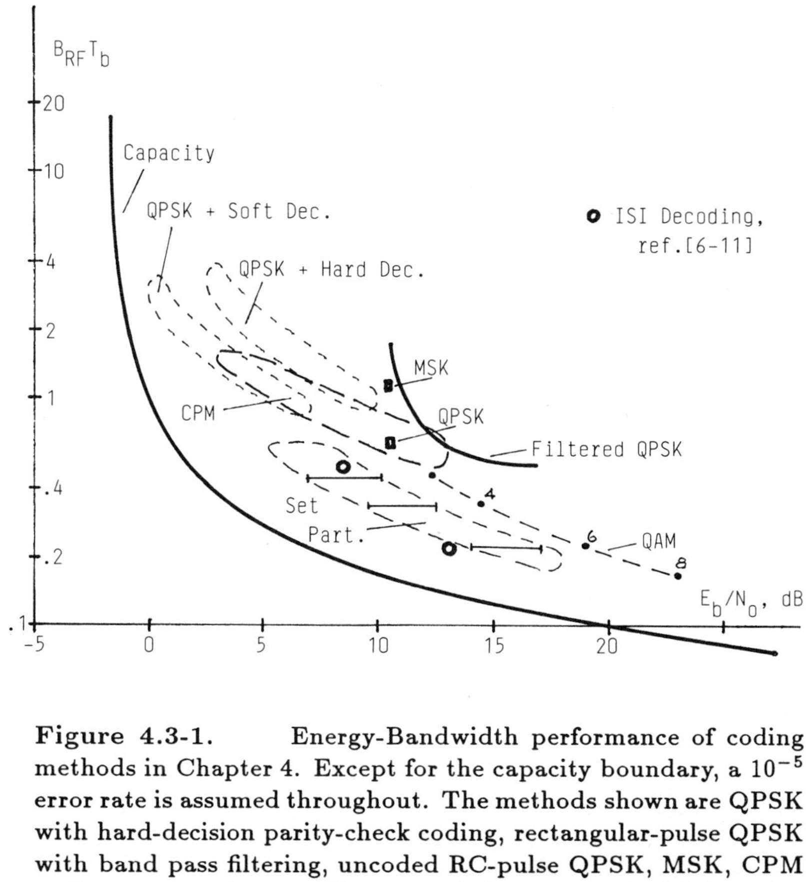


**Figure 4.3-1.** Energy-Bandwidth performance of coding methods in Chapter 4. Except for the capacity boundary, a $10^{-5}$ error rate is assumed throughout. The methods shown are QPSK with hard-decision parity-check coding, rectangular-pulse QPSK with band pass filtering, uncoded RC-pulse QPSK, MSK, CPM coding, set-partition coding, uncoded QAM, and ISI decoding from ref. [6-11]. All QPSKs assume 35% excess bandwidth pulses except the rectangular QPSK; so also does the set-partition coding. For CPM, the 99% power bandwidth is used. Note that all schemes here have envelope variation except CPM and MSK.

they may not even carry an integer number. We need a uniform way to assess competing schemes, in terms of the bandwidth and energy that they need per unit of information carried. The dimensions of $E_b/N_0$ are $(\text{bit})^{-1}$ and of $B_{RF}T_b$ are cycles/bit†.

†In SI units $B_{RF}T_b$ has units Hz-s/bit, which is cycles/bit if one interprets Hz as cycles/s; the dimension Hz/bit also appears in the literature, but this is incorrect.

In the foregoing work on parity-check codes we took no notice of bandwidth, but it is possible to locate parity-check codes in the energy-bandwidth plane by coupling the codes with a modulator. The bandwidth of this combination is $B_{RF}T_b/R$ cycles/data bit, where $B_{RF}T_b$ accounts for the modulator bandwidth per input bit to the modulator and $R$ is the code rate in data bits/code bit. We see that the code expands the modulator bandwidth by the factor 1/R. As for energy consumption, eqs. (4.2-15) to (4.2-17) show that $10\log_{10}(Rd_{\text{free}}/2)$ less energy in dB is needed by the combination to achieve the error benchmark, compared to the modulator alone.

Figure 4.3-1 shows the approximate energy-bandwidth location of practical parity-check codes combined with a QPSK modulator. As the rate $R$ declines, both coding gain and bandwidth grow and the code location tracks off to the upper left. Using soft decision creates another region about 3 dB to the left of the hard decision region.

The figure assumes a Nyquist-pulse QPSK with 35% excess bandwidth, the modulation described in Example A-2 in Appendix A. This scheme is typical of practical applications. (It will also be employed in the set-partition codes of Section 4.4). Other types of practical QPSK exist and can be used here but their bandwidth is wider, which shifts the performance of the coding combination considerably upward in the plot. The uncoded performance of one of these appears for comparison under the heading "Filtered QPSK". Here a rectangular-pulse constant-envelope QPSK, an infinite bandwidth scheme, is filtered down to various bandwidths; as the filter pass band shrinks, the signals need more energy to overcome intersymbol interference effects, and the energy-bandwidth performance follows a curved trajectory downward to the right.

Before leaving energy and bandwidth, we would like to locate the capacity of the Gaussian channel on Fig. 4.3-1. A point there represents an attempt to signal at an energy and bandwidth of $E_b/N_0$ and $B_{RF}T_b$, normalized per bit. Is this enough energy and bandwidth to create a channel of capacity one bit? If not, no intensity of coding can drive the error probability close to zero; if so, good performance is at least possible. The capacity formula thus creates a boundary in the plane between possible and impossible energy–bandwidth combinations.

The boundary for Gaussian noise appears in the figure. We can derive it from two facts that appear in Appendix A. First, the number of signal space dimensions available in time $T$ and RF bandwidth $B_{RF}$ is $2B_{RF}T$. Second, the capacity of this channel is

$$C = B_{RF}T \, \log_2\left(1 + \frac{2E_d}{N_0}\right) \quad \text{bits,} \tag{4.3-1}$$

where $E_d$ is the average signal energy per signal space dimension and $N_0$ is the spectral density of the in-band noise. From (4.3-1), the capacity per second is

$$C_t = \frac{C}{T} = B_{RF} \log\left(1 + \frac{2E_d}{N_0}\right) \quad \text{bits/s.}$$

If we divide $B_{RF}$ by $C_t$, we have the bandwidth per bit carried by this capacity-achieving system, which is the ordinate axis of Fig. 4.3-1; the ordinate point is

$$\frac{B_{RF}}{C_t} = \frac{1}{\log\left(1 + \dfrac{2E_d}{N_0}\right)} \quad \text{Hz-s/bit.} \tag{4.3-2}$$

It remains to replace $E_d$ with the energy per data bit, $E_b$. We have

$$\begin{aligned} E_d &= (\text{Total energy})/(\text{Total dimensions}) \\ &= \frac{E_b C}{2B_{RF}T} \\ &= \frac{E_b C_t}{2B_{RF}} \end{aligned} \tag{4.3-3}$$

and substituting this into (4.3-2) gives

$$\frac{B_{RF}}{C_t} = \frac{1}{\log\left[1 + \dfrac{E_b/N_0}{B_{RF}/C_t}\right]}. \tag{4.3-4}$$

The capacity boundary in Fig. 4.3-1 is the relation between the coordinates $B_{RF}/C_t$ and $E_b/N_0$ given in (4.3-4). Measured by energy, the hard-decision parity-check — QPSK combination lies 5 to 10 dB away from capacity at the $10^{-5}$ error benchmark; in bandwidth it lies perhaps 10-fold higher. In the rest of the chapter we will try to close these gaps.

## Continuous Phase Modulation Codes

CPM codes carry data in the continuous, coded phase of an RF sinusoidal carrier. They were invented for the digital satellite and mobile radio channels, which work best with constant-envelope signaling and often must run at low signal energy. These facts motivate a coded, phase-only system running at a relatively low rate in bits/signal interval. The requirement of phase continuity was added because it reduces bandwidth and provides coding gain.

Our plan will be to describe the signals and then show how the coding concepts of trellis, distance, and decoding work with them. CPM as a method of coding originated with refs. [4-16] and [4-17] in the late 1970s. Alternatives to our presentation exist in several texts [4-22]–[4-24], and ref. [4-18] is an advanced text devoted mostly to CPM. Tutorials and journal special issues are another source of information [4-19]–[4-21].

A constant-envelope phase-varying signal has the form

$$\sqrt{\frac{2E}{T}}\cos\big(\omega_o t + \phi(t)\big) \tag{4.3-5}$$

and in a CPM signal the phase $\phi(t)$ is

$$\phi(t) = 2\pi \sum_{i \le n} u_i h q(t - iT) \tag{4.3-6}$$

during the $n$-th symbol interval $nT < t \le (n+1)T$. We need to explain each of these variables. $E$ is the energy in the signal (4.3-5) during a symbol interval of length $T$ and $\omega_o$ is the RF carrier frequency in rads/s. Eq. (4.3-6) consists of three variables, the data symbols $u_i, i = 0, 1, \ldots$, a proportionality constant $h$, and a *phase response pulse* $q(t)$. The modulating phase $\phi(t)$ is the convolution of recent data symbols with this phase pulse. We think of the $b$-ary data as the plus-minus integers $\pm1, \pm3, \ldots, \pm(b-1)$; for example, a binary CPM encoder accepts data $\pm1$, while a quaternary coder accepts the 4-ary symbols with levels $(\pm1, \pm3)$. We will continue this symbol convention through the rest of the chapter.

The heart of a CPM code is the phase response pulse. Some of these are shown in Fig. 4.3-2. The pulse represents the total response of the modulating phase to each data symbol: The response function begins at zero, varies over $\nu - 1$ intervals, and settles down to $1/2$ after time $(\nu - 1)T$. Each data symbol $u_i$ eventually causes a phase shift of $\pi u_i h$ radians, which arrives at this value through a pattern set by $q(t)$. $q(t)$ plays the same role here as the convolutional code generator sequences defined in eq. (4.1-2). CPM codes are based on continuous-time phase convolution while the codes in Section 4.1 employ discrete-time symbolic convolution, and in either case, the trick of the design lies in the generator sequence. Just as with the symbolic codes, CPM codes have a constraint length $\nu$ which is the number of active intervals in the generator response plus one.

The proportionality constant $h$ in (4.3-6) is called the *modulation index.* It plays a role similar to the index $\beta$ in analog FM in that a larger index generally leads to better error performance but a wider spectrum. A CPM code's spectrum also depends critically on the phase response

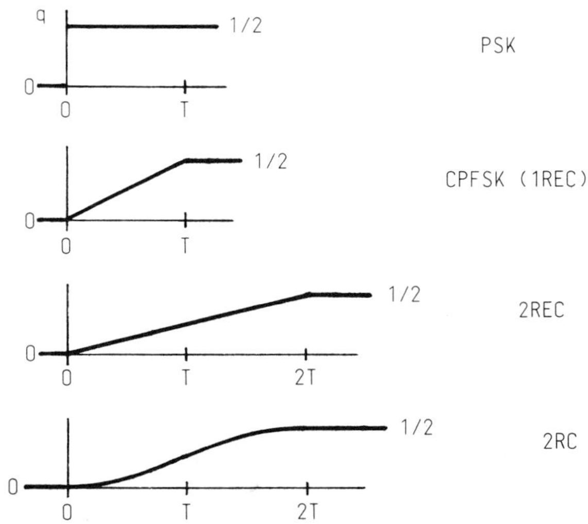


**Figure 4.3-2.** **Some CPM phase response pulses.**

shape. The region of the highest slope in $q(t)$ affects the width of the main lobe in the spectrum, while the spectral sidelobes are set by the number of continuous derivatives in the pulse (see [4-18], p. 168). Some CPM spectra appear at the end of the section. A final influence on the spectrum is the data alphabet size $b$, since the integer-valued data scale the total phase shift. All of these factors also affect the free distance of the code in a complex way.

The first phase response pulse in Fig. 4.3-2 is the instantaneous response of rectangular-pulse PSK†, which we can think of as a trivial case of CPM. Technically, this is differentially encoded PSK, since as in all CPM schemes the data are carried in the change of phase rather than in the value of the phase. The next pulse is the CPFSK (continuous-phase frequency-shift keying) pulse; it is a linear increase of phase spread over one interval. We could also spread the increase over two intervals, which gives the third case. This would be a more complicated code in which the phase at any one time changes in a pattern set by the last *two* data symbols. The CPM names for these last two cases are 1REC and 2REC; the digit refers to the spread of the pulse (CPM schemes are named after

†Rectangular PSK is also called NRZ (non-return to zero) PSK and pure PSK; it should be distinguished from pulse-shaped PSK, as described in Appendix A.

the spread, not the constraint length) and the name REC originated in the rectangular shape of the pulse derivative. The last case in Fig. 4.3-2 is a scheme called 2RC, in which a raised cosine pulse shape spreads over two intervals. RC pulses contain two more continuous derivatives than do REC pulses and, therefore, lead to lower spectral sidelobes.

We next turn to tree and trellis structures in CPM codes. In CPM coding there are two notations of trellis, the first being the usual progression of states versus time and the second being a plot of all *phase trajectories* against time. Of the two, the second is the more useful, and sometimes a tree plot is the most useful of all. Phase has an obvious physical significance, and signal phase is closely related to the signal state, in any case.

The standard state description [4-18] for a CPM signal at time $nT$ consists of a phase $\theta$ and the last $\nu - 2$ data symbols. $\theta$ represents the accumulated phase of all previous symbols up to the $(n - \nu + 2)$th one $u_{n-\nu+1}$, i.e., the phase in (4.3-6) that is no longer changing with time. If the modulation index $h$ is $j/k$, $j$ and $k$ integers, then it can be shown that $\theta$ is always a multiple of $2\pi/2k$. There are thus $2k$ phases and $b^{(\nu-2)}$ symbol combinations in this state description, for a total of $2kb^{(\nu-2)}$ states. More efficient descriptions exist.

A simple example of trees and trellises is given by the CPFSK case.

**Example 4.3-1: CPFSK (1REC) Binary CPM with h = 1/2.**

> This trellis code is created by the second pulse in Fig. 4.3-2. The different data sequences cause $\phi(t)$ to trace out the phase tree of Fig. 4.3-3, a structure built of diamond shapes. It is easy to see the memory and continuity of $\phi(t)$ across the symbol interval boundaries. The highlighted path through the tree is the one created by the data $+1, +1, +1, -1$. Because $h$ is 1/2, the phase separation between two paths that split is $(1/2)2\pi$ after one symbol interval (the length of the pulse) goes by. Descendants of these paths can merge again after another interval; for instance, the paths $+1, +1, +1, -1$, and $+1, +1, -1, +1$ split at $2T$ and merge at $4T$.
>
> Since phase is modulo $2\pi$, the phase tree eventually wraps around onto itself and certain nodes become indistinguishable. In fact, all CPM schemes with rational $h$ do this, and what we have is a natural definition for a CPM phase trellis: It is the progression of phase for all data sequences plotted modulo $2\pi$. The $h = 1/2$ CPFSK phase trellis appears below the phase tree in Fig. 4.3-3, and it is clear that many nodes have merged. We can think of the tree as wrapped around the cylinder in Fig. 4.3-4.

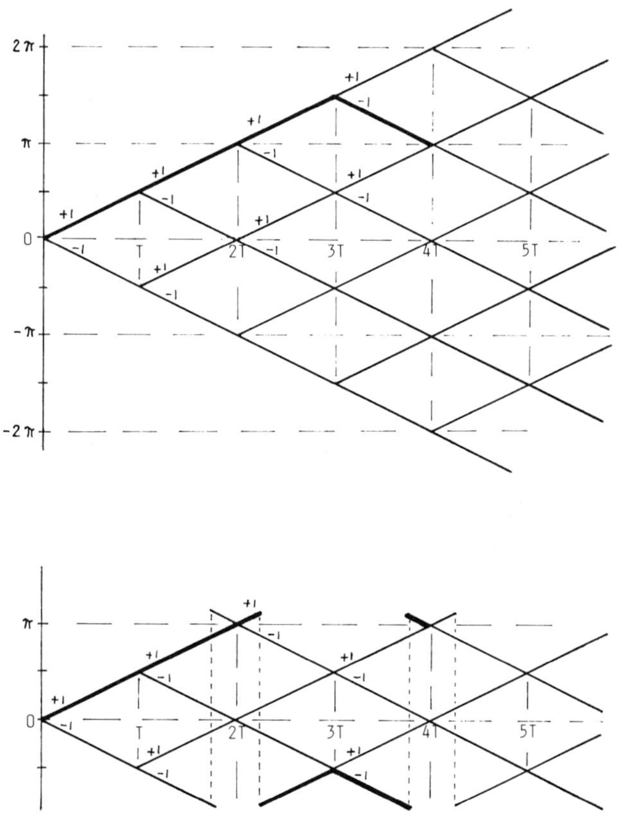


**Figure 4.3-3:** Phase tree for CPFSK with $h = 1/2$, showing some of the early phase trajectories. The tree is redrawn as a phase trellis (below) by merging nodes with the same phase mod $2\pi$. Selected data symbols are marked by +1 and -1.

A true state trellis for $h = 1/2$ CPFSK appears in Fig. 4.3-5, drawn in the style of the convolutional trellises of Section 4.2. In a length-1 CPM scheme the previous data symbol does not contribute to the definition of the present state, so that the state is defined by the cumulative phase alone, and this is shown as a multiple of $\pi/2$. The present symbol plus this phase determines which state comes next. We can see that even and odd states alternate in this scheme. The state trellis gives a poorer feel for the scheme than does the phase trellis, but sometimes a phase trellis shows a crossing of phase where a true state merge does not occur; only the state trellis records true merges. In CPFSK all the phase crossings are true state merges.

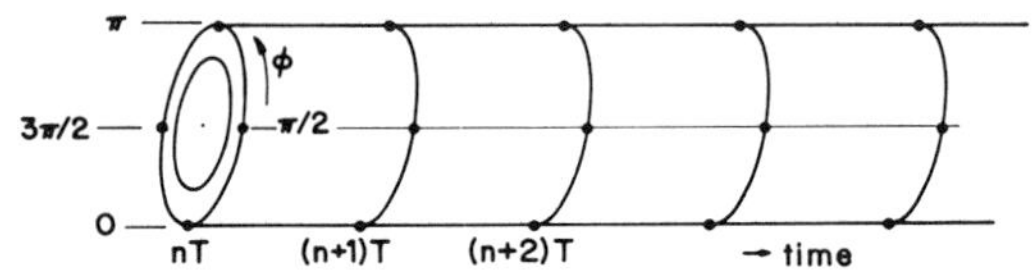


**Figure 4.3-4.** The phase cylinder beginning at time $nT$.

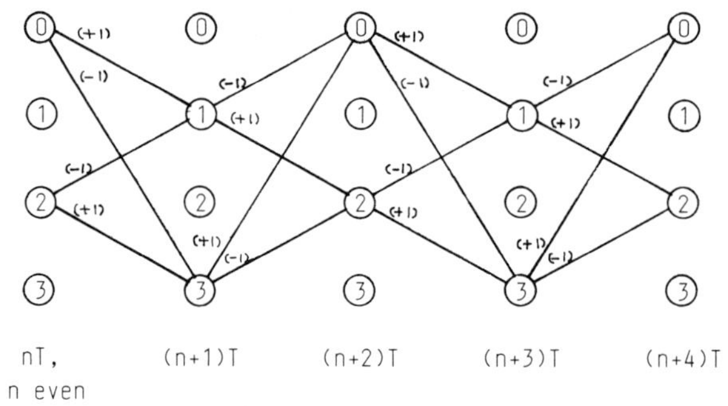


**Figure 4.3-5** The state trellis for $h = 1/2$ CPFSK. Even and odd states occur after even and odd multiples of $T$. The state designator is the cumulative phase as a multiple of $\pi/2$. Two complete cycles are shown. Symbol $u_n$ arrives just after time $nT$.

The CPFSK in the example was first proposed by Doelz and Heald [4-25] who called it minimum-shift keying (MSK) and thought of it as a single-interval simple modulation; much later, de Buda [4-26] observed that the phase continuity was a kind of coding memory and that the optimal detector had to observe two intervals. Several others studied CPFSK with other modulation indices in succeeding years (see [4-27]–[4-29]); a full analysis of CPFSK as a coding scheme appeared finally in the 1979 Ph.D. thesis of T. Aulin (published as [4-17]).

*Multi-h codes* are an extension of CPFSK in which the modulation index that applies in an interval rotates cyclically through a small set $\{h_1, ..., h_Q\}$ [4-16]. The rotating index leads to a more complex trellis and

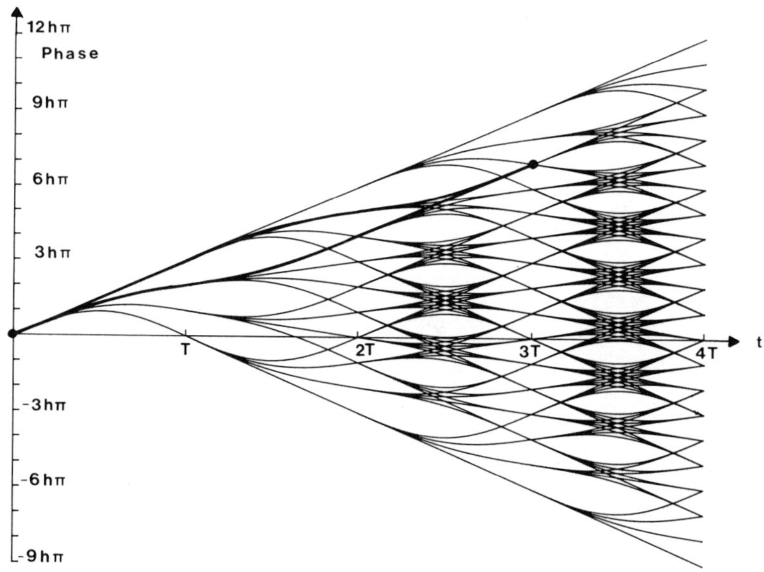


**Figure 4.3-6** Phase tree of quaternary $2RC$ with index $h$. The plot assumes that symbols $\ldots, u_{-2}, u_{-1}$ are all $+3$. An example of a length-3 error event is highlighted between the two dots. Adapted from ref. [4-18], p.79, ©Plenum Press 1986, used by permission.

a larger free distance. The rotation idea may also be combined with the longer phase response functions [4-18].

Next, we give an example of a more complex CPM code.

Example 4.3-2: 2RC Quaternary CPM with h = 1/4.

A phase tree for this scheme appears in Fig. 4.3-6 and a state trellis in Fig. 4.3-7. The accumulated phase and one previous data symbol appear in the state description. The phase tree here is much more complex than the CPFSK one in Fig. 4.3-3, but phase transitions are smoother, which reduces signal bandwidth, and merges occur after three intervals instead of two, which increases the free distance. The tree assumes that the symbol previous to the root node was a $+3$. Merges are harder to spot in this trellis; the highlighted example shows the paths with data symbols $+3, +1, +3$ and $+1, +3, +3$, which split at the root node and merge at the round dot.

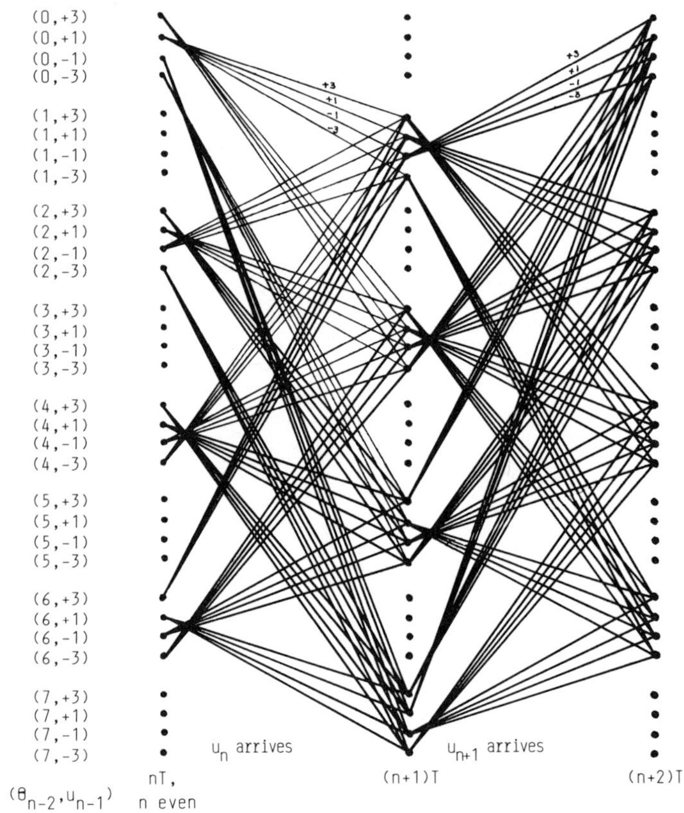


**Figure 4.3-7** State trellis for quaternary $2RC$ with $h = 1/4$, showing one cycle. The state at time $nT$ is made up of previous data symbol $u_{n-1}$ and the phase accumulated to time $(n-2)T$, shown as a multiple of $\pi/8$. The present symbols on the transitions are mostly suppressed for clarity. The trellis has the same even-odd character as Fig. 4.3-5.

## CPM Distance Properties

Just as with parity-check codes, the error performance of a CPM code depends primarily on the distance from the transmitted code word to its nearest neighbor. In the Gaussian white-noise channel that we now assume, this is the distance between two signals in the signal space

$$D^2[s_1(t), s_2(t)] = \int \left[s_1(t) - s_2(t)\right]^2 dt, \tag{4.3-7}$$

where the integral extends over the time that the signals differ. (The derivation of (4.3-7) is reviewed in Appendix A). Suppose $s_1$ and $s_2$ are sinusoids of the form (4.3-5), with phases $\phi_1(t)$ and $\phi_2(t)$. It is not hard to show that (4.3-7) becomes

$$\frac{2E}{T}\int\left[1-\cos(\phi_1(t)-\phi_2(t))\right]dt \tag{4.3-8}$$

in the limit of large carrier frequency. Only phase difference matters in this formula.

Eq. (4.3-8) is still not in the most convenient form. By convention, signal space distances are normalized by twice the data bit energy in order to remove the artificial effects of alphabet size and signal energy; that is, we seek

$$d^2\left[s_1(t),s_2(t)\right]=\frac{D^2\left[s_1(t),s_2(t)\right]}{2E_b}.$$

Here $E_b$, the energy per data bit, is $E/\log_2 b$, the CPM signal energy divided by the log of the alphabet size. So (4.3-8) becomes

$$d^2\left[s_1(t),s_2(t)\right]=\frac{\log_2 b}{T}\int\left[1-\cos(\phi_1(t)-\phi_2(t))\right]dt \tag{4.3-9}$$

with the integral extending over the time the phases differ. On this normalized scale, QPSK has square distance 2, and two parity-check-coded QPSK signals with Hamming distance $d_H$ will lie at signal space square distance $2d_H$.†

Now we can find the normalized free distance of a CPM code. If a transmitted signal $s_i(t)$ carries data $...u_{-1},u_0,u_1...$ the nearest neighbor to it in signal space is the one at distance

$$d_m^2=\min_{s_j(t)} d^2\left[s_i(t),s_j(t)\right],$$

where the index $j$ runs over all signals that differ from $s_i(t)$ for the first time in symbol $u_0$. In the state trellis, these are all the paths that split from the transmitted path after time 0. For the convolutional codes of Section 4.2, $d_m^2$ is the free distance because every transmitted code word has the same distance structure, but this is often not the case with bandwidth-efficient codes. CPM codes, in particular, need a second minimization over what signal is sent. We can write

†Distances are almost always quoted in "square" terms, but the word square is often dropped.

$$d_{\text{free}}^2 = \min_{s_i(t)} \min_{s_j(t)} d^2\left[s_i(t),\; s_j(t)\right] \tag{4.3-10}$$

with the condition that $s_j(t)$ differs from $s_i(t)$ for the first time in interval 0. The probability that an ML receiver decides $s_j$ when $s_i$ was sent is $Q(d_{\text{free}}\,\sqrt{E_b/N_0})$.

These concepts do not really differ from the Hamming distance manipulations in Section 4.2, except for the double minimization. An efficient dynamic program to find (4.3-10) was first proposed in ref. [4-16]; the scheme is explained in [4-18] or Mulligan and Wilson [4-30].

The 2RC code in Example 4.3-2 will illustrate free distance and the related concepts of decision depth and column distance in CPM codes. In general, these parameters are complicated functions of the phase pulse length, the alphabet size and the modulation index. Figure 4.3-8 plots square minimum distance against the index for the particular pulse and alphabet of quaternary 2RC. There are a number of curves here for different decision depths (or observation windows) in the trellis decoder. As the decoder observes longer trellis paths, the worst-case distance between any transmitted path and neighbor becomes larger, until a greater length does not improve the distance; this length is the decision depth function $L_D(d_{\text{free}})$ at $d_{\text{free}}$ for this index and $d_{\text{free}}$ is the best obtainable distance. For instance, at about $h = 0.55$ the free distance of this 2RC is about 4.2 and the required decision depth is 15 trellis levels. An observation of only 4 levels gives a minimum distance of 2.9 while single-level observation yields just 0.5.

It can be seen in Fig. 4.3-8 that there are several valleys in the distance picture, modulation indices at which distance is much worse than it is at nearby indices. These are called weak modulation indices. In quaternary 2RC, they occur at $h = .5,\ .67,\ 1.0$ and some larger indices. To obtain the very largest free distance we must carefully choose $h$; the best $h$ appears to be near $h = .95$ (a slightly better one occurs at $h$ about 1.74).

Precisely which error events lead to $d_{\text{free}}$ varies in a complex way with the modulation index. At small $h$ it is the same event for all CPM schemes, the one for which the difference between the transmitted and neighbor data symbols is $\ldots 0,\ +2,\ -2,\ 0, \ldots$ or $\ldots 0,\ -2,\ +2,\ 0,\ \ldots$ The highlighted event in Fig. 4.3-6 is one such event. This event defines the initial rise of distance in Fig. 4.3-8.

Some peak values of $d_{\text{free}}^2$ that occur in CPM codes are given in Table 4.3-1. The modulation index is the one in the range $0 < h < 1$ giving

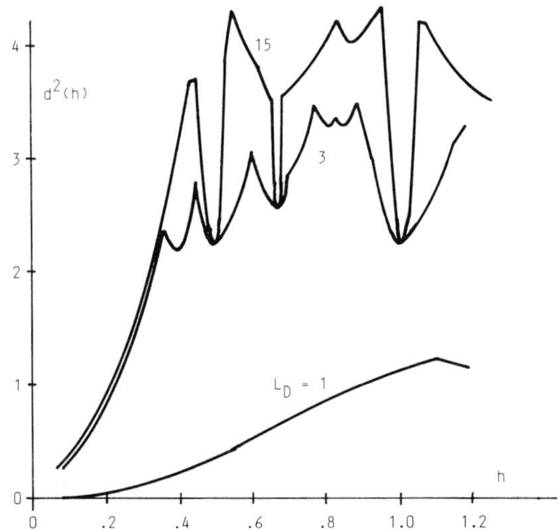


**Figure 4.3-8.** Minimum distance of quaternary 2RC as a function of modulation index $h$, for decision depths 1, 3, and 15. The curve for depth 15 is the free distance except near weak modulation indices at $h = 0.5, 0.67$, and 1.0. Data taken from Aulin [4-31].

the peak distance. Also given is the decision depth, the path length required in a trellis decoder to achieve the $d_{\text{free}}$ level of error performance. The table shows that $d_{\text{free}}$ improves with the alphabet size and to a lesser degree with the phase pulse length.

## Energy-Bandwidth Performance of CPM Codes

Figure 4.3-9 plots some typical spectra of CPM codes using the bandwidth scale $B_{RF}T_b$ defined at the beginning of the section; only the right half of the spectrum is shown. The two codes quaternary 3RC, $h = 0.3$ and binary 1REC, $h = 0.5$ (i.e., MSK) are shown along with the QPSK spectrum. All three of these schemes have $d^2_{\text{free}}$ about 2, and thus have about the same error rate. The phase smoothness of a CPM code reduces spectral sidelobes; this is already clear with 1REC and is strongly apparent with 3RC. The main lobe width of 3RC is reduced as well, because of its long pulse and quaternary data alphabet; these characteristics also bring $d^2_{\text{free}}$ up to 2.

The maximum likelihood decoder for CPM over a good, white noise channel is the Viterbi algorithm. This is just the decoder for convolu-

**Table 4.3-1.** A Selection of Best CPM Square Free Distances at Modulation Indices Below Unity, Including the Required Index and Decision Depth. QPSK Has Square Distance 2.

| Phase Response | Code Rate, log b | Index, h | $d^2_{free}$ | Decision Depth |
|---|---|---|---|---|
| CPFSK (1REC) | 1 | .72 | 2.4 | 3 |
| | 2 | .91 | 4.2 | 9 |
| | 3 | .96 | 6.1 | 41 |
| 2RC | 1 | .76 | 2.7 | 5 |
| | 2 | .94 | 4.4 | >15 |
| 3RC | 1 | .95 | 3.5 | >15 |
| | 2 | .75 | 5.3 | 12 |
| 3REC | 1 | .93 | 3.6 | 10 |

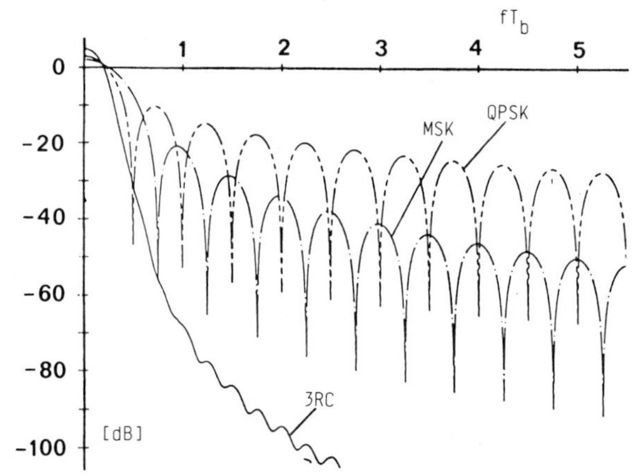


**Figure 4.3-9.** Power spectra of 3RC with $h = 0.3$, MSK (which is CPFSK with $h = 0.5$) and rectangular-pulse uncoded QPSK. All three schemes have about the same free distance. From ref. [4-18], p.173, ©Plenum Press 1986, used by permission.

tional codes in Section 4.2, with a CPM trellis and with signal space distance instead of Hamming distance. Some attractive CPM schemes, such as binary 1REC, have small trellises, but others have quite large trellises, so that sequential decoding is preferred to Viterbi decoding.

Some error measurements are given for both decoders in Chapter 6 and we will not repeat them here; see particularly Example 6.2-2.

Figure 4.3-1 shows the performance of the best CPM codes in the energy-bandwidth plane. Larger alphabet size and longer pulse length give codes with larger energy and smaller bandwidth. At first glance it appears that CPM extends the soft-decision parity-check performance down and to the right into a lower bandwidth region, and that the set-partition codes in the next section perform considerably better than CPM. However, RF envelope variation plays a critical role here. CPM coding has constant envelope while set-partition coding does not, and the performance of the parity-check coding would be much worse than shown if it had constant envelope. The real key to energy-bandwidth performance lies in envelope variation, provided that the channel allows it.

# 4.4 SET-PARTITION CODING

Now we turn to yet another kind of trellis coding. In this method *sets* of data-bearing signals are associated with an encoder state transition, rather than individual signals, and the signals are points in a two-dimensional signal space like that of QPSK. The design of a code consists of choosing sets of points and then assigning the sets to trellis transitions. Like CPM codes, these codes work by imposing a pattern directly on the signal phases and amplitudes, and the distance between code words is measured by Euclidean signal space distance. The code words have explicit bandwidth, but unlike CPM coding, it is set a priori and is not part of the code design.

## QAM Signal Point Constellations

Set-partition coding is mostly based on a quadrature amplitude modulation, or QAM, model of signaling. In this model we represent a pulse-modulated RF carrier as a point in a two-dimensional signal space. How to locate this point is reviewed in Appendix A. For now, we will simply state that the modulated signal is

$$s(t) = \sqrt{\frac{2E_s}{T}}\left[I(t)\cos\omega t - Q(t)\sin\omega t\right], \tag{4.4-1}$$

in which $T$ is the time devoted to a transmitted symbol and $E_s$ is its energy on the average. $\omega$ is the RF carrier frequency. $I(t)$ and $Q(t)$ are sequences of modulating pulses, the *in-phase* and *quadrature* sequences, respectively, and are defined by

$$I(t) = \sum_n c_n^I g(t - nT)$$
$$Q(t) = \sum_n c_n^Q g(t - nT) \qquad (4.4\text{-}2)$$

in which certain values of the weights $c$ represent the data symbols. A new pulse $g(t)$ appears each $T$ seconds in both the $I$ and $Q$ waveforms.

QAM imposes certain orthogonality conditions on the pulse shape (see App. A), and under these the signal space representation of (4.4-1) is a point in a two-dimensional space, one such space for each symbol interval. All the different signals that can appear in the $n$-th interval $[nT, (n+1)T]$ form a *constellation* of points, the pairs $(c_n^I, c_n^Q)$ that can occur there. Figure 4.4-1 shows a constellation of 8 points around a circle. These represent the signals of octal phase-shift keying, in which the signal phase in the $n$-th interval is $\tan^{-1}[c_n^Q/c_n^I]$; these 8 points carry 3 data bits. Figure 4.4-4 shows at the top an array of 32 points on a rectangular grid that carries 5 data bits per symbol interval. This scheme is called 32 cross QAM because of the cross-like arrangement of points, and we can generalize the idea of points on a rectangular grid to an $m$-point array that carries $\log_2 m$ bits. The term QAM is also used in a generic sense to refer to any modulation based on an $I$-and-$Q$ point constellation, whether on a rectangular grid or not.

It can be shown that the power spectrum of a generic QAM signal (4.4-1) has the same shape as that of the basic pulse $g(t)$, no matter what the arrangement of signal space points. The calculation appears in digital communications texts like refs. [4-18], [4-23], [4-24], or [1-12]. It is thus clear that whatever coding we do with the constellation points will not affect the signal spectrum: If the pulse has double-sideband bandwidth $B_{RF}$ and the scheme after coding carries $R$ data bits per symbol interval, then $T_b$ is $T/R$ and the bandwidth consumption of the scheme is $B_{RF}T_b$ per data bit.

A variety of pulses may be used to create a QAM constellation. The smallest bandwidth of these pulses is $B_{RF} = 1/T$. The favored class of pulses for QAM is the raised-cosine class, which is the subject of Example A-2 in Appendix A. The bandwidth of these pulses is $(1+\alpha)/T$, where $\alpha$, the excess bandwidth factor, satisfies $0 \le \alpha \le 1$. A typical, practical value for $\alpha$ is 0.35, and we will assume this value is the case for all the schemes in this section. The normalized bandwidth consumption is consequently

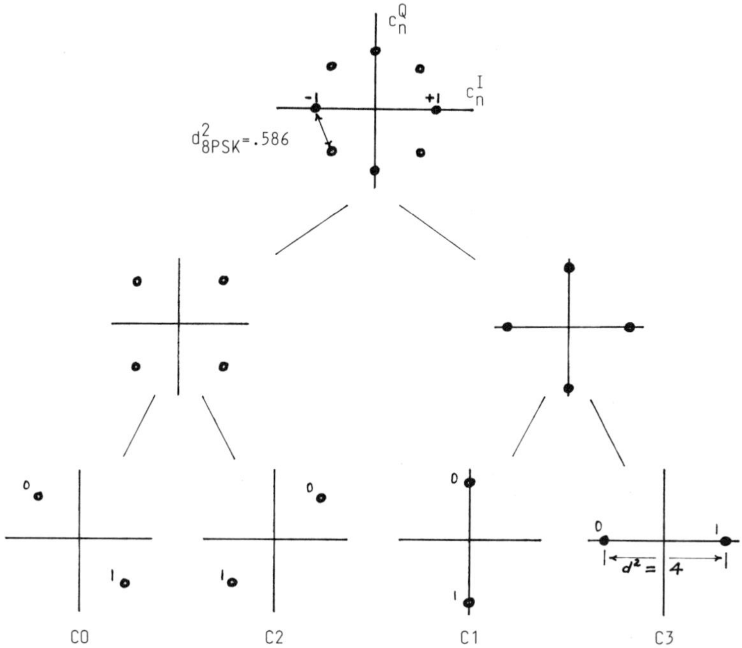


**Figure 4.4-1.** Rate 2 set-partition code: Partitioning of 8PSK master constellation to form subsets. Data bits are labeled in the bottom partition.

$$B_{RF}T_b = \frac{1+\alpha}{R} = \frac{1.35}{R} \text{ cycles/bit} \tag{4.4-3}$$

for a scheme carrying $R$ bits per symbol interval.

Just as it did for CPM signals, the Euclidean distance between two QAM signals stems from (4.3-7). We need to find the normalized distance $d^2[s_1(t),\ s_2(t)]$ between two signals; then the probability of mistaking one for the other in a maximum likelihood receiver is the usual expression $P_e = Q(d/\sqrt{2N_0})$. The first step is to compute (4.3-7):

$$D^2\left[s_1(t),\ s_2(t)\right] = \int \left[s_1(t) - s_2(t)\right]^2 dt$$

$$= \frac{E_s}{T} \int [(\Delta I)^2 + (\Delta Q)^2]\, dt$$

$$= [(\Delta c_n^I)^2 + (\Delta c_n^Q)^2] \frac{E_s}{T} \int g^2(t)\, dt. \qquad (4.4\text{-}4)$$

The integrals here extend over the time that the integrands are nonzero. The notation $\Delta I$ means the difference between the $I_1(t)$ for $s_1(t)$ and the $I_2(t)$ for $s_2(t)$, and similarly $\Delta c_n^I$ is the difference between the $n$th interval weights in the two signals. The second line holds in the limit of large carrier frequency (see Problem 4-11). In the third line the expression $(E_s/T) \int g^2(t)\, dt$ is a constant for a given QAM system.

To normalize (4.4-4), we need to divide it by twice the average energy per data bit, $2E_b$. The average energy of a single QAM RF pulse is

$$\mathcal{E}\left[(c_n^I)^2 + (c_n^Q)^2\right] \frac{E_s}{T} \int g^2(t)\, dt, \qquad (4.4\text{-}5)$$

using steps similar to those that produced (4.4-4), and the expectation here is over which point in the constellation is sent. From the orthogonality condition on $g(t)$, it can be shown that this is also the average energy of a train of pulses. The standard assumption is that all the constellation points are equiprobable, and under this condition it will be convenient to define

$$c^2_{\mathrm{ave}} \triangleq \mathcal{E}\left[(c_n^I)^2 + (c_n^Q)^2\right]. \qquad (4.4\text{-}6)$$

The average energy per data bit in a scheme carrying $R$ data bits/interval is

$$E_b = \frac{1}{R} c^2_{\mathrm{ave}} \frac{E_s}{T} \int g^2(t)\, dt. \qquad (4.4\text{-}7)$$

From (4.4-4)–(4.4-7) we can now write the normalized Euclidean distance:

$$d^2\left[s_1(t), s_2(t)\right] = \frac{R}{2} \frac{\left(\Delta c_n^I\right)^2 + \left(\Delta c_n^Q\right)^2}{c^2_{\mathrm{ave}}}. \qquad (4.4\text{-}8)$$

The numerator in (4.4-8) is just the square distance between two points in the constellation.

Set-partition coding will use signals that differ across more than one interval, and so we need to extend (4.4-8) a little to cover this case. By retracing the steps it can be seen that $(\Delta c_n^I)^2$ needs to be replaced by

$$(\Delta c^I)^2 = \sum_n (\Delta c_n^I)^2, \tag{4.4-9}$$

where the sum is over the intervals that the signals differ, and similarly for the term $(\Delta c_n^Q)^2$. With this substitution, $d^2$ is the sum of the square $I$ and $Q$ contributions over all the affected intervals.

## Definition of Set-Partition Coding

We start from a QAM-style constellation of signal points. To create a set-partition code, we must do three things:

i) Define the master QAM constellation in the first place;
ii) Split this constellation into subsets;
iii) Devise a finite-state machine that selects a subset for each succeeding interval.

A simple splitting appears in Fig. 4.4-1. There are three separate design challenges here, that all contribute to the quality of the code. A good master constellation is one that has its points packed in efficiently, with a low average consumption of energy. The union of all the subsets should be the master constellation, while at the same time the points in each subset should be well separated from each other. Finally, we need a good finite-state machine. Just as in the previous sections, the progress of this machine is described by a trellis, and we want paths in the trellis to stay separate as long as possible. Almost all set-partition codes in fact employ *convolutional code* trellises of the type in Section 4.2.

A set-partition code carries data in two ways. Part of the data stream drives the machine that selects the subsets; if we use a $b$-ary branching trellis, $\log b$ bits will be needed for this task. The remaining bits are carried in the subset, once it is selected. If each of these is an $m$-ary sub-constellation, $\log m$ bits are carried in the subset. In total, the rate of the code is

$$R = \log b + \log m \quad \text{data bits/interval.} \tag{4.4-10}$$

This requires $bm$ points out of the master constellation, $b$ subsets of $m$ points, and which $bm$ it will be depends on the data stream.

The idea of set-partition coding is due to Ungerboeck [4-32], who also proposed an effective heuristic way to partition the sets.

Before going deeper, we should practice with a simple but important code based on an octal PSK master constellation. The code is called an 8PSK/4PSK-type code because the master constellation is 8PSK and the subsets resemble QPSK constellations.

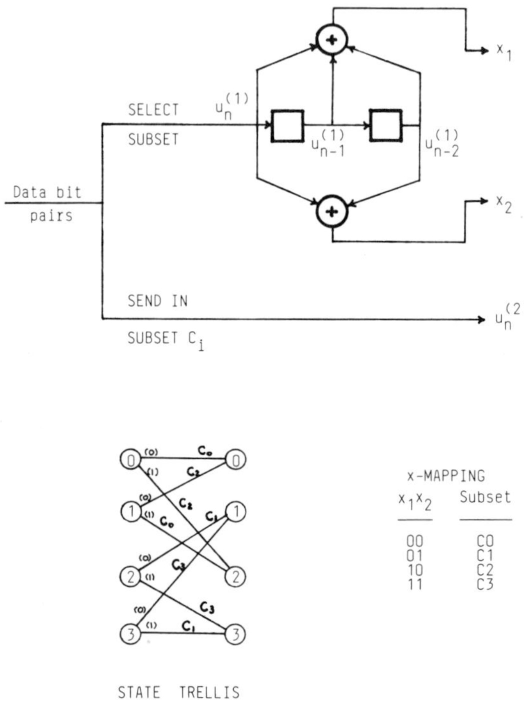


**Figure 4.4-2** Rate-2 set-partition code: Encoding of data bits, with subset selection via the rate-1/2 code of Fig. 4.2-5. The coder output selects the present subset by the table shown. The state is $(u_{n-1}^{(1)}, u_{n-2}^{(1)})$ as a base 10 integer. $u_n^{(1)}$ is shown in parentheses on the state trellis.

### Example 4.4-1: 8PSK/4PSK Rate-2 codes.

The master constellation for these codes appears in Fig. 4.4-1, together with two levels of splitting into subsets. The splitting follows Ungerboeck's procedure of selecting alternate constellation points for each subset in a split. After the second splitting, four subsets of two points each are formed. A rate-2 code may be formed by sending 1 bit in the subset and 1 bit in the choice of the subset; thus $\log b = 1$ and $\log m = 1$. A system to do this appears in Fig. 4.4-2 along with a trellis that shows how the "choice" bit selects the subset. The finite-state machine that does this selection is simply the rate-1/2 convolutional encoder of Fig. 4.2-5; its two output bits di-

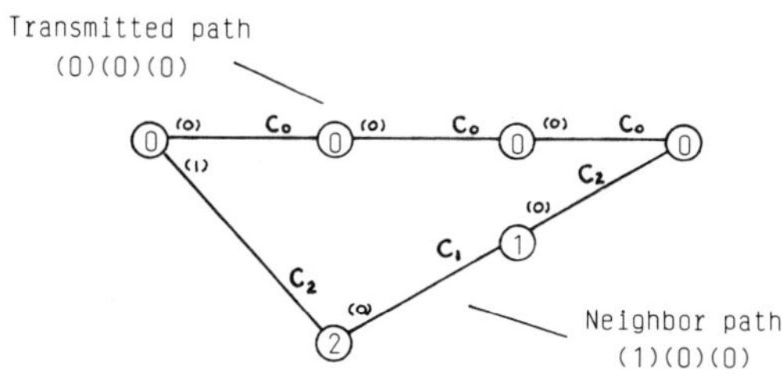


**Figure 4.4-3.** Trellis transitions for a transmitted and neighboring sequence in an 8PSK rate-2 set-partition code. Subset selection bit in ().

rectly select the subset according to the numbering convention in Fig. 4.4-1.

What is the free distance of this code? To answer this question we need to consider two kinds of neighbors of a transmitted sequence, those within the same subset and those in other subsets. From Fig. 4.4-1, we can see that every point in the two-element subsets has a neighbor in the same subset at square-distance 4 on the scale of the figure. So based on same-subset calculations alone, the minimum distance is 4; the overall free distance cannot exceed this.

It remains to consider the case where the transmitted sequence lies along one state transition path and a neighbor lies along another. One such example is shown in Fig. 4.4-3. Between the split and merge nodes of these trellis paths there are three symbol intervals, two in which the transmitted and neighbor sequences have points in subsets $C_0$ and $C_2$ respectively, and one interval in which the subsets are $C_0$ and $C_1$. In the $C_0/C_1$ interval, the closest approach of a signal point in $C_0$ to a point in $C_1$ is 0.586, which is the minimum square-distance of the (unnormalized) 8PSK. In the $C_0/C_2$ intervals, points cannot lie closer than square-distance 2. Thus, the worst case total square-distance for all three intervals is $2 + 2 + .586 = 4.586$. It turns out that none of the different-trellis-path cases have distances less than this, and we conclude that the free distance calculation is dominated by the same-subset distance, 4.

By using (4.4-8) to normalize this distance, we can compare it to the distance of uncoded QPSK and 8PSK. The code rate is 2 bits/interval and $c^2_{\text{ave}} = 1$ because all master constellation points have unit energy. Thus (4.4-8) gives $d^2_{\text{free}} = 4$. This compares with a QPSK normalized square minimum distance of 2 and an 8PSK distance of 0.879 (see Problem 4-12). We conclude that the set-partition code has a 3 dB energy advantage over QPSK, a scheme with the same $B_{RF}T_b$; it has a 6.6 dB advantage over octal PSK, but it needs 50% more bandwidth.

In the 8PSK example we can see a number of issues in set-partition code design. These will become more apparent in Example 4.4-2 which follows shortly.

*The subset size.* Perhaps the most basic decision is how to split the data bits between subset selection and subset size. The rate $R$ is split between log $m$ bits carried in the subsets and log $b$ bits that select the subset. As log $m$ drops, the subsets contain fewer points at a greater minimum distance from each other; this is good, because the free distance cannot exceed this same-set minimum distance. On the other hand, as log $b$ grows the rate of the convolutional subset selector in Fig. 4.4-2 grows along with it, and experience shows that this makes it harder to keep the subset trellis paths sufficiently apart. The best split is a compromise between these two effects.

In Example 4.4-1 we split the data bits one each. With a rate-2 code and an 8PSK master constellation, there is no other choice. If no bits drive the subset selection, we have QPSK with $d^2_{\text{free}} = 2$. If both bits drive the subset, we really have a rate-2/3 convolutional code connected to an 8PSK modulator. While simple designs of this sort have been found with $d^2_{\text{free}}$ above 5, (see [4-18], Chap. 11) such a scheme is not a set-partition code, and in fact it is known that no rate-2 set-partition code on the 8PSK constellation exceeds $d^2_{\text{free}} = 4$. In Example 4.4-2 we will start with a larger constellation and see that several valid data splits are possible; the codes will be more powerful, too.

*Master constellation size and shape.* In Example 4.4-1 the master constellation had $m'$ points, with $\log m' = 3$, while the code rate was $R = 2$ bits/interval. We could restate this by saying that the rate of the uncoded constellation, 3, exceeds the code rate by 1. Experience shows that this is a good rule in general, and most good set-partition codes follow the rule

$$\log m' = R + 1. \tag{4.4-11}$$

The shape of the master constellation has an effect on the code free distance, and we can see this from the normalized distance definition (4.4-8). Suppose we fix the non-normalized minimum distance of a constellation: This fixes the numerator of (4.4-8). Now we can improve the normalized minimum distance by reducing the average energy factor

$c^2_{\text{ave}}$. One way to do this is to pack the points more efficiently; rectangular packing, as in Fig. 4.4-4, is a little less efficient than hexagonal packing. Another way is to remove the corners of the constellation and relocate them so as to form a more circular constellation; the *cross constellation* [4-33], [4-34] is such an attempt. It is easier to understand these ideas by thinking about a simple QAM constellation, but they apply equally to trellis coded QAM.

*The subset trellis.* Most set-partition codes use a trellis created by a convolutional code finite-state machine. Lengthening the convolutional constraint length creates a more subtle machine, keeps the trellis paths apart for longer, and increases the distance between signals along differing paths. Eventually, the smallest such distance becomes larger than the same-subset minimum distance, after which improving the machine does not improve the free distance.

A critical aspect of the subset selection is the mapping between the convolutional coder output and the subsets. For example, the mapping in Fig. 4.4-2 follows no sensible order; if the set numbering follows the natural binary ordering, which is 0, 1, 2, 3, a much inferior code results. Often the best mapping must be found by trial and error.

*Finding free distance.* Although we found $d_{\text{free}}$ by an ad hoc method, the same distance algorithms that work on other trellis codes will work for set-partition codes. These explore the trellis as usual but the distance between two paths is now the least distance over the points in the subsets along the way. As always, $d_{\text{free}}$ is the lesser of any distance found by this trellis search and the same-set minimum distance. The concepts of column distance and decision depth apply to set-partition codes, but little is presently known about these functions.

We will now explore another important example of set-partition coding, this one based on a cross QAM master constellation. The rate-4 code here finds application in 9.6 kb/s wireline modems.

**Example 4.4-2: 32QAM/16QAM Rate-4 code.**

We begin with the 32-point cross constellation shown in Fig. 4.4-4. By the same ad hoc method as in Example 4.4-1, we can form four 8-point subsets or eight 4-point subsets, as shown in the figure. In the first case, the subsets carry $\log m = 3$ data bits, and one bit is left over to select the subset; there being four to choose from, a rate-1/2 convolutional code will do the selection. This 3:1 split of the data bits turns out to be suboptimal because the same-subset distance is not large enough. From eq. (4.4-8), $d^2_{\min}$ inside a subset

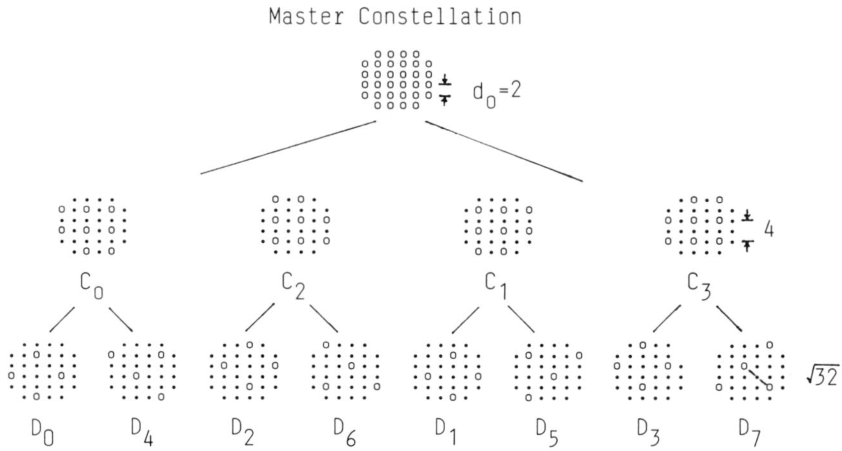


**Figure 4.4-4.** Rate-4 set-partition code: Partitioning of the 32-point cross QAM constellation into size 8 and size 4 subsets.

is 1.6. (We compute this as follows: The rate $R$ is 4; the same-subset square distance is 16 from Fig. 4.4-4; $c^2_{\text{ave}}$ is 20, the average energy of the 32-point constellation in the figure, since the code uses all 32 points equally over the long run. See also Problem 4-13). On the other hand, the worst-case distance between signals along differing trellis paths exceeds 1.6 with just a 4-state trellis. This tells us that our subsets should have fewer points and greater minimum distance.

The optimal rate-4 arrangement is the 2:2 split in Fig. 4.4-5, two bits carried by the subsets and two bits selecting the subset via a rate-2/3 convolutional code. Now the same-subset $d^2_{\text{min}}$ is 3.2. Using the 8-state subset selector in the figure, we find that $d^2_{\text{free}}$ is 2, with the trellis structure now the limiting factor (See Problem 4-13). Can $d^2_{\text{free}}$ be raised to 3.2? The answer is yes, if the rate-2/3 selector has 128 or more states.

It remains to calculate the coding gains for these codes. An uncoded square 16QAM constellation has $d^2_{\text{min}} = 0.8$. The code in Fig. 4.4-5 has $d^2_{\text{free}} = 2$, a gain over this of 3.98 dB. Codes with the 2:2 split can range up to 6.02 dB gain, the limit being the same-subset distance. Codes with a 3:1 split are limited to a gain of 3.01 dB.

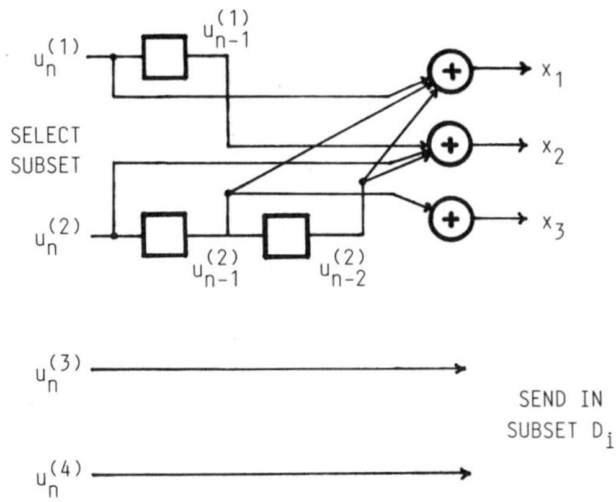


| $x_1x_2x_3$ | Subset |
|---|---|
| 000 | D0 |
| 001 | D1 |
| 010 | D2 |
| 011 | D3 |
| 100 | D4 |
| 101 | D5 |
| 110 | D6 |
| 111 | D7 |

**Figure 4.4-5.** Rate-4 set-partition code: Subset selection method based on rate-2/3 convolutional code. The coder output selects the present subset by the table shown. In the notation of Section 4.2, the generators here are $g_{1,1} = (1\ 0\ 0), g_{1,2} = (0\ 1\ 0), g_{1,3} = (0\ 0\ 0); g_{2,1} = (0\ 1\ 1), g_{2,2} = (1\ 0\ 1), g_{2,3} = (0\ 1\ 0)$.

## Extensions of Set-Partition Coding

The 1980s saw massive research in set-partition coding and related techniques. We can mention but a few advances here. Researchers have found better codes by departing from the traditional convolutional-code subset selector in Figs. 4.4-2 and 4.4-5. A feedback, rather than feedforward, architecture will sometimes generate the same transitions in fewer states; the addition of nonlinear elements and other circuitry to the shift-register structure can lead to better free distance and other improvements. A readable summary of these improvements appears in Ungerboeck [4-33]. An innovation of another type was the invention by Wei [4-35] of codes with rotational symmetry. These codes can withstand errors of a multiple of 90 degrees in the receiver phase reference. Set-partitions of master constellations with more than two dimensions were introduced by Gallager in [4-36]. Much recent work appears in the special journal issue [4-20].

Perhaps the greatest theoretical advance was the interpretation of set-partition coding in terms of algebraic cosets by Calderbank and Sloane [4-37] and Forney [4-34]. In this much more sophisticated view, the master constellation is a lattice and the subsets are cosets of the lattice.

Much of this research extends techniques that predate set-partition coding. Coding with lattice points dates back at least to the 1960s; some practitioners were Leech, Sloane, deBuda, and Shannon [4-38], [4-39]. Another older technique is using a binary-symbol code like a convolutional code to encode a smaller data alphabet onto a larger set of phase and amplitude points. In the language of set-partition coding, none of the data bits are carried in the subset. The design of a good code of this type consists of finding a good binary code and a good mapping from its output to the constellation points; goodness is measured by free Euclidean distance in the constellation, not Hamming distance in the binary code. These codes can be much better than true set-partition codes, although they may be harder to implement. An example of older work is Zetterberg [4-40]. Parity-check codes may be combined with CPM codes to good effect. Some references to such work appear in [4-18], Chap. 11.

## Decoding

The decoding of set-partition codes does not differ greatly from that of other trellis codes. In a Viterbi decoder, trellis paths are extended from level to level, with only a single survivor path kept at the new node. We must keep in mind that a whole subset of signals attaches to each trellis branch, and that just the closest of these is kept with the path; then the algorithm eliminates all but the best path into each node, finding the closest of the close.

The complexity of decoding thus contains two factors, finding the closest point in a branch subset to the received signal and finding the survivor into a trellis node. Recall that a subset has $m$ points, a constraint-$\nu$ trellis has size $b^{\nu-1}$, and the rate satisfies $R = \log b + \log m$. We can assume that a large subset search costs less than a large trellis, so that a set-partition code should have relatively large subsets, if possible. But as we have seen, large subsets limit free distance.

Table 4.4-1 lists some of the best set-partition codes known at this writing. The codes here have the best coding gain at the convolutional code rate and state size shown. All use some form of convolutional subset selection. As a rule, these codes obtain high gain at quite low complexity.

**Table 4.4-1.** A Selection of Good Set-Partition Codes and Their Normalized Free Distances. The Convolutional Subset Selector Code Rate Is $(\log b)/(1 + \log b)$. The Bandwidth of These Codes Is $2B_{\text{QPSK}}/R$ Hz-s/data bit, Where $B_{\text{QPSK}}$ Is the Bandwidth of QPSK that Uses the Same I and Q Pulses.

| Master Constellation | Code Rate, R | Subset: Selection Split log m: log b | $d^2_{free}$ | States |
|---|---|---|---|---|
| 8PSK | 2 | 1:1 | 4.0 | 4 |
| 16PSK | 3 | 2:1 | 2.0 | 4 |
| | 3 | 2:1 | 3.0 | 64 |
| | 3 | 1:2 | 3.1 | 256 |
| Rect. 16QAM (Fig. A-1) | 3 | 2:1 | 2.4 | 4 |
| | 3 | 1:2 | 3.0 | 8 |
| | 3 | 1:2 | 4.8 | 128 |
| Cross 32QAM (Fig. 4.4-4) | 4 | 3:1 | 1.6 | 4 |
| | 4 | 2:2 | 2.0 | 8 |
| | 4 | 2:2 | 3.2 | 128 |

# 4.5 INTERSYMBOL INTERFERENCE DECODING

When data-bearing pulses affect other, nearby symbol intervals, the result is intersymbol interference, or ISI. Interference can be intentional or not. Channel coding schemes relate successive data symbols to each other and these are examples of ISI by design. Band limiting a data signal with a filter produces ISI that is undesired but still intentional. This kind of filter creates ISI that is usually fixed and known and can be designed against, however unwelcome the interference may be. An example of unintentional ISI is multipath propagation in a mobile channel; here the interference is time-varying and identifying its nature is a major problem.

An *equalizer* is a device that mitigates ISI. We can think of an equalizer as having two parts, one that identifies what the ISI is like and one that removes or reduces its effect. Sometimes the device is really one circuit or algorithm that performs both functions. The identification function is perhaps the more challenging, but it is the ISI removal function that relates to coding and so we will concentrate on it. Furthermore, we will discuss only maximum likelihood removal. Full discussion of equalizers may be found in digital communications texts; see especially the text by Proakis [4-24] or the survey by Qureshi [4-41].

Assume, then, that the nature of the ISI – be it a filter response or a channel model – is known at the receiver. In these and many other cases, we can view the interference as a convolution of a channel impulse response with the data-bearing signal. In fact, we can coalesce the channel response with the data pulse and any other system responses to form one composite impulse response $h(t)$ to a data symbol; thus symbol $u_n = \pm 1$, or $\pm 3$, ... produces at the receiver $\pm h(t - nT)$ or $\pm 3h(t - nT)$ etc. The total received signal is the superposition

$$s(t) = \sum_n u_n h(t - nT). \tag{4.5-1}$$

We will show in this section that when $h(t)$ has finite duration, $s(t)$ can be graphed on a trellis structure just like the ones earlier in the chapter, and that the data can be "detected" from $s(t)$ by a trellis decoder. Here is yet another manifestation of the trellis structure. It is not unreasonable to discover trellises in the ISI problem: We found them in convolutional codes and ISI of the type we study is another kind of convolutional coding that differs only in its arithmetic field.

## Maximum Likelihood ISI Receivers

At the outset we assume additive white Gaussian noise (AWGN) in the channel and an $N$-symbol message sequence $\mathbf{u} = (u_0, u_1, ..., u_{N-1})$. As shown at the top of Fig. 4.5-1, $s(t)$ plus noise produces $r(t)$ at the receiver input.† The maximum likelihood (ML) detector for $\mathbf{u}$ finds the hypothetical message sequence $\hat{\mathbf{u}}$ that maximizes the conditional probability of $r(t)$ given $\hat{\mathbf{u}}$; i.e., for the $r(t)$ observed, it performs

$$\begin{aligned} \text{Find:} &\quad \max_{\hat{u}_0, \hat{u}_1, \ldots} P\{r(t) \mid \hat{u}_0, \hat{u}_1, \ldots\} \\ \text{Put out:} &\quad \hat{\mathbf{u}} \text{ achieving the max.} \end{aligned} \tag{4.5-2}$$

Appendix A derives the ML detector in more detail, and by following the steps there it can be shown that (4.5-2) for AWGN boils down to

†In many practical situations, thermal noise in the receiver input circuit dominates other noises and becomes the "channel" noise; then AWGN precisely models the channel.

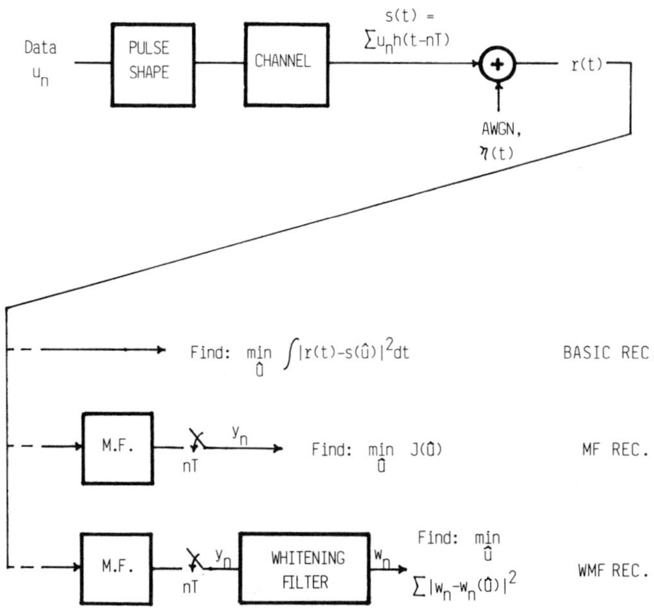


**Figure 4.5-1.** Continuous-time ISI channel model plus models of the three equivalent receivers discussed in Section 4.5. $s(u)$ is the signal produced by data sequence $u$.

$$\text{Find:} \quad \min_{\widehat{u}_0, \widehat{u}_1, \ldots} \int \left| r(t) - s(t) \right|^2 dt. \tag{4.5-3}$$

In words, the detector output is the legal $s(t)$ that lies closest to $r(t)$ in Euclidean distance. We have arrived at the top receiver in Fig. 4.5-1. The integral here extends over the time time that $r(t)$ is observed.

A few more reductions give the simpler *matched filter* receiver in the middle of the figure. First we reduce the integral in (4.5-3) to

$$\int \left| r(t) \right|^2 dt - 2\,\mathrm{Re} \int r(t) s^*(t)\, dt + \int \left| s(t) \right|^2 dt.$$

The first term is a constant during the minimization and may be ignored. We may express the remaining terms as the *metric* $J(\mathbf{u})$, with the definitions shown:

$$J(\mathbf{u}) \triangleq -2\,\mathrm{Re} \left[ \sum_i u_i^* y_i \right] + \sum_i \sum_j u_i^* u_j \epsilon_{i-j}, \tag{4.5-4}$$

where

$$y_i \triangleq \int r(t)h^*(t-iT)\,dt \tag{4.5-5}$$

is the *matched filter response* and

$$\epsilon_{i-j} \triangleq \int h^*(t-iT)\;h(t-jT)\,dt \tag{4.5-6}$$

are called the *elements* of the channel $h(t)$. The $y_i$ here are conventional matched filter outputs, from a filter matched to the overall channel impulse response followed by a sampling at time $iT$. We use complex variables to cover the case when the signals are bandpass, in which case the real part of $u_i$ is an in-phase data symbol and the imaginary part is a quadrature symbol, and the impulse response $h(t)$ and the noise are complex. However, all these will be real in what follows.

The effect of the matched filter receiver is to recast the continuous-signal receiver (4.5-3) as a *discrete-time* receiver. It is defined by

$$\begin{aligned} &\text{Find:} && \min_{\widehat{u}_1,\widehat{u}_2,\ldots} J(\widehat{\mathbf{u}}) \\ &\text{Put out:} && \widehat{\mathbf{u}} \text{ achieving the min;} \end{aligned} \tag{4.5-7}$$

here $J(\widehat{\mathbf{u}})$ is a function of discrete-time variables. We reiterate that these receivers make the same decision, because (4.5-3) and (4.5-7) carry out the same calculation. The matched filter (or MF) receiver for ISI grew out of a number of papers in the early 1970s, including MacKechnie [4-43], Ungerboeck [4-44], Messerschmitt [4-45], and Falconer and Magee [4-46], among others.

Figure 4.5-1 shows a third receiver called the *whitened matched filter*, or WMF, receiver. It consists of an MF receiver followed by a certain filter $1/F^*(z^{-1})$, the whitening filter, whose purpose is to produce an output with white noise components. The filter's input sequence, $y_o$, $y_1$, ..., contains correlated noise components because of the effect of $h(t)$ in eq. (4.5-5) on the white noise component of $r(t)$, and the filter is designed in just the right way so that its outputs $w_0$, $w_1$, ... contain *independent* noise components.

The WMF receiver was proposed by Forney [4-42] in 1972. Its advantage is that its distance calculation is simple additive Euclidean distance. Let $w_n$ be the actual filter output and $w_n(\widehat{\mathbf{u}})$ be the output if data $\widehat{\mathbf{u}}$ are carried noiselessly through the channel. Then the advantage means that a maximum likelihood WMF receiver is defined by

$$\text{Find:} \qquad \min_{\widehat{u}_0, \widehat{u}_1, \ldots} \sum_n \left| w_n - w_n(\widehat{\mathbf{u}}) \right|^2 \tag{4.5-8}$$
$$\text{Put out:} \qquad \widehat{\mathbf{u}} \text{ achieving the min.}$$

The disadvantage of the WMF receiver is that a stable realization of the whitening filter $1/F^*(z^{-1})$ may not exist. This is often the case, for instance, in a mobile radio channel. Also, it may be hard to solve for $F$ mathematically. In these cases we may be limited to the basic or the MF receiver. We will solve for the whitening filter next.

## Discrete-time Gaussian Channel Models

The foregoing discrete-time receivers imply discrete-time models for our continuous-time Gaussian channel, to which we now turn. We assume that the channel response $h(t)$ is finite with span $LT$ seconds, and begin with the WMF receiver, which gives the simplest model. The key to the model is the whitening filter.

We arrive at the filter through some rather mathematical steps. It will be easiest to express all sequences as z-transforms; the matched filter/sampler output, for instance, is

$$Y(z) = y_0 + y_1 z^{-1} + y_2 z^{-2} + \cdots$$

and similarly $U(z)$ is the transform of the data. Let's take a closer look at $Y(z)$. Since the received signal is $r(t) = \sum u_n h(t - nT) + \eta(t)$, we know from definition (4.5-5) that

$$y_k = \int \left[ \eta(t) + \sum u_n h(t - nT) \right] h^*(t - kT)\, dt.$$

In terms of z-transforms this relation is

$$Y(z) = \epsilon(z) U(z) + \xi(z) \tag{4.5-9}$$

where $\epsilon_k$ is given by definition (4.5-6), $-L \le k \le L$, and

$$\xi_k \; = \; \int \eta(t) h^*(t - kT)\; dt. \tag{4.5-10}$$

This last variable is a sample of white noise filtered by a filter with response $h^*(-t)$; it is a sample of *colored noise.*

The WMF receiver filters $Y(z)$ one more time in order to make this noise component white. How might this be done? Our procedure will be to find the autocorrelation of $\xi_k$, find a set of filters that make it white, and then extract the right one. From stochastic process theory (see Proakis [4-24], Chap. 3), the autocorrelation is

$$\mathcal{E}[\xi_i^* \xi_j] = 2\ N_0\ \epsilon_{i-j}, \tag{4.5-11}$$

where $N_0/2$ is the power spectral density of the white channel noise. Next we must form a *spectral factorization* of $\epsilon(z)$: This entails finding a transform $F(z)$ such that

$$\epsilon(z) = F(z)F^*(z^{-1}). \tag{4.5-12}$$

Such a factorization has the property that a filtering of a Gaussian $\xi(z)$ by $1/F^*(z^{-1})$ (or by $1/F(z)$) produces a white Gaussian sequence; that is,

$$N(z) = \frac{\xi(z)}{F^*(z^{-1})}$$

is the transform of a white sequence. So any $1/F^*(z^{-1})$ that satisfies (4.5-12) is a candidate whitening filter.

There are many factorizations of $\epsilon(z)$ and we need to look more closely to find an acceptable one. Since $\epsilon(z)$ is the polynomial

$$\epsilon(z) = \sum_{k=-L}^{L} \epsilon_k z^{-k}$$

with $\epsilon_k = \epsilon^*_{-k}$, it has $2L$ roots that occur in conjugate reciprocal pairs, $\rho_\ell$ and $1/\rho^*_\ell$; this follows from the observation that $\epsilon(z) = \epsilon^*(z^{-1})$. At the same time the $L$ roots of $F(z)$ must be conjugate reciprocals of the L roots of $F^*(z^{-1})$. We can conclude that if $\rho_1, \ldots \rho_L$ are the roots of $F(z)$, then $F^*(z^{-1})$ has roots $1/\rho^*_1, \ldots 1/\rho^*_L$. There are $2^L$ such sets of roots.

Which sets are acceptable combinations? If we apply the whitening filter $1/F^*(z^{-1})$ to the sampler output (4.5-9), we get

$$\frac{Y(z)}{F^*(z^{-1})} = \frac{U(z)\epsilon(z)}{F^*(z^{-1})} + N(z),$$

which by (4.5-12) is

$$U(z)F(z) + N(z). \tag{4.5-13}$$

Clearly, this is a *discrete-time channel model*, consisting of a length-L response discrete-time filter $F(z)$ and some additive white noise sequence.

Figure 4.5-2 illustrates this model. Of all the $F(z)$'s that solve (4.5-12), we should choose one that creates a reasonable model. Often it is obvious from the channel which one this is; see for instance Example 4.5-1.

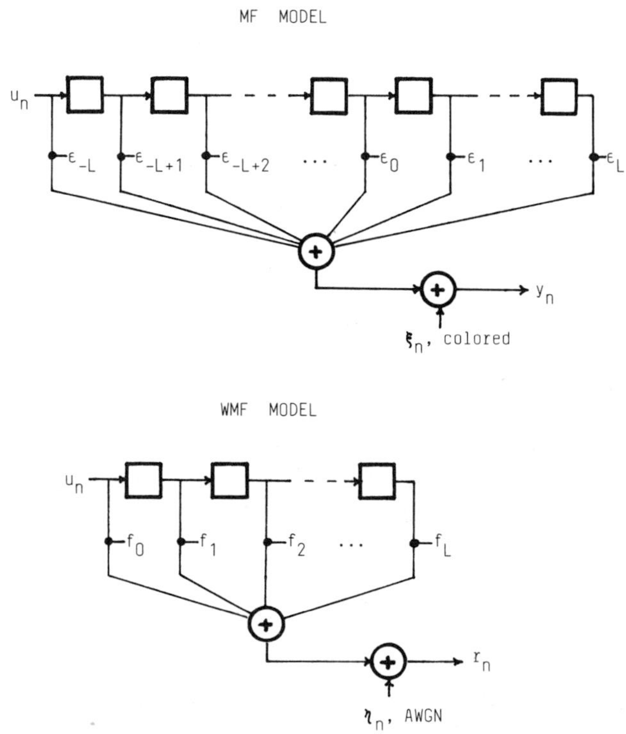


**Figure 4.5-2.** Matched filter and whitened matched filter discrete-time models for the continuous-time channel with response $h(t)$. $L$ is the ISI memory.

The whitening filter must be the matching $1/F^*(z^{-1})$, since this filter whitened the noise in $Y(z)$. If any zeroes of $F^*(z^{-1})$ lie on or outside the unit circle, $1/F^*(z^{-1})$ has no stable realization, and because of the reciprocal root property, this occurs whenever the model $F(z)$ has a root *inside* the unit circle. Whether or not this presents a difficulty depends on one's point of view. An unstable whitening filter in Fig. 4.5-1 means that quantization noise or other disturbance at the filter input will lead to divergence in the output. On the other hand, Forney [4-42] shows that the overall filter $H^*(z^{-1})/F^*(z^{-1})$ that operates on the channel noise $\eta(t)$ is always stable, and of course the term $F(z)U(z)$ is stable. Perhaps a good policy is to accept the WMF model when $1/F^*(z^{-1})$ is stable, and at least consider the MF model otherwise.

Figure 4.5-2 also shows the discrete-time model created by the MF receiver. This is directly (4.5-9), the data filtered by $\epsilon(z)$ plus the colored noise $\xi(z)$, defined by (4.5-10). The model is twice the length of the WMF model, although this may effectively revert to L when the receiver is constructed.

**Example 4.5-1: A Two-Ray Multipath Channel.**

This is a common way to model the multipath propagation that occurs in a digital mobile radio channel. Let the channel response to a unit data symbol at $t = 0$ be

$$h(t) = \delta(t) + a\delta(t - 2T). \qquad (4.5\text{-}14)$$

Here $\delta(t)$ is the impulse function and the model is in effect a discrete-time model of a direct component and a component delayed two symbol intervals. More sophisticated models add a phase factor to each term. Let us find the WMF model (4.5-13) for this channel. From (4.5-6), the $z$-transform of the elements is,

$$\epsilon(z) = \sum_{k=-2}^{2} \epsilon_k z^{-k} = a^* z^2 + (1 + |a|^2) + az^{-2} \qquad (4.5\text{-}15)$$

which factors into

$$(az^{-2} + 1)(a^* z^2 + 1).$$

Suppose we try $F(z) = az^{-2} + 1$ and $F^*(z^{-1}) = a^* z^2 + 1$ as the spectral factorization (4.5-12). Then the channel model (4.5-13) turns out to be $U(z)(az^{-2} + 1) + N(z)$, which in the time domain says that a channel output $r_n$ satisfies

$$r_n = u_n + au_{n-2} + \eta_n. \qquad (4.5\text{-}16)$$

This fits the physical reality (4.5-14) perfectly.

What do the maximum likelihood receivers look like for this channel? Eq. (4.5-16), however it might have been derived, really suggests the basic ML receiver (4.5-3); that is, it suggests that the receiver should compute the distance between discrete-time channel outputs $r_0, r_1, \ldots$ and a hypothetical signal $\widehat{s}_0, \widehat{s}_1, \ldots$, for which $\widehat{s}_n = \widehat{u}_n + a\widehat{u}_{n-2}$, and find

$$\min_{\widehat{\mathbf{u}}} \sum |r_n - \widehat{u}_n - a\widehat{u}_{n-2}|^2. \tag{4.5-17}$$

As for the MF receiver, it filters the sequence $r_0, r_1, \ldots$ according to (4.5-5) and produces the stream $y_n = r_n + a^* r_{n+2}$, which is the anticausal filtering $Y(z) = R(z)F^*(z^{-1})$. To make this causal, we need to replace $F^*(z^{-1})$ with the 2T-delayed filter

$$z^{-2}F^*(z^{-1}) = a^* + z^{-2}. \tag{4.5-18}$$

This together with the metric (4.5-4) will give the MF receiver (4.5-7), which uses delayed $y_0, y_1, \ldots$ to compute $J(\widehat{\mathbf{u}})$ for hypothetical data sequences. Compared to the simplicity of (4.5-16), it seems hardly worth the effort.

To create a WMF receiver, we add the whitening filter to filter (4.5-18). The whitener is $1/F^*(z^{-1}) = 1/(a^*z^2+1) = z^{-2}/(a^*+z^{-2})$, a filter with poles outside the unit circle whenever $|a| < 1$. Since many multipath propagations have $|a| < 1$, this is troubling. Were we actually to construct a whitener in such a case, it would be unstable. However, the concatenation of the whitener and the matched filter (4.5-18) has $z$-transform unity (with perhaps a delay factor), which is certainly stable. Perhaps we should conclude that the MF and WMF receivers have no real application and be content with the basic ML receiver based on (4.5-16).

**Example 4.5-2: Two-Ray Channel with Raised-Cosine Pulses.**

We will see more differences in the receivers if we replace the discrete-time impulses in Example 4.5-1 with true pulses. We will use the raised-cosine pulses in Example A-2, which featured in Section 4.4 and in the QPSK comparison in Section 4.3. Now the response of the channel to a unit data symbol at time 0 is

$$h(t) = g(t) + ag(t - 2T), \tag{4.5-19}$$

with $g(t)$ the raised-cosine pulse. *We find that $\epsilon(z)$ is still (4.5-15)*: The reason is the orthonormality of $g(t)$ and its time shifts $g(t-kT)$. We may as well factor $\epsilon(z)$ in the same way and end up with model (4.5-16), a discrete-time model that mimics the continuous-time channel (4.5-19).

We can still build the basic ML receiver (4.5-3), but it now integrates continuous-time signals and this is a disadvantage. Our discrete-time receiver options are the MF and WMF receivers.

We cannot do without the matched filter/sampler combination because this is the means of conversion to discrete time. The filter is a continuous-time one with impulse response $h^*(-t)$ suitably truncated and made causal. The MF receiver executes (4.5-7) with the sampler outputs $y_n$. The WMF receiver appends a discrete-time whitener $1/F^*(z^{-1})$ as in Example 4.5-1, which will be an unstable filter when $|a| < 1$. We should consider the MF receiver in that case. If we do use the WMF receiver, we employ the discrete-time model (4.5-16), in which a hypothetical signal sample will be $\widehat{s}_n = \widehat{u}_n + a\widehat{u}_{n-2}$, and the received signal sample $r_n$ is the whitener output.

## Trellises and Trellis Decoders

For finite response channels, the MF and WMF channel models provide trellis models for the ISI that are much like those in the previous sections. The WMF model (4.5-13) is particularly useful. In that one, the channel response to a data sequence $U(z)$ is $F(z)U(z)$, in which $F(z)$ is a finite polynomial. In the time domain the model output at time $n$ is

$$s_n = \sum_{i=0}^{\nu-1} u_{n-i} f_i \tag{4.5-20}$$

plus white noise. It is easy to see that this model is a finite-state machine: Its output at time $n$ depends on the $L = \nu - 1$ previous $b$-ary data symbols plus the present $n$-th one, just as with symbolic convolutional codes. The machine has $b^{\nu-1}$ states.

A trellis maps the progress of a finite-state machine over time, and Fig. 4.5-3 shows the ISI trellis for Examples 4.5-1 and -2 with binary data. Since the ISI memory is $\nu - 1 = L = 2$, the trellis has $2^2$ states. On each branch appears the model output and in parentheses the data symbol that drives that transition. Except for the output values this trellis is identical to the convolutional one in Fig. 4.2-3, and this is no accident because the structure in both cases derives from a shift register, Fig. 4.1-1 in one case and the WMF model of Fig. 4.5-2 in the other. Both trellises have constraint length 3.

Trellises are also easy to construct for the MF and basic receiver models. From Fig. 4.5-2 we see that the MF model has twice the shift register length and therefore $b^{2(\nu-1)}$ states in its trellis. We remark that the use of even an infinite-duration data pulse $g(t)$ (see Example 4.5-2)

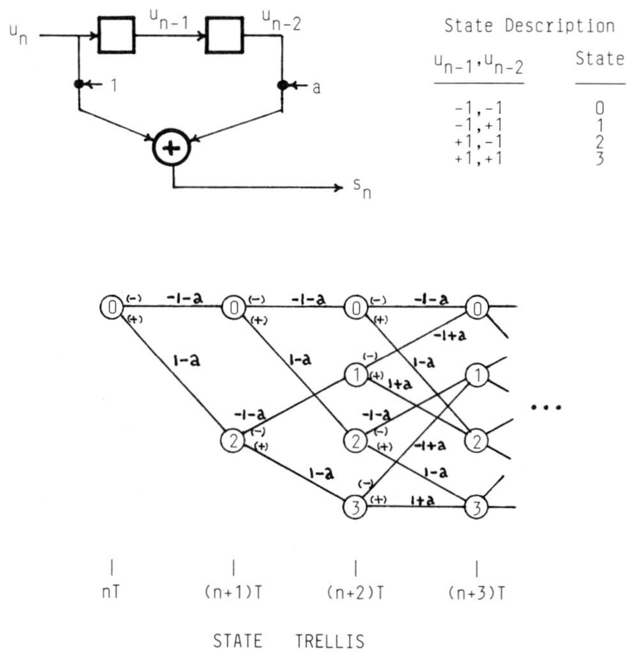


**Figure 4.5-3.** Finite-state machine and trellis for the ISI $s_n = u_n + a u_{n-2}$ in Examples 4.5-1 and -2. The initial register contents are assumed to be -1s. Data symbols in ().

does not affect the state size, so long as the pulses are time-orthogonal. As for the basic receiver, a trellis model for this is a little more abstract but is still not difficult. One must replace the model output like (4.5-20) with a continuous-time formula that shows how a finite number of data symbols produce $s(t)$ in (4.5-3).

The Viterbi algorithm finds the least-metric path through a trellis and in so doing performs an ML decoding of the transmitted signal. We should now design a practical sliding-block Viterbi decoder like the ones in earlier sections. As before, some important issues are the path metric, the path memory length (or decision depth), the signal free distance and the consequent error probability.

*Path metrics.* What is needed is a recursive formula to compute the metric or distance of a path at time $n$, given its value at time $n-1$, and (4.5-3), (4.5-4) and the WMF receiver output may be adapted to this task. Taking the basic receiver first, we may express the distance to time $n$ in (4.5-3) as

$$\int_0^{nT} \left|r(t) - \widehat{s}(t)\right|^2 dt = \int_{(n-1)T}^{nT} \left|r(t) - \widehat{s}(t)\right|^2 dt + \int_0^{(n-1)T} \left|r(t) - \widehat{s}(t)\right|^2 dt \qquad (4.5\text{-}21)$$

Here we assume that $\widehat{s}(t)$ is the response to a certain sequence $\widehat{u}_0, ..., \widehat{u}_{n-1}$ and that the channel impulse response is causal. Consequently, the rightmost term does not depend on $\widehat{u}_n$, although the $(n-1)T$ – to – $nT$ "increment" term does depend on a full span of $\nu$ symbols. We thus have a recursion in the sense that the new distance is the old one at time $(n-1)T$ plus an increment that depends on variables known at the old and new trellis states. Amidst the intricacies of modeling ISI, it is good to keep in mind that the basic formula (4.5-21) always works and is the best that one can do with the information available to time $nT$.

For the MF receiver, a naive metric recursion would be to take the terms in the outer sum of (4.5-4) one at a time to obtain

$$J(\widehat{u}_0, \ldots \widehat{u}_n) = -2\,\mathrm{Re}[\widehat{u}_n^* y_n] + \sum_{j=n-L}^{n+L} \widehat{u}_n^* \widehat{u}_j \epsilon_{n-j} + J(\widehat{u}_0, \ldots \widehat{u}_{n-1}). \qquad (4.5\text{-}22)$$

But this is non-causal in the sense that it uses symbols $\widehat{u}_{n+1}, \ldots, \widehat{u}_{n+L}$ which are not yet available. A simple fix is to insert an $L$ delay in the calculation so that the increment term becomes

$$-2\,\mathrm{Re}[\widehat{u}_{n-L}^* y_{n-L}] + \sum_{j=n-2L}^{n} \widehat{u}_{n-L}^* \widehat{u}_j \epsilon_{n-L-j}.$$

Other increment expressions are possible.

For the WMF receiver, the noise reaching the Viterbi algorithm is white and the proper metric is simply the sum of the square differences at each trellis level; that is, the distance $D^2$ at level $n$ is

$$D^2(\widehat{u}_0, ..., \widehat{u}_n) = \left| w_n - \sum_{i=0}^{L} \widehat{u}_{n-i} f_i \right|^2 + D^2(\widehat{u}_0, ..., \widehat{u}_{n-1}), \qquad (4.5\text{-}23)$$

where $w_n$ is the WMF receiver output. This is basically a discrete-time version of (4.5-21).

While all these recursions amount to the same thing when applied to an entire message $\mathbf{u}$, they will give varying assessments of a message known only to time $n$. Which metric to use in an application depends on channel characteristics like group delay and whether the channel is minimum phase.

*Distance, length, and error probability.* The free distance of a signal set subjected to ISI is in principle easy to compute from a basic Euclidean distance expression like (4.3-7). As with any trellis code, we want the worst case inter-signal distance between two signals. A particularly simple way to carry this out is through the WMF channel model. Let the sequence $\Delta = \mathbf{u}^{(1)} - \mathbf{u}^{(2)}$ be the difference between two indefinitely long data sequences; then we seek the minimum over all nonzero $\Delta$ of

$$\sum_k \Big| \sum_i u_{k-i}^{(1)} f_i - \sum_i u_{k-i}^{(2)} f_i \Big|^2 = \sum_k \Big| \sum_i \Delta_{k-i} f_i \Big|^2 . \tag{4.5-24}$$

Efficient algorithms exist to find the minimum of (4.5-24). Next, (4.5-24) must be normalized as usual by twice the energy per data bit, which is the scaled element $\epsilon_0 \sigma_u^2$, where $\sigma_u^2$ is the variance of the data symbol values.

An example of free distance is shown in Fig. 4.5-4, which shows the normalized free distance of the channel $F(z) = 1 + a_1 z^{-1} + a_2 z^{-2}$ with binary input data $\pm 1$. The figure gives free distance contours in terms of degradation in dB from $d_{\text{free}}^2 = 2$, the distance of the straight antipodal signaling channel with $F(z) = 1$. We can see that over most of the tap combinations, the Viterbi decoder attains the same free distance as the antipodal case, that is, it fully defeats the ISI. The largest unavoidable loss is a little more than 2 dB, which occurs at about $F(z) = 1 + 1.5z^{-1} + z^{-2}$. Note that the cases in Example 4.5-1 occur along the horizontal axis, and here we see that all cases of ISI are fully defeated.

Once the free distance is found, the error probability of a Viterbi ISI decoder is approximately $Q(d_{\text{free}}\sqrt{E_b/N_0})$ at high SNR, just as it is for other trellis codes. Decision depths may also be found for ISI decoders in the same way that they are found for convolutional and CPM codes; see ref. [4-47].

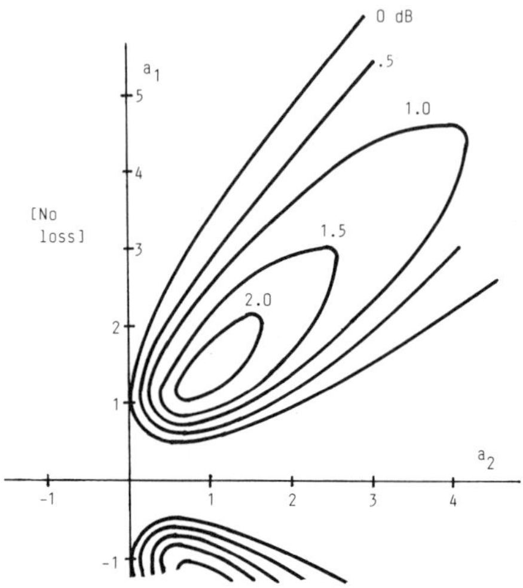


Figure 4.5-4. A contour plot showing the energy loss in dB of maximum likelihood decoding, compared to an antipodal signaling scheme without ISI. The channel model is $F(z) = 1 + a_1 z^{-1} + a_2 z^{-2}$ with binary data. The plot is symmetrical about the horizontal axis; losses are zero for any channel in the second and third quadrants. Data from ref. [4-47].

## Equalizers and the Viterbi Decoder

The Viterbi decoder provides the ultimate decoder in a channel with good SNR, and we have seen that it often entirely defeats the ISI. It is known, however, that not all the ISI trellis states must be searched in order to obtain essentially ML performance. Chapters 5 and 6 explore several ways to reduce trellis searches, and ISI examples are featured in Sections 5.1 and 6.2. Chapter 6, in particular, shows that huge reductions are possible.

We can think of ordinary equalizers as machines that pursue just a single path through the ISI trellis. A great many of these have been proposed, and their design is as much an art as a science. Some standard references on equalizer design were given at the beginning of the section, and we cannot go further into the subject in this book. We can mention, however, that equalizers fall very generally into two classes, the zero-forcing and the decision feedback classes. The first of these consists simply of the filter $1/F(z)$ and nothing more; we have seen that this filter may not be stable, and in such a case the zero-forcing equalizer is useless (see Problem 4-16). A decision feedback equalizer recreates the ISI from its data decisions and subtracts the result from the incoming signal; in effect this is a single-path search of the ISI trellis. Messerschmitt [4-45] shows how to compare the performance of these schemes to the ML receiver and demonstrates that the basic ML receiver is always better than the decision feedback equalizer, which in turn is always better than the zero-forcing kind. The differences are small in some channels but very large in others.

## Infinite Response Channels

It often happens that channels and/or pulses have an infinite impulse response and in such a case a truly accurate trellis ISI model has infinitely many states. Effective band limitation filters in channels need poles in their transfer functions and these are the classic source of infinite response channels. Recent advances have shown several ways to deal with this case, and we can review them briefly.

A first approach proposed by Xiong (ref. [6-25]), is simply to extend the WMF receiver design to channels with poles: Now $\epsilon(z)$ and the channel model $F(z)$ have poles and an infinite response. We can then either truncate the model $F(z)$, which will produce a trellis model, or leave it as it is, which leads to a *tree* model of the ISI, that is, a branching structure that expands indefinitely. To search for a good decoding in a tree ISI model, we certainly cannot employ a Viterbi algorithm, but a tree search algorithm of the kind featured in Chapter 6 will work well. In thinking about such a tree decoder it is useful to remember that the number of states grows without limit but the number of *state variables* that define the response is small and fixed and usually equal to the number of poles. We can implement the decoder by keeping track of the system state through its state variables.

Another state variable approach, this one for the basic ML receiver (4.5-3), is proposed in [6-11]. The distance-calculating recursion that replaces (4.5-21) is based on a simple state-variable description of the channel. It is shown that a small tree search essentially achieves the ML performance for some practical channel band limitation filters.

The concept of truncating an infinite model in order to obtain a finite one is another approach that we will take up in Section 5.1.

# PROBLEMS

4-1 a) Consider the rate-$\frac{1}{3}$ convolutional code with $g_1 = (1\ 1\ 1), g_2 = (1\ 1\ 1), g_3 = (1\ 0\ 1)$.
i) Draw the block diagram of the code generator.
ii) If the input sequence is (0 1 1 1 0 1) determine the output sequence.
b) Consider the rate-$\frac{2}{3}$ convolutional code with $\mathbf{g}_{11} = (1\ 1\ 1), \mathbf{g}_{12} = (0\ 0\ 1), \mathbf{g}_{13} = (1\ 0\ 0), \mathbf{g}_{21} = (0\ 1\ 0), \mathbf{g}_{22} = (1\ 0\ 1)$, and $\mathbf{g}_{23} = (1\ 1\ 1)$.
i) Draw the block diagram of the code generator.
ii) Determine the code word generated if $\mathbf{u}_1 = (1\ 1\ 0\ 1)$ and $\mathbf{u}_2 = (1\ 0\ 0\ 1)$.

4-2 Consider the rate-$\frac{1}{3}$ convolutional code of Problem 4-1(a).
a) Draw the tree corresponding to this code.
b) Draw the trellis diagram generated by this code.
c) Verify that both the tree and trellis generate the same output sequence if the input sequence is (0 1 1 1 0 1).

4-3 For each of the convolutional codes of Problem 4-1 determine
a) the generator polynomial, and
b) the code polynomial generated by the input sequences given in Problem 4-1.

4-4 Consider the convolutional code of Problem 4-1(a). Using the Viterbi algorithm, decode the received sequence $\mathbf{y} = (000\ 110\ 110\ 110\ 001)$.

4-5 Consider the rate-$\frac{1}{3}$ convolutional code of Example 4.1-1.
a) Verify that its transfer function $T(D, I, \beta)$ is given by (4.2-4).
b) Determine $T(D, I, \beta)$ for the rate-$\frac{1}{3}$ code of Problem 4-1(a).
c) Determine a bound on the error probability for both these codes.
d) Show the bound on $P(E, d)$ given in (4.2-7).

4-6 Draw the phase trellis and the state trellis of CPFSK with modulation index $h = 1/3$.

4-7 Using eq. (4.3-10), find the free distance of binary CPFSK (or 1*REC*) as a function of the modulation index $h$. Assume a small $h$ so that a length-2 error event leads to the free distance; this is a diamond-shaped one in Fig. 4.3-3. Evaluate your expression for $h = 1/2, 1/3, 1/4$.

4-8 Draw the phase tree of $h = 1/4$ quaternary 2*REC* (i.e., the linear phase response pulse over 2 intervals in Fig. 4.3-2). Compare your tree to the 2RC tree in Fig. 4.3-6. Find the minimum distance error event, using (4.3-9) and restricting yourself to paths that split out of the root node and merge after 3 levels. For small enough $h$ you have found the free distance of this coded modulation.

4-9 Show that eq. (4.3-7), the distance between two CPM signals, becomes (4.3-8) in the limit of large carrier frequency.

4-10 Consider CPFSK error events whose lengths are 3 intervals, rather than the 2 intervals in Problem 4-7.

a) For $h = 1/2$, find the worst-case such event and its distance (use eq. (4.3-9) to measure distance).

b) Above this $h$, length-3 error events eventually lead to smaller distances than length-2 events. For which $h$ does this first occur?

4-11 Show that the second line in (4.4-4) holds in the limit of large carrier frequency, if the signals are as defined in (4.4-1) and (4.4-2).

4-12 a) Show that the normalized square minimum distance of uncoded octal PSK is 0.879.

b) What loss is this compared to uncoded QPSK?

c) The normalized square-distance of the set-partition code in Ex. 4.4-1 is 4. Compute the gain of this over uncoded QPSK and over uncoded octal QPSK.

4-13 a) Find the quantity $c^2_{\mathbf{ave}}$ for the 32QAM cross constellation of Ex. 4.4-2.

b) From this, find the normalized square free distance of the second set-partition code in the example (the one with the 2:2 split).

c) Find the normalized square minimum distance of 16QAM, the scheme in Fig. A-1 with a $4 \times 4$ square block of points centered on the origin. Then compute the coding gain of this rate 4 scheme over the rate 4 code in (b).

4-14 a) Draw a section of the trellis of the scheme in Fig. 4.4-5. Assume that the state variable is $(u^{(1)}_{n-1}, u^{(2)}_{n-1}, u^{(2)}_{n-2})$, and list the states in the order $(000), (001), \ldots, (111)$.

b) Consider one case of a transmitted and neighbor path that split and merge 4 levels later, and find the distance between them. From this result, what can you say about the normalized square free distance?

4-15 By starting from definition (4.5-10), show that eq. (4.5-11) is the autocorrelation of the matched filter output colored noise variable $\xi_k$. (This problem requires knowledge of stochastic processes).

4-16 Model a 3-ray multipath channel by the discrete-time response $F(z) = 1 + az^{-1} + z^{-2}$ to a unit data symbol. Here the first and third rays have unit strength but the second varies.

a) For which $a$ is the whitening filter $1/F^*(z^{-1})$ stable?

b) A simple way to equalize this channel is the zero-forcing equalizer, a scheme based around a filter with response $1/F(z)$. When is this filter stable?

c) At which $\omega$ does the spectrum $F(j\omega)$ have zeroes, for a given second ray size $a$? Note that for these sinusoidal inputs the equalizer in (b) will fail.

4-17 Suppose a channel has continuous-time impulse response $h(t) = 1,\ 0 < t < 3T/4$, and 0 otherwise. This models a channel with a constant time dispersion of length $3T/4$. Using the whitened matched filter method, find a discrete-time model $F(z)$ for this channel as in (4.5-13).

4-18 a) Find the matched filter receiver metric (4.5-4) for the channel in Problem 4-17.
b) How would you implement the basic receiver (4.5-3)?

# *REFERENCES*

[4-1] P.Elias, "Error-free coding," *IRE Trans. Information Theory*, IT-4, pp. 29–37, 1954.

[4-2] G.D. Forney, Jr., "Convolutional codes I: Algebraic structure," *IEEE Trans. Information Theory.,* Vol. IT-16, pp. 720–738, Nov. 1970.

[4-3] A.J. Viterbi, "Error bounds for convolutional codes and an asymptotically optimum decoding algorithm," *IEEE Trans. Information Theory,* Vol. IT-15, pp. 177–179, Jan. 1979.

[4-4] L.R. Bahl, C.D. Cullum, W.D. Frazer, and F. Jelinek, "An efficient algorithm for computing free distance," *IEEE Trans. Information Theory,* Vol. IT-18, pp. 437–439, May 1972.

[4-5] K.J. Larson, "Comments on 'An efficient algorithm for computing free distance'," *IEEE Trans. Information Theory,* Vol. IT-19, pp. 577–579, July 1973.

[4-6] J.J. Bussgang, "Some properties of binary convolutional code generators," *IEEE Trans. Information Theory*, IT-11, pp. 90–100, 1965.

[4-7] J.P. Odenwalder, *Optimal decoding of convolutional codes,* Ph.D. Thesis, University of California, Los Angeles, 1970.

[4-8] L.R. Bahl and F. Jelinek, "Rate-$\frac{1}{2}$ convolutional codes with complementary generators," *IEEE Trans. Information Theory,* Vol. IT-17, pp. 718–727, 1971.

[4-9] K.J. Larson, "Short convolutional codes with maximal free distance for rates 1/2, 1/3, 1/4, " *IEEE Trans. Information Theory*, Vol. IT-19, pp. 371–372, 1973.

[4-10] E. Passke, "Short binary convolutional codes with maximal free distance for rates 2/3 and 3/4," *IEEE Trans. Information Theory*, Vol. IT-20, pp. 683–689, 1974.

[4-11] D.J. Costello, "A construction technique for random error correcting codes," *IEEE Trans. Information Theory*, Vol. IT-15, pp. 631–639, 1969.

[4-12] A.J Viterbi and J.K. Omura, *Principles of Digital Communications and Coding*, McGraw-Hill, 1979.

[4-13] S. Mason and H. Zimmermann, *Electronic Circuits Signals, and Systems,* Wiley, NY 1960.

[4-14] J. Hagenauer, N. Seshadri, and C.-E.W. Sundberg, "The performance of rate-comaptible punctured convolutional codes for digital mobile radio," *IEEE Trans. Communications*, vol. 38, pp. 966–980, July 1990.

[4-15] J. Hagenauer, "Rate-compatible punctured convolutional codes (RCPC codes) and their applications," *IEEE Trans. Communications*, vol. 36, pp. 389–400, April 1988.

[4-16] J. B. Anderson and R. deBuda, "Better phase-modulation error performance using trellis phase codes," *Electronics (Letts.)*, vol. 12, pp. 587-588, 28 Oct. 1976; J. B. Anderson and D. P. Taylor, "A bandwidth-efficient class of signal space codes," *IEEE Trans. Information Theory*, vol. IT-24, pp. 703–712, Nov. 1978.

[4-17] T. Aulin, C.-E. Sundberg and N. Rydbeck, "Continuous phase modulation (Parts I and II)," *IEEE Trans. Communications*, vol. COM-29, pp. 196–225, March 1981.

[4-18] J. B. Anderson, T. Aulin and C.-E. Sundberg, *Digital Phase Modulation*, Plenum, New York, 1986.

[4-19] Special Section on Combined Modulation and Encoding, *IEEE Trans. Communications*, vol. COM-29, pp. 185–297, March 1981.

[4-20] Issues on Bandwidth and Power Efficient Coded Modulation, *IEEE J. Sel. Areas Communs.*, vol. SAC-7, pp. 873–1016 and 1277–1469, Aug. and Dec. 1989.

[4-21] C.-E. Sundberg, "Continuous phase modulation," *IEEE Communications Magazine*, vol. 24, pp. 25–38, April 1986.

[4-22] J. D. Gibson, *Principles of Digital and Analog Communications*, Macmillan, New York, 1989.

[4-23] R. E. Ziemer and R. L. Peterson, *Digital Communications and Spread Spectrum Systems*, Macmillan, New York, 1985.

[4-24] J. G. Proakis, *Digital Communications*, 2nd Ed., McGraw-Hill, New York, 1989.

[4-25] M. L. Doelz and E. T. Heald, Minimum-shift data communication system, U. S. Patent 2,977,417, March 28, 1961.

[4-26] R. deBuda, "Coherent demodulation of frequency-shift keying with low deviation ratio," *IEEE Trans. Communications*, vol. COM-20, pp. 429–436, June 1972.

[4-27] M. G. Pelchat, R. C. Davis and M. B. Luntz, "Coherent demodulation of continuous phase binary FSK signals," *Proc.*, International Telemetering Conf., Washington, D.C., pp. 181–190, 1971.

[4-28] W. P. Osbourne and M. B. Luntz, "Coherent and noncoherent detection of CPFSK," *IEEE Trans. Communications*, vol. COM-22, pp. 1023–1035, Aug. 1974.

[4-29] T. A. Schonhoff, "Symbol error probabilities for M-ary coherent continuous phase frequency-shift keying (CPFSK)," *Conf. Record*, Intern. Conf. on Communications, San Francisco, Calif., pp. 34.5–34.8, June 1975.

[4-30] M. G. Mulligan and S. G. Wilson, "An improved algorithm for evaluating trellis phase codes," *IEEE Trans. Information Theory*, vol. IT-30, pp. 846–851, Nov. 1984.

[4-31] T. Aulin, "Three papers on continuous phase modulation (CPM)," Ph.D. Thesis, Telecommunication Theory Dept., Tech. Univ. of Lund, Sweden, Nov. 1979.

[4-32] G. Ungerboeck, "Channel coding with multilevel/phase signals," *IEEE Trans. Information Theory*, vol. IT-28, pp. 55–67, Jan. 1982.

[4-33] G. Ungerboeck, "Trellis-coded modulation with redundant signal sets - Parts I and II," *IEEE Communications Magazine*, vol. 25, pp. 5–21, Feb. 1987.

[4-34] G. D. Forney, Jr., "Coset codes - Parts I and II," *IEEE Trans. Information Theory*, vol. IT-34, pp. 1123–1187, Sept. 1988.

[4-35] L.-F. Wei, "Rotationally invariant convolutional channel coding with expanded signal space - Parts I and II," *IEEE J. Sel. Areas Communs.*, vol. SAC-2, pp. 659–686, Sept. 1984.

[4-36] G. D. Forney, Jr., R. G. Gallager, *et al.*, "Efficient modulation for band-limited channels," *IEEE J. Sel. Areas Communs.*, vol. SAC-2, pp. 632–647, Sept. 1984.

[4-37] A. R. Calderbank and N. J. A. Sloane, "New trellis codes based on lattices and cosets," *IEEE Trans. Information Theory*, vol. IT-33, pp. 177–195, Jan. 1987.

[4-38] J. Leech and N. J. A. Sloane, "Sphere packings and error-correcting codes," *Can. J. Math.*, vol. 23, pp. 718–745, 1971.

[4-39] R. deBuda, "The upper bound of a new near-optimal code," *IEEE Trans. Information Theory*, vol. IT-21, pp. 441–445, July 1975; R. deBuda, "Some optimal codes have structure," *IEEE J. Sel. Areas Communs.*, vol. SAC-7, pp. 893–899, Aug. 1989.

[4-40] L.-H. Zetterberg, "A class of codes for polyphase signals on a band-limited Gaussian channel," *IEEE Trans. Information Theory*, vol. IT-11, pp. 385–395, July 1965.

[4-41] S. U. H. Qureshi, "Adaptive equalization," *Proc. IEEE*, vol. 73, pp. 1349–1387, Sept. 1985.

[4-42] G. D. Forney, Jr., "Maximum-likelihood sequence estimation of digital sequences in the presence of intersymbol interference," *IEEE Trans. Information Theory*, vol. IT-18, pp. 363–378, May 1972.

[4-43] L. R. MacKechnie, "Maximum likelihood receivers for channels having memory," Ph.D. Thesis, Dept. Elec. Eng., Univ. of Notre Dame, Notre Dame, Ind., Jan. 1973.

[4-44] G. Ungerboeck, "Adaptive maximum-likelihood receiver for carrier-modulated data transmission systems," *IEEE Trans. Communications*, vol. COM-22, pp. 624–636, May 1974.

[4-45] D. G. Messerschmitt, "A geometric theory of intersymbol interference - Parts I and II," *Bell Syst. Tech. J.*, vol. 52, pp. 1483–1539, Nov. 1973.

[4-46] D. D. Falconer and F. R. Magee, Jr., "Adaptive channel memory truncation for maximum likelihood sequence estimation," *ibid.*, pp. 1541–1562.

[4-47] C. C.-W. Wong and J. B. Anderson, "Optimal short impulse response channels for an MLSE receiver," *Conf. Record*, Intern. Conf. on Communications, Boston, Mass., pp. 25.3.1–25.3.5, June 1979.

[4-48] N. Seshadri and J. B. Anderson, "Decoding of severely filtered modulation codes using the (M,L) algorithm," *IEEE J. Sel. Areas Communs.*, vol. SAC-7, pp. 1006–1016, Aug. 1989.

# CHAPTER 5

# Reduced Trellises

At this point in our story we have seen many good source and channel codes. The trellis has given us a conceptual framework for many of them and the Viterbi algorithm is an optimal way to traverse a trellis. We have good codes and a way to think about them.

But we do not have practical coded communication unless these codes and trellises are small enough. In Chapter 2 the really effective DPCM source codes often have thousands of states in their trellises. In Chapter 4 the outlook was mixed: Useful convolutional and set-partition codes exist with only 8-32 states, but really powerful codes exist, too, with hundreds and thousands of states, for which a Viterbi trellis traversal is not practical. Intersymbol interference decoding presents a particular challenge. Although Section 4.5 describes problems like multipath resolution that are nicely solved by the Viterbi algorithm, the classic ISI problem – the infinite-impulse-response channel filter – has an infinite trellis model.

We must distinguish between an algorithm that stores a code and produces a needed trellis branch on demand, and one that searches a trellis for a good source encoding or a good channel decoding. The foregoing chapters solve the first problem by generating code branches from a short generator sequence rather than storing an entire tree or trellis. The second problem remains unsolved and we turn to it in this and the next chapter.

One can divide the problem of efficient code searching into two general approaches, reducing the *size* of the code trellis and reducing the *search* of a given trellis to a small region. For example, there may be states that make insignificant change in code symbols or values compared to other states, and perhaps these can be isolated and deleted from the trellis; this illustrates reducing the size of a trellis. It is also likely that a decoding (or encoding) that ends up in one part of a trellis precludes decoding in far away parts, and it seems therefore that not all of a trellis needs searching. In this chapter we will explore the first approach, size reduction, and in Chapter 6 we will study limiting the search, a much larger subject. Both chapters emphasize channel coding because the majority of applications are here.

Generally, the trick of any reduction scheme is to maintain intact the error or distortion performance of the unreduced coder. About this we should state some ground rules. The performance of a code is a quality of the code, and a size reduction of a proper code inevitably reduces its performance too. We can hope the performance loss is minimal. A reduced search is another matter, for Chapter 6 shows that great reduction is possible without affecting the inherent performance of a code. A goal now is to find the minimum such search.

## 5.1 TRUNCATION OF THE IMPULSE RESPONSE

An infinitely long impulse response is surely one that needs truncating and this section is a study of code generators with this property. Truncation also applies to finite responses that are too long. Thus far two cases have appeared of code generators with an infinite response. The first was predictive the quantizer class of source codes in Section 2.4, and particularly the DPCM code generator (2.4-8), whose $z$-transform was

$$\widehat{Z}(z) = \frac{\widehat{U}(z)}{1 - A(z)}. \tag{2.4-8}$$

Recall that $\widehat{U}(z)$ is a driver sequence of simple quantizer levels, $A(z)$ is computed from the source autocorrelation by (2.4-3), and $\widehat{Z}(z)$ is the sequence of code word letters, or "reproducer values". An alternate realization of this all-pole filter is $\widehat{Z}(z) = C(z)\widehat{U}(z)$, where $C(z) = c_0 +$

$c_1 z^{-1} + c_2 z^{-2} + \cdots$ is an infinite sequence, and a time-domain expression for the sequence $\widehat{Z}(z)$ is

$$\widehat{z}_n = \sum_{i=0}^{\infty} c_i \widehat{u}_{n-i}. \tag{2.4-9}$$

The second case, infinite response ISI, was just discussed in Section 4.5. A discrete-time WMF model for this is

$$R(z) = F(z)U(z) + N(z) \tag{4.5-13}$$

in which $U(z)$ is now the sequence of transmitted data symbols, $N(z)$ is white noise, $R(z)$ is the channel model output sequence, and $F(z) = f_0 + f_1 z^{-1} + f_2 z^{-2} + \cdots$ represents the infinite channel response. In the time domain this is

$$r_n = \sum_{i=0}^{\infty} f_i u_{n-i} + \eta_n. \tag{5.1-1}$$

In either of these cases, (2.4-9) or (5.1-1) is a convolution of a low-precision sequence with an infinite "generator" sequence. A trellis model of either process, like the ones in the preceding chapter, will have infinitely many states. What can be done about this? Can we still construct a trellis decoder?

The key to this question is to realize that for (2.4-9) or (5.1-1) to generate a finite-energy process, the sequence $C(z)$ or $F(z)$ must have finite energy, and in particular its terms must converge to zero. There must be some point at which we may truncate these sequences and thus create a finite trellis coding model, without much affecting the code.

## Direct Truncation

The direct approach to truncation is to ignore the terms in $F(z)$ or $C(z)$ after the $L$-th, and arrive at a finite-state code model like that in (4.5-20) or Fig. 4.5-2. This trellis code model will have $b^L$ states. It will consist of the first $L + 1$ terms of (2.4-9) or (5.5-1). We will explore two examples of response truncation, one for channel coding and one for source coding.

**Example 5.5-1: The One-Pole Channel Filter.**

A continuous-time one-pole channel is one with response to a unit data symbol

$$h(t) = \begin{cases} Ae^{-at}, & \text{for } t \geq 0 \\ 0, & \text{otherwise.} \end{cases} \tag{5.1-2}$$

(see definition (4.5-1)). Carrying out the WMF model derivation in Section 4.5, we get (see Problem 5-1)

$$\varepsilon_k = \frac{A^2}{2a} e^{-a|k|T}, \tag{5.1-3}$$

$$\varepsilon(z) = \frac{A^2(1-e^{-2aT})}{2a} \frac{1}{1-z^{-1}e^{-aT}} \frac{1}{1-ze^{-aT}}. \tag{5.1-4}$$

We obtain (5.1-4) through the identity

$$\frac{1}{1-qz^{-1}} = 1 + (qz^{-1}) + (qz^{-1})^2 + \cdots. \tag{5.1-5}$$

For the discrete-time model $F(z)$ of $h(t)$, we make the obvious spectral factorization

$$\begin{aligned} F(z) &= \frac{A\sqrt{(1-e^{-2aT})/2a}}{1-z^{-1}e^{-aT}} \\ &= A\sqrt{\frac{(1-e^{-2aT})}{2a}} \left[1 + e^{-aT}z^{-1} + e^{-2aT}z^{-2} + \cdots\right]. \end{aligned} \tag{5.1-6}$$

In the time domain the whole channel model is

$$r_n = A\sqrt{\frac{(1-e^{-2aT})}{2a}} \sum_{i=0}^{\infty} \left(e^{-aT}\right)^i u_{n-i} + \eta_n. \tag{5.1-7}$$

Figure 5.1-1 plots the factor $(e^{-aT})^i$ against $i$ when $T = 1/a$ and the factor at the front of (5.1-7) is unity, and it is clear in this case that the taps in the model $F(z)$ after $f_4$ or so contribute virtually nothing. For binary data, the model $F(z) = 1 + .37z^{-1} + .14z^{-2} + .05z^{-3} + .02z^{-4}$ is an accurate 16-state trellis model of the channel response. When the factor $e^{-aT}$ is close to one, as it is in a slowly decaying channel, this kind of model will be unacceptably large.

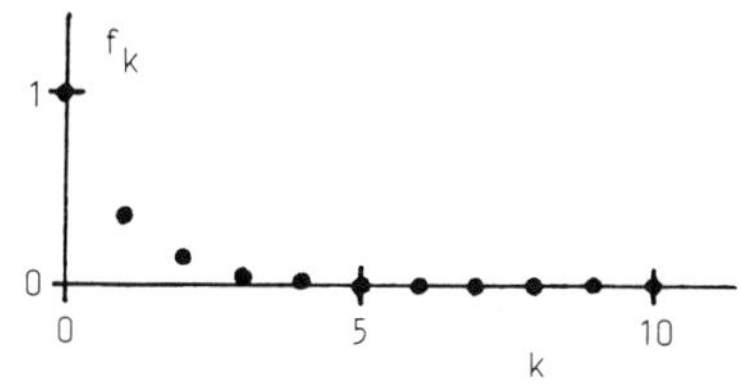


**Figure 5.1-1.** Impulse response of the exponential channel model in Ex. 5.1-1 for $T = 1/a$. The first five values are 1, 0.37, 0.14, 0.05, 0.02.

**Example 5.1-2: A Fixed DPCM Source Code.**

In Section 2.4, the standard DPCM code matched to the long-term statistics of speech had generator $1/[1-A(z)]$, with $A(z) = 1.75z^{-1} - 1.22z^{-2} + .30z^{-3}$. Eq. (2.4-7) gives the time-domain expression and Fig. 2.4-2 shows a partial code tree for this three-pole infinite response. The generator impulse response $C(z)$ appears in Fig. 5.1-2. We see that this response is quite a bit longer than the one in Ex. 5.1-1, but $A(z)$ is after all only a rough approximation, matched to an estimate of speech statistics, and we can truncate its impulse response to perhaps $c_9$ without much effect. The result at rate 1 is a 512 state trellis code. At rate 2, a more important rate, the trellis has 262144 states.

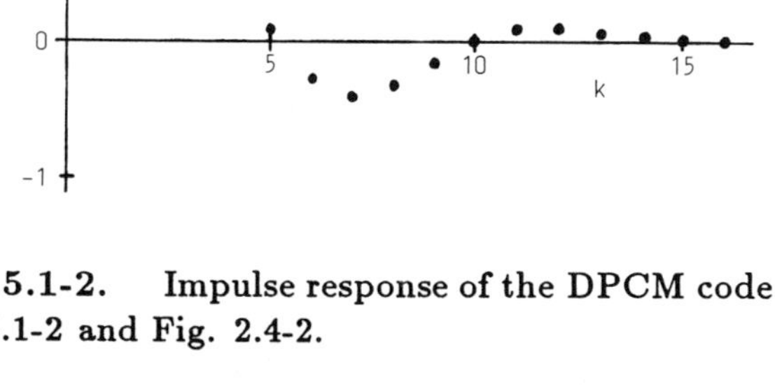


**Figure 5.1-2.** Impulse response of the DPCM code generator in Ex. 5.1-2 and Fig. 2.4-2.

## Truncation with Tail Memory

As the last example shows, the simple truncation of an infinite-response generator sometimes fails to produce a suitable trellis code. In fact simple coding structures often still exist and the key is to change the way that we carry along the memory of past symbols. Suppose we break an infinite response like Fig. 5.1-1 or -2 into a strong, main part and a long, weaker *tail response.* A generic code generator might then look like

$$s_n = \sum_{i=0}^{L} c_i u_{n-i} + \tau(u_0, \ldots, u_{n-L-1}) \tag{5.1-8}$$

in which the tail response is gathered together into

$$\tau(u_0, \ldots, u_{n-L-1}) \triangleq \sum_{i=L+1}^{n} c_i u_{n-i}.$$

The path map symbols $u_0, u_1, \ldots$ here are data symbols in channel coding and basic quantizer levels in source coding. The ones near time $n$ each strongly affect the code word letter $s_n$, while the tail symbols have a possibly strong cumulative effect but a weak individual effect. In other words, the code generation needs individual path map symbols in the main response but only a cumulative measure of the tail.

Several papers in the late 1980s [5-1]–[5-5] introduced the tail method of response truncation. We can return to Exs. 5.1-1 and -2 for an illustration. For the one-pole ISI channel example, eq. (5.1-7) contains the code generator formula, which we can reduce to

$$s_n = \sum_{i=0}^{n} q^i u_{n-i}, \qquad q = e^{-aT}. \tag{5.1-9}$$

(We have set the lead factor in (5.1-7) to unity). Code letter $s_n$ is given by

$$s_n = \sum_{i=0}^{L} q^i u_{n-i} + \tau(u_0, \ldots, u_{n-L-1}) \tag{5.1-10}$$

in which

$$\tau(u_0, \ldots, u_{n-L-1}) = \sum_{i=L+1}^{n} q^i u_{n-i}.$$

But the tail response satisfies the convenient relation

$$\tau(u_0, \ldots, u_{n-L-1}) = q\tau(u_0, \ldots, u_{n-L-2}) + q^{L+1}u_{n-L-1}, \tag{5.1-11}$$

which means that we may generate a new tail from the previous one times $q$ plus the increment $q^{L+1}u_{n-L-1}$. A Viterbi-like trellis decoder would work as follows: The decoder searches a $b^L$-state trellis, the states being defined by $(u_{n-L}, \ldots, u_{n-1})$, and associated with each state is the tail variable defined in (5.1-10); from the state, the tail variable and the next symbol $u_n$, eq. (5.1-10) gives the trellis branch letter; at each $(n+1)$-level node, a survivor path and its tail variable are kept and the others destroyed.

Tests of this one-pole channel decoder show that its error performance can be almost as good as that of a very large trellis decoder (see [5-1] and [5-2]), even when the truncation $L$ is just one or two. The only real disadvantage of the tail method is that the Viterbi search can force a decision about an output after only $L$ symbols are viewed, and $L$ is generally much shorter than the decision depth $L_D$ of the ISI code (for a discussion about $L_D$, see Section 4.2). The limit to performance is thus often the short decision depth. By limiting the *search* of the trellis rather than the trellis depth, as in Chapter 6, we remove this disadvantage.

Tail-memory decoders have been applied to ISI, CPM and set-partition decoding. Methods exist to calculate the minimum distance of all these cases, with and without a tail memory, and from these distances may be estimated the large-SNR error performance.

Turning next to the DPCM code in Ex. 5.1-2, we can write the code generation in the time domain as eq. (2.4-7),

$$\widehat{z}_n = \sum_{i=1}^{3} a_i \widehat{z}_{n-i} + \widehat{u}_n;$$

here 3 is the order of the autoregressive speech model and the formula represents a three-pole filter function. Two iterations of (2.4-7) will yield a version of (5.1-8) that defines a two-symbol-observing tail memory generator:

$$\widehat{z}_n = (a_1^2 + a_2)\widehat{u}_{n-2} + a_1\widehat{u}_{n-1} + \widehat{u}_n + \tau(\widehat{u}_0, \ldots, \widehat{u}_{n-3})$$

with tail response

$$\begin{aligned}\tau(\widehat{u}_0, \ldots, \widehat{u}_{n-3}) = (a_1^3 + 2a_1a_2 + a_3)\widehat{z}_{n-3} + (a_1^2a_2 + a_2^2 + a_1a_3)\widehat{z}_{n-4} \\ + (a_1^2a_3 + a_2a_3)\widehat{z}_{n-5}.\end{aligned} \tag{5.1-12}$$

Here we represent the effect of $\widehat{u}_0, \ldots, \widehat{u}_{n-L-1}$ by the three variables $\widehat{z}_{n-3}, \widehat{z}_{n-4}$ and $\widehat{z}_{n-5}$, which in reality are *state variables* of the three-pole generator filter. By using state variables we can store the tail component with three variables rather than $n - L$.

A trellis-like DPCM encoder will now work as follows. The encoder searches a $b^2$-state trellis in which the states are defined by the $b$-ary levels $\widehat{u}_{n-1}$ and $\widehat{u}_{n-2}$; associated with each trellis state are the three filter state variables $\widehat{z}_{n-3}, \widehat{z}_{n-4}$ and $\widehat{z}_{n-5}$, different values for each trellis node; from these values, the trellis state, and the next quantizer level $\widehat{u}_n$, eq. (5.1-12) gives the next branch letter $\widehat{z}_n$.

The state variable approach introduced here is a powerful, general method for dealing with infinite-response coding, although we cannot take the space to pursue it further. The method was actually used in the channel filter example too, since $\tau$ in (5.1-11) functions as the single state variable for the one-pole filter there. We also use state variables to generate the infinite-response codes in Chapter 6. A more general treatment of the method for ISI decoding appears in ref. [5-2], with applications to Butterworth channel filters.

# 5.2 PARTITION OF THE PATH MAP ALPHABET

We have seen that code generators often have an impulse response that decays more or less to zero over a long period. Consequently, data symbols or source code quantizer levels that are well in the past do not much affect the present code letter. Recall that we gave the name path maps to these symbols or levels because they map a path through the code tree, and that the precise effect of each map symbol is given by convolutions like eqs. (2.4-9) or (5.1-1) in Section 5.1. Since early symbols have little effect, they need not be stored very accurately and they may as well be quantized into rough, low-precision symbols. This is the idea of partitioning the path map symbol alphabet.

The result is a simpler trellis. To see this, let us somewhat idealize the channel decoding process and assume that the object is to decode the first symbol $u_0$ after viewing channel outputs $r_0, r_1, \ldots, r_L$. (We could equally well consider a source encoder.) The path maps here are various collections of hypothetical data symbols $\hat{u}_0, \ldots, \hat{u}_L$. Suppose we assign the precisions $b_0, b_1, \ldots, b_L$ to map symbols at tree depths $0, 1, \ldots, L$, meaning that at depth $L$ a symbol $\hat{u}_L$ quantized into one of $b_L$ bins (see

Section 2.2). The precisions satisfy $b_0 \leq b_1 \leq \ldots \leq b_L$ and $b_\ell \leq b$ for each $\ell$, where $b$ is the branching factor of the code tree. We wish to extend the search of the code structure from depth $L-1$ to depth $L$. While all path maps leading to branches at depth $L-1$ are retained in full precision, only the reduced precisions are used to generate code letters on depth-$L$ branches, and if more than one path reaches depth $L$ with the same reduced-precision expression, all but the minimum-distant one are dropped. It is clear that no more than the product $b_0 b_1 \ldots b_L$ paths survive to depth $L$. The decoder releases the lead symbol $\hat{u}_0$ (in full precision) of the best path as the decoder output.

This decoder is really the Viterbi algorithm, at work on an implicit trellis of $b_0 b_1 \ldots b_L$ states. It contains the principle of optimality set out in Section 4.2 and it selects survivor paths. Without the map-alphabet partitioning, the code generation will contain $b^{L+1}$ states at depth $L$, and since some $b_\ell$ can be much smaller than $b$, there is much scope for reduction here.

Chevillat and Elephtheriou [5-3] and Eyuboglu and Qureshi [5-4], [5-5] proposed this idea for decoding set-partition codes in an ISI channel; some of the earlier ISI-tail symbols in (5.1-1) are quantized to a smaller alphabet. The second paper also reduces the precision of the set-partition specification; that is, it considers a constellation point $(c_n^I, c_n^Q)$ to be placed in one of the fewer, larger subsets in a partitioning scheme like Fig. 4.4-4 rather than in the lower, more plentiful subsets actually used by the transmitter. Both papers report heavy reduction in trellis size without much loss in error performance, but the precise relation between size and loss remains unknown.

## Average-Matched Filtering

A related scheme of trellis reduction is much older and stems from the idea of matched filtering, which is described in Appendix A and Sections 4.3–4.5. The idea of a matched filter receiver is to find the correlation between a hypothetical section of a signal and the received noisy signal; this is repeated for the different hypothetical signals, and the one with the best correlation is the receiver output. In a coded modulation for which a data symbol affects a long run of transmitter outputs, there will need to be very many matched filterings. The idea of average-matching is to match a recent signal section precisely but to match only the *average* signal in the earlier period, given what is present in the recent section. The idea is not very different from the ISI truncation with a tail memory in Section 5.1, although the effect of the recent signal section is reckoned in a new way. In both methods the structure that remains has size

$b^{L+1}$. Note that matched filtering is equivalent to the minimum-distance decoding in Sections 4.5 and 5.1.

Although the idea applies generally, the average-matched filter (or AMF) receiver has been applied mostly to CPM coding. We will study that application in more detail. Consider the signal (4.3-5).

$$s(t) = \sqrt{\frac{2E}{T}} \cos(\omega_0 t + \phi(t)),$$

in which $\phi(t)$ is the data bearing phase (4.3-6). Figure 5.2-1 illustrates $\phi(t)$ for the case of binary CPFSK; the solid line is an assumed phase in the interval $(0, T)$ while the dashed phases represent all the possible antecedent phases before time 0. $\phi(0)$ is some value $\phi_0$. A pure AMF receiver consists of two filters. The first is matched to the solid phase in the figure during $(0, T)$, corresponding to data $+1$, and to the average of all the dashed phases before time 0; the second is matched to data $-1$ during $(0, T)$, which causes phase $\phi_o - h\pi t/T$, and to the same average of antecedents before time 0. The receiver decides in favor of the receiver with the most positive output at time $T$.

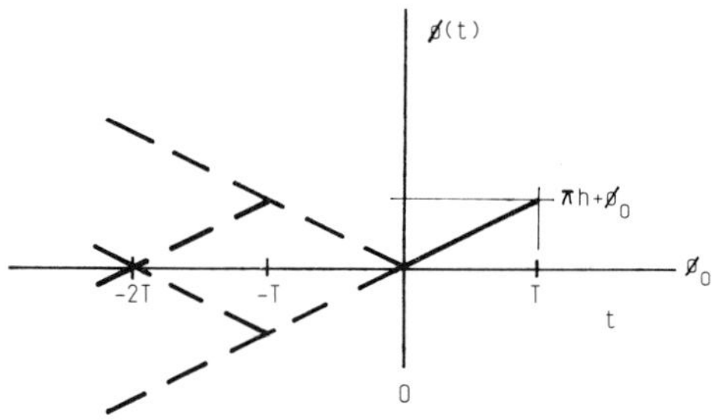


Figure 5.2-1. Example of AMF receiver design for binary CPFSK signals. CPM variables are defined in Section 4.3. Averaging is over the signals with the dashed phase trajectories, with each equally weighted.

The real work of designing an AMF receiver is computing the average signal to which the earlier part of the filters matched. For instance, during interval $(-T, 0)$ one must find

$$\frac{1}{2}\cos\left(\omega_0 t - \frac{h\pi t}{T}\right) + \frac{1}{2}\cos\left(\omega_0 t + \frac{h\pi t}{T}\right),$$

ignoring the constant $\sqrt{2E/T}$, and during $(-2T, -T)$ one must average four signals. Problem 5-4 computes our CPFSK example. In general this averaged signal declines to zero for increasingly negative time and eventually it may be truncated and ignored. This is a statement that the distant past provides no information about the present. In the recent past, however, only certain signals are compatible with the present assumed signal and these provide for an approximate matched filter. It is an interesting fact that the AMF receiver (not the Viterbi algorithm!) is the maximum likelihood receiver for full response CPM and PSK at *low SNR*.

A generalization of the AMF idea is the *multiple-symbol* AMF receiver. In this receiver the assumed and perfectly matched section of the signal extends back $L_A$ intervals before the present, before which the matching is to the average antecedent as usual. The receiver so designed needs $b^{L_A}$ matched filters and, of course, performs better over the high-SNR channel. The multiple-symbol receiver is especially needed for partial response (i.e., long pulse) CPM and for coded QPSK with an extended constraint length.

Osborne and Luntz [5-6] introduced the AMF receiver which was later extended by others [5-7], [5-8]. Studies have shown (see [5-8]) that for CPM with a short pulse and small $h$ the AMF receiver suffers little loss with 1/2 to 1/4 the number of filters compared to the full matched filter receiver. The saving carries over to trellis decoder versions of the receiver.

## 5.3 PULSE RESPONSE SIMPLIFICATION

The complexity of the coding schemes in this book stems mostly from their generator impulse responses – from the response length or from the slowness with which the response decays. If a way can be found to reduce the pulse complexity without reducing the code free distance, then the code will be less complex and so will its decoder. Once again, CPM coding has seen the greatest application of pulse simplification, although the idea applies to other codes. The CPM application contains one further subtlety: Only the decoder and its trellis need to assume a simpler pulse, and the transmitter may go on using the complex pulse. This is important because the original pulse may lead to a narrower bandwidth or some other advantage. The savings we seek are in the receiver.

An example of CPM pulse simplification at the receiver appears in Fig. 5.3-1, in which the phase trees are drawn for binary 3RC and binary 3REC. Since intersignal distance is roughly proportional to phase difference in this plot, it is clear that the minimum distance of both schemes is about the same. Yet the 3RC code has considerably narrower bandwidth in the spectral tail region, which will be an advantage in a crowded mobile radio application. According to Section 4.3, the trellis decoder for binary 3RC has $2^2$ states while the decoder for 2REC has only $2^1$ states, where we have ignored the common effect of the index $h$ on the state size. Thus, we save 50% on the decoder size by pretending that the code was 2REC. The free distance loss is miniscule (see [5-8], Chap. 8).

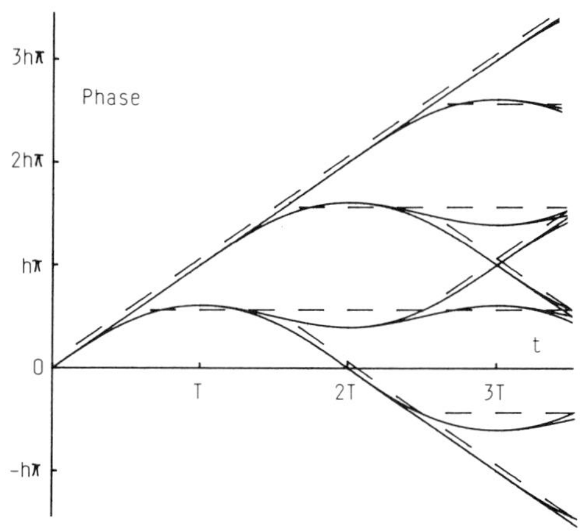


**Figure 5.3-1.** A comparison of the phase trees of binary 2REC (dashed) and 3RC, showing the close fit. From ref. [5-8], p. 281, ©Plenum Press 1986, used by permission.

Sometimes the pulse assumed at the decoder needs to have a time offset in order to achieve a better phase tree fit. An outright numerical optimizing of the simpler pulse is proposed in ref. [5-8]. Generally, the trellis reduction versus loss in free distance tracks that given in Section 5.2 for AMF receivers: Reduction to 1/2 to 1/4 are possible with little loss. Pulse simplification for CPM coding originated with the text [5-8] and especially with the thesis [5-9] and papers of A. Svensson. These works show how to estimate the error performance when the decoder expects a different signal than the transmitter sends, a more general calculation called the *mismatched receiver* problem.

# PROBLEMS

5-1 Derive $\epsilon_k$ for the 1-pole channel in Ex. 5.1-1 and show that $\epsilon(z)$ is given by (5.1-4).

5-2 Suppose a channel has 2-pole discrete-time model 1/[1-A(z)] where $A(z) = a_1 z^{-1} + a_2 z^{-2}$.
a) Give the time-domain formula for this channel in terms of code letters $s_n$ and data symbols $u_n$.
b) Derive a tail memory formula like eq. (5.1-10) for this model with truncation at L=2.

5-3 Figure 2.4-2 gives some of the depth-5 code tree for the binary-branching code tree whose generator impulse response appears in Fig. 5.1-2.
a) Plot several paths of length 8 for the quaternary-branching ($b = 4$) tree with the same generator, showing the code letters along the paths. Use the quantizer input levels and bins in Table 2.2-2, scaled upward by a factor 1/.80 to reflect the scaling in Fig. 2.4-2. Assume that map symbols are 0 before time 0.
b) Repeat, but quantize the map symbols other than the latest five into two bins only; i.e., into levels $\pm 1$ depending on whether the 4-level symbol is above or below 0. How close are the paths to those in part (a)?

5-4 Consider binary CPFSK CPM coding with index $h = 1/4$. Assume that the signal in $(0, T)$ starts from zero phase and ramps linearly up to phase $h\pi/T$, and assume a unit magnitude for the transmitted cosine.
a) Compute the average of the signals that can lead up to the assumed signal. Do this for several previous intervals. You have designed the average-matched section of an AMF receiver filter.
b) To what should this section be matched if the assumed signal starts at 90 degrees.

5-5 Consider the exponential channel in Ex. 5.1-1, whose discrete-time model is eq. (5.1-7). Let the data $\mathbf{u}$ be the quaternary symbols $\pm 1, \pm 3$; but in generating the code letters

$$s_n = \sum_{i=0}^{\infty} q^i u_{n-i}, \qquad q = e^{-aT},$$

quantize $u_{n-1}$ and $u_{n-2}$ to +2 if they are positive and -2 otherwise, and set all the earlier $u_{n-i}$ to 0. Construct a section of the 16-state trellis for this code generation.

# REFERENCES

[5-1] A. Duel-Hallen and C. Heegard, "Delayed decision-feedback sequence estimation," *IEEE Trans. Communications*, vol. COM-37, pp. 428–436, May 1989.

[5-2] N. Seshadri and J. B. Anderson, "Decoding of severely filtered modulation codes using the (M,L) algorithm," *IEEE J. Sel. Areas Communs.*, vol. SAC-7, pp. 1006–1016, Aug. 1989.

[5-3] P. R. Chevillat and E. Eleftheriou, "Decoding of trellis-encoded signals in the presence of intersymbol interference and noise," *IEEE Trans. Communications*, vol. COM-7, pp. 669–676, July 1989.

[5-4] M. V. Eyuboglu and S. U. Qureshi, "Reduced-state sequence estimation with set partitioning and decision feedback," *IEEE Trans. Communications*, vol. COM-36, pp. 13–20, Jan. 1988.

[5-5] M. V. Eyuboglu and S. U. Qureshi, "Reduced-state sequence estimation for coded modulation on intersymbol interference channels," *IEEE J. Sel. Areas Communs.*, vol. SAC-7, pp. 989–995, Aug. 1989.

[5-6] W. P. Osborne and M. B. Luntz, "Coherent and noncoherent detection of CPFSK," *IEEE Trans. Communications*, vol. COM-22, pp. 1023–1036, Aug. 1974.

[5-7] W. Hirt and S. Pasupathy, "Suboptimal reception of binary continuous phase shift keying signals," *Proc. IEE*, Part F, vol. 128, pp. 125–134, June 1981.

[5-8] J. B. Anderson, T. Aulin and C.-E. Sundberg, *Digital Phase Modulation*, Plenum Press, New York, 1986.

[5-9] A. Svensson, "Receivers for CPM," Ph.D. Thesis, Telecommunication Theory Dept., Technical University of Lund, Lund, Sweden, May 1984.

CHAPTER 6

# Sequential Coding Algorithms

In the foregoing chapters we have seen tree and trellis schemes that are more or less brute force in their approach to coding and decoding. The Huffman code generation algorithm in Section 2.2 creates an efficient code, but the source encoding algorithm that uses this code must store the entire code and must compare an incoming sequence to the whole code. Practical DPCM-type codes have thousands of tree branches. The rate-distortion codes in Sections 2.4 and 2.5 have similar drawbacks: They are good but potentially very large codes and we have not yet proposed an efficient method to search their code books.

Much the same applies to the trellis channel codes in Chapter 4. We used the Viterbi decoder there to search the trellises. The trellises of convolutional, CPM, and set-partition codes are a much more efficient representation of the codes than their code trees, but these trellises can still be uncomfortably large. Intersymbol interference trellises can be gigantic. Are there effective schemes that search only a part of these trellises? Is it necessary to search the whole trellis as the Viterbi algorithm does? How can a partial search be made efficient? This chapter addresses these questions.

It will be useful to review once again the distinction between *storing* tree code words or branches and *encoding* a source output or *decoding* a channel output. We have essentially solved the problem of storing large codes in Chapter 4. Rather than store whole codes, we generate code branches as needed from a short generator sequence. The codes in Chapter 4 (and some of those in Section 2.4) are all generated by a convolution process. But our encoding and decoding algorithms, whose job it is to find a successful coding, still search the entire tree or trellis. Our coding effort is thus still large.

We will begin the chapter with the basics of sequential coding and turn then to some of the better known algorithms. The chapter emphasizes channel decoding because this is the chief application at present. Sections 6.3-6.6 present a formal analysis of sequential decoding for the binary symmetric channel, the only case for which a complete analysis is known. Experimental tests for all the trellis codes in Chapter 4 closely reflect the BSC analysis. The chapter ends with a section on sequential source coding.

The major tool in sequential coding analysis is the random code. Random codes were introduced in Section 2.3 as a means to prove coding theorems, and we find them useful once again in this chapter. The new codes will be binary channel codes, like those in Section 3.1 except that they will now be tree codes. Every symbol in the tree code is selected independently, with probability 1/2 for a 0 and 1/2 for a 1. At first glance random tree codes may seem artificial and technical. But convolutional codes, the important practical class of binary tree codes, have strongly pseudorandom properties. The random-code properties that we use here, which have to do with minimum distance and numbers of tree paths, are followed closely by practical codes. An important fact about convolutional codes – and perhaps the key to their usefulness – is that they create pseudorandom trees, just as maximal shift registers create pseudonoise sequences.

## 6.1 BASICS OF SEQUENTIAL CODING

A sequential channel decoder detects information in small symbol groups. A sequential source encoder encodes source outputs in small groups. Both repeat a small, basic routine for each group. The algorithm design principle here is divide and conquer: Reduce a gross coding job to many small, identical steps.

Tree and trellis codes carry information in a way that makes sequential coding easier. The words of a tree code lie along a perfectly regular tree graph in which $b$ branches stem from each node and a group of $a$ binary code symbols appears on each branch. The symbol group provides a natural output unit for the repeating routine and the tree nodes form a sequence of $b$-fold decision points. To recapitulate Chapters 2 and 4, a channel encoder associates a $b$-ary data symbol with each $b$-fold tree branching and transmits the corresponding tree path; the channel decoder must decide which path was sent by comparing the path as received to the code tree. A source encoder searches a source tree code for a path close to the source outcome; the source decoder simply converts the path map back into a code word. In source coding the encoding is the challenge, while in channel coding it is the decoding.

A trellis structure is similar to a tree except that its paths sometimes merge. At the local level, a large trellis looks and acts like a tree. Since the *raison d'etre* for a sequential coder is that it looks at just a small part of a trellis, the coder "sees" a tree even when it works in a trellis. For this reason, and for simplicity of analysis, we discuss only trees.

How may we measure the effectiveness of a sequential coder? An obvious measure in channel decoding is the probability of error. In source coding the equivalent is the reproducer sequence distortion. Another measure is the coding delay, that is, the depth of tree needed to encode or decode a symbol. The *cost* of the algorithm can be measured in several ways. How many paths on the average must the coder search per unit output? In case the search varies a lot, what maximum storage allocation must be set aside? Finally, some ways of searching paths require more algorithm steps than others. How many steps are needed per output?

## The First-Symbol Decoder

We now turn to channel coding, and describe a slightly specialized sequential channel decoder called a *first-symbol* decoder. It provides the standard framework for analysis. It contains the important properties of practical sequential decoders and makes them easier to derive and understand. Real decoders differ in only a few common-sense ways.

The object of the first-symbol decoder is to decode the first code tree branch only. When it finishes this job, it releases the branch as the decoder output and purges all the tree paths left in storage. It then begins decoding anew, with the node at the end of the released branch as the root node, and the basic cycle repeats until the whole received sequence is decoded. Our analysis will focus on one of these cycles. A real decoder might not release output branches one at a time, and it would not purge paths such as descendants of the released branch that it was certain to view again. We will encounter other small variations when we consider specific tree search methods.

It will be convenient to divide the code tree paths into subsets. The correct, or transmitted, branch out of the root and all its descendants form the *correct subset*, abbreviated CS. Each incorrect branch together with all its descendants forms an *incorrect subset*, or ISS. There are $b$-1 of these. We can also talk about incorrect subsets of later nodes along the correct path. All of this is shown in Fig. 6.1-1 for a binary branching tree. The decoder's job is to identify which subset is the correct one.

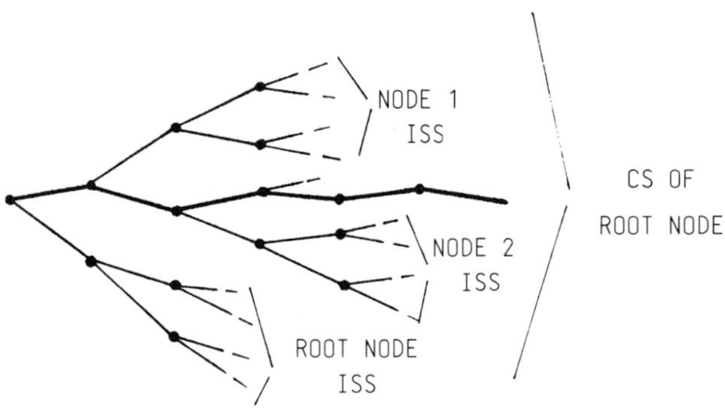


**Figure 6.1-1.** Correct subset and some incorrect subsets in a binary-branching code tree. The heavy path was transmitted.

The statistical behavior of the CS paths is completely different from that of ISS paths. The correct path is the same as the received path except for occasional errors. Consider the BSC. If we plot the correct path Hamming distance to the received path versus tree depth, we obtain a random walk like that in Fig. 6.1-2. The walk is defined by:

$$
\begin{aligned}
P\{0 \text{ step}\} &= (1-p)^a \\
P\{+1 \text{ step}\} &= a\,(1-p)^{a-1}\,p \\
&\vdots \\
P\{+k \text{ step}\} &= \binom{a}{k}\,(1-p)^{a-k}\,p^k \\
&\vdots \\
P\{+a \text{ step}\} &= p^a.
\end{aligned}
\tag{6.1-1}
$$

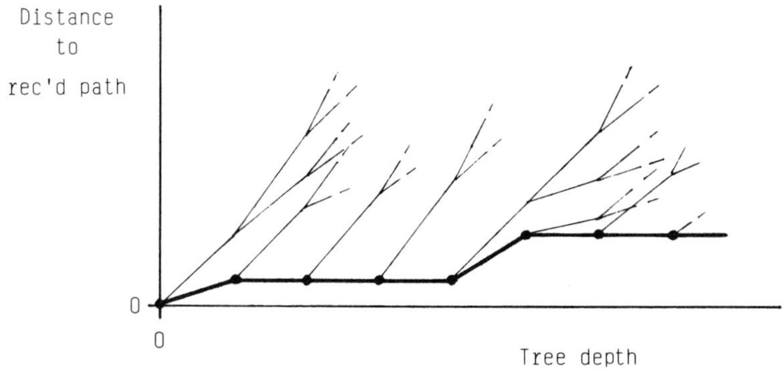


**Figure 6.1-2.** An example of a correct path random walk. A single and a double bit error per branch are shown. Some random walks of incorrect paths are shown for comparison.

Ordinarily the upward drift of the walk is quite small. From each correct path node, $b-1$ incorrect paths branch off and start random walks of their own, but these walks tend rapidly upward. If the tree code is a random one, an incorrect path takes an upward step of size $k$ with probability $\binom{a}{k}2^{-a}$. This means an average drift up of 1/2 per tree path bit, compared with only $p$ per bit along the correct path.

An incorrect subset consists only of incorrect paths. Each one of these drifts up at the 1/2-per-bit rate, but our interest now is not in the typical path, but in the *least* path, since it is this one that threatens correct decoding. A sensible decoder, when confronted with two paths of the same length, must surely favor the least distant. If an ISS path falls below the correct path random walk at decision time, a sequential decoder makes an error. Actually, our attitude is slightly conservative here, because another path in the correct subset, besides the transmitted one, could fall below the least ISS path and save the day. But this is unlikely. Largely, sequential decoding analysis is the study of two random processes, the correct path random walk and the ISS minimum path. These are the focus of Section 6.3.

## Path Metrics

The Hamming distance of a path from the received path is an example of a path metric, or measure of goodness, for paths. Unfortunately, Hamming distance breaks down as a metric when paths of several lengths need to be compared. A long path, for example, could have a little larger Hamming distance than a short path but still be a better path because its distance per unit length is less. Another metric, called the *Fano metric*, provides a solution to this problem.

The Fano metric can be justified in two ways. The first is to show that it minimizes the effort in some important kinds of decoding, and we will do this in Section 6.4. The second is to show that it stems directly from the principle of maximum likelihood, which underlies our decoders. From Appendix A, a maximum likelihood decoder is one that finds the code word $\mathbf{x}_i$ that maximizes

$$\log\left[\frac{P\{\mathbf{y}|\mathbf{x}_i\}P\{\mathbf{x}_i\}}{P\{\mathbf{y}\}}\right], \tag{6.1-2}$$

under the assumption that the code words are equally likely. Here $\mathbf{y}$ is the received sequence and $\mathbf{x}_i$ is a *partial* code word. Expression (6.1-2) is the likelihood of this partial word. The lengths of $\mathbf{y}$ and $\mathbf{x}_i$ are $\ell$ in terms of tree branches and $n = \ell a$ in terms of bits.

A brief evaluation of (6.1-2) for the BSC and random codes is as follows. Because of the randomly selected code, any $\mathbf{y}$ may be received with equal probability, so $P\{\mathbf{y}\} = 2^{-a\ell}$. There are $b^\ell$ equally likely code words. Because of the BSC,

$$P\{\mathbf{y}|\mathbf{x}_i\} = p^k(1-p)^{n-k},$$

where $k$ is the number of bit errors. Substituting these quantities into (6.1-2), we get

$$\mu(\mathbf{y}) = k\log p + (n-k)\log(1-p) + n(1-R). \tag{6.1-3}$$

Eq. (6.1-3) is usually expressed on a per-code-bit basis, which for bit $j$ is

$$\begin{cases} \log_2 2p - R, & \text{if } y_j \neq x_{ij} \\ \log_2 2(1-p) - R, & \text{if } y_j = x_{ij}. \end{cases} \quad \text{(The Fano Metric)} \tag{6.1-4}$$

R. M. Fano [6-1] first proposed (6.1-4) and Massey [6-2] showed its maximum likelihood optimality. Our argument can be extended to convolutional codes with the binary input DMC [6-3].

From the form of (6.1-4) we can see that long paths receive good (i.e., positive) metrics so long as they do not contain too many errors. Each new received bit adds the increment $\log_2 2(1-p)$ if it agrees with a code word bit, from which is subtracted $R$. The quantity $R$ is called the *bias* of the metric. It is deducted from the metric total once for each arriving bit, while the other term in (6.1-4) depends on whether the bits agree. The bias term is the contribution to the metric from the path's length, as opposed to its correctness. The form of (6.1-4) – a correctness term plus a length term – is a basic form that applies well beyond the BSC. In Section 6.3 we will study coding with an arbitrary bias $B$ not necessarily equal to the code rate.

Example 6.1-1.

> Suppose a rate 1/2 code is used with a BSC whose crossover is $p =$ 0.01. The per-bit Fano metric (6.1-4) is
>
> $$\begin{array}{ll} -6.14 & \text{(bits disagree)} \\ +.49 & \text{(bits agree).} \end{array}$$
>
> Consider two received sequences, one of length 16 bits with one error, and one of length 2 bits with no errors. The first has total metric 1.21 and the second has .98, and so the longer path is rated better by the Fano metric, even though it has one error.
>
> We could create a new metric by increasing the bias to 0.55 from 0.50. This will reward forward motion less. Now the long sequence has metric .34 and the short one has metric .87. The short one is now "better".

To compute metrics in a practical decoder, we see from the example that an algorithm need only increment a total by one of two numbers, depending on whether or not the bits in two sequences agree. It is customary to scale and round off these numbers to simple integers in order to save arithmetic; the increments in the example might be converted to $-12$ and $+1$. In more complicated channels than the BSC, the table of numbers will be larger because the channel outputs and the code symbols may not be binary, but the same ideas apply.

## When Will Decoding Stop?

We need a *stop criterion*, a rule that tells when the search of the code tree has gone on long enough. The first-symbol decoder views paths and their lengths and errors relative to the received path. From these it must decide whether to stop, and having stopped, it must estimate which is the correct subset of the root node.

Several stop criteria have been suggested for sequential decoding and several others for sequential encoding of sources. For decoding we use mostly the *probability of error criterion*, which states that decoding may stop when it is evident that the decoder error probability is no worse than a fixed number $P_e$. There are two failure modes for a decoder. In searching out some paths and dropping others, it may drop the paths in the correct subset; lacking any vestige of the correct subset, it cannot possibly decode correctly. In the other mode, CS paths are present at stop time but an ISS path seems better, at least in the sense of the path metric. The decoder selects the best path it sees and makes an error. Section 6.3 shows how to control the search so that these modes are never more probable than $P_e$.

Whatever the definition of the stop criterion, two boundaries limit the search of the tree. The first, called the *stop line*, is the locus of all path lengths and metrics that stop the search. There is no search beyond this line because decoding has stopped. The second boundary, called the *drop line*, is the line beyond which paths are not good enough for further scrutiny. Some simple sketches of these lines follow in Fig. 6.2-1. We will define them more carefully in Section 6.3.

In Section 6.5 we will discuss two other stop criteria. One of these stops the search when all combinations of $t$ or fewer errors have been corrected. If more than $t$ errors occur within the depth of search, correct decoding is not guaranteed. This criterion does not guarantee a particular error probability and, in fact, it takes no particular note of the channel error mechanism. It is used in applications where the error count is more important than the error probability.

## How Should the Search Evolve?

After a decoder extends a code tree path to a new depth and computes its new metric, it must decide which path to extend next. The decision may depend on the metric values of the candidates. Their values, in turn, depend on which metric and which bias the search uses. The search may allow backtracking to a shorter path. If so, it can take a look at long but risky paths. These are examples of the organization and pattern of the tree search prior to its stopping, which we call the search *mode*. We will discuss some basic modes in Section 6.2, when we describe the important decoder algorithms.

For every search mode a decoder will explore a different total number of tree paths. An obvious aim of decoder analysis is to find out which mode views the fewest paths. Some modes search more paths but have other advantages, such as synchronism or a less deep search, that outweigh efficiency.

The method of counting searched tree paths varies a little between theory and practice. In practical sequential decoding, one simply counts the number of paths (or branches) viewed per branch of output. In the theoretical analysis, we count just the ISS paths per output branch. Specifically, we count the ISS paths in one cycle of the first-symbol version of an algorithm. These two methods really count the same thing because every path viewed by the practical version except for the correct path is in the incorrect subset of some node along the correct path (see Fig. 6.1-1).

An important aspect of searching is the variability that is encountered from output to output. Wide variation is a disadvantage, since sufficient storage and computing ability must be allocated for the worst case. There are some basic rules about the variation in the number of ISS paths out of a correct path node. Variability will be small if the search of the ISS paths is over a fixed region of the tree, that is, over paths whose metrics and depths are subject to a priori fixed limits. Variability will be large if the size of this region depends on an observable like the behavior of the correct path. This is because the number of ISS paths, while fixed in a fixed region, varies rapidly with the region size. In several search modes that we still discuss, the size of the ISS search depends on the worst excursion of the correct path metric, which itself varies widely. These searches, thus, have wide variation.

## 6.2 IMPORTANT CHANNEL DECODING ALGORITHMS

Searching and sorting are an important subject in computer science. An everyday example is accessing files on a disk. In this chapter we are searching for an object somewhere deep in a tree or trellis structure, at the end of a path whose distance and depth combination meet a criterion. A classic work on searching and sorting from a computer science perspective is Knuth [6-4]. In sequential decoding each search type has different advantages and a different analysis. We will start with these types and illustrate each with an important practical decoder.

A first basic method is the *metric-first* search. Here a measure of goodness, the metric, evaluates how close a node in the structure is to the objective; the next penetration into the structure is in front of the node already found that has the best metric. In sequential decoding the nodes lie at the end of partially explored code tree paths. Long and short paths with similar metrics will cause the search to track forward and backward many times as it finds and extends the best path.

Figures 6.2-1(a) and (b) illustrate metric-first searches of a code tree. For clarity, only the incorrect subsets are shown (stippled) and one other path, the correct one. The axes are Hamming distance and tree depth. The search lies within a bounded region, as we will see in Section 6.3. Any path hitting the stop line boundary stops the search, and the correct path does so in the figure. The first figure shows a metric-first search controlled by the simple Hamming distance metric. This means that the search proceeds layer by layer, as symbolized by the horizontals in the figure, until the stopping condition is fulfilled. The Fano metric (6.1-4) yields the pattern of slopes in the second figure, in which each successive line upward represents a more negative (i.e., poorer) metric value. The slope depends on the bias.

The figure shows a basic property of metric-first searching, that *the metric along all paths must descend to the worst value along the correct path*, if the correct path is to be found. Otherwise, the correct path is never fully extended and some other path stops the search. For the Fano metric, all paths that hit a sloping line are extended before any that touch the next line upward; if any of the progeny touch a lower line these are extended first. In a successful search, the end result is that all of the ISS paths lie at a metric equal to the minimum correct path value, and the correct path (and a few near relatives) lie by themselves at a better metric at the stop boundary. The correct path wanders out to a poorer metric and then in again to a better one. All of this is illustrated in Fig. 6.2-1(b). A more concrete example of metric-first search progress will be given shortly in our description of the stack algorithm.

Another approach to searching is the *depth-first*, or "greedy" search. Now the priority is to proceed to the end of the structure as rapidly as possible. Some sort of limits to this greed for progress are set up. In sequential decoding there are several. A path that hits certain combinations of distance and depth will be dropped forever; this path has hit the drop line. Another limit is the stop line; a path hitting this terminates the search. A third type of limit halts a path's extension only temporarily. Algorithms with this type of limit typically extend paths depth-first until they hit a certain metric value. If the search does not reach a satisfactory end, the algorithm expands this metric limit. Depth-first searches and limit expansions alternate until some path hits the stop line and the search concludes.

We can illustrate a depth-first search against a metric limit with Fig.

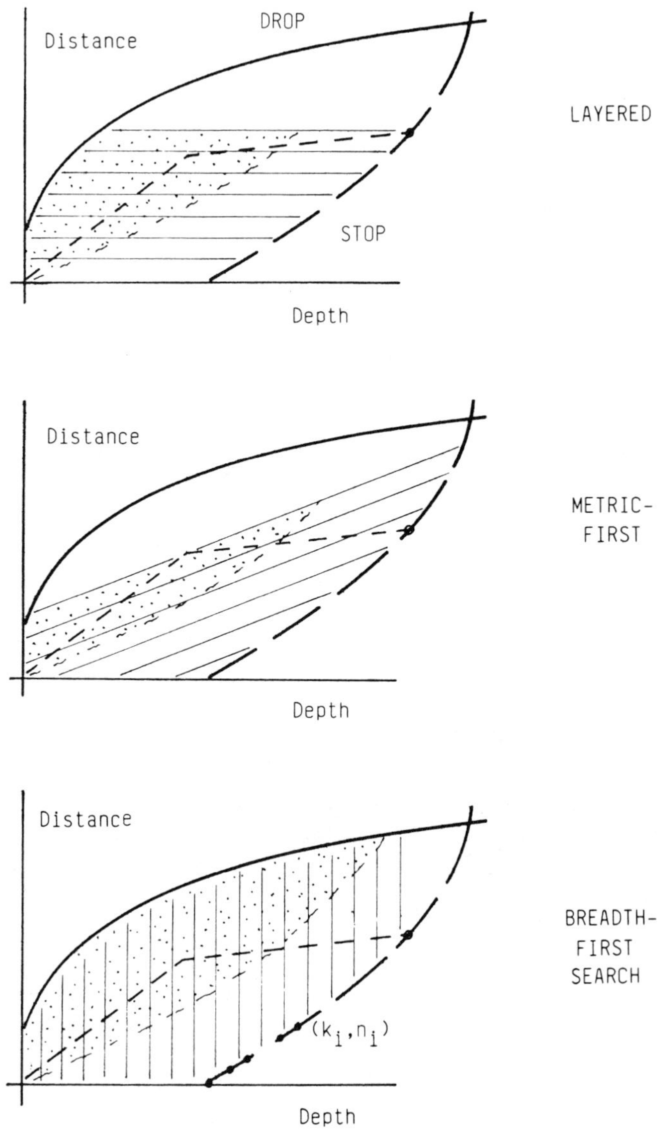


**Figure 6.2-1.** Some examples of search modes; (a) Layered search, (b) Metric-first, (c) Breadth-first. The layered search is a metric-first search with Hamming distance as the metric. Stop and drop lines are indicated; the short dashed line is the correct path. Stippling shows the region of ISS paths. The axes are Hamming (or Euclidean) distance and path depth.

6.2-1(b). The search extends one path after another until all the paths below a certain sloping line are filled in. The order of extending does not matter. If none of these paths fulfill the stopping condition, the metric limit is raised to the next sloping line (which corresponds to a poorer metric) and all paths are extended once again. Here, again, is the property that we first saw with metric-first searches: The correct path cannot reach the stop boundary until the metric limit reaches the poorest metric along the correct path. As a consequence, both metric-first searches and depth-first searches with a metric limit visit about the same number of ISS paths.

Depth-first searches can and probably should employ other search limits than a metric limit, but thus far only sequential decoders with a metric limit have been proposed. As we will see in Section 6.6, depth-first decoders are important because in terms of *algorithm steps*, rather than paths visited, they are the most efficient decoders known.

A third type of search is the *breadth-first* search. This type explores all nodes at one depth of penetration that it will ever explore and then moves on to the next depth. The search never backtracks. A breadth-first search must be more conservative about rejecting paths since it cannot return for a second look, and consequently it looks at more paths than the other searches, but the continuous forward motion is often an advantage. Figure 6.2-1(c) shows a breadth-first decoder at work on the same case as in the other parts of the figure. The vertical lines symbolize the depth-by-depth search. It is clear that more ISS paths are explored.

While we have discussed three basic search types here, it should not be assumed that the only good decoders are ones that are purely these types. Good mixed-mode decoders, no doubt, exist. We will turn now to descriptions of some well known decoders, and one of these, the multiple stack algorithm, is of a mixed type.

## The M-Algorithm: A Breadth-First Search

The idea of the $M$-algorithm is to look at the $M$ best paths at each depth of the code tree. It is a breadth-first search, the breadth being $M$. Because of the $b$-fold branching in the tree, $bM$ paths are created if each of the $M$ present paths is fully extended to the next depth. The $M$-algorithm drops all but the best $M$ of these, in terms of distance. Since all paths are of the same length, this is the best $M$ in terms of the Fano metric, too. The $M$-algorithm has been widely applied in both channel and source coding; it finds use in CPM, intersymbol interference, and other types of coding, in which the distance is Euclidean rather than Hamming distance. It first appeared in source coding [6-6].

Here is a standard first-symbol version of the algorithm. Searching starts from a root node.

**The M-Algorithm**

Repeat **Steps 1–4** for each tree level.

**Step 1.** (*Path extension*). Extend all stored paths to the next depth, creating $b$ new paths from each stored path.

**Step 2.** (*Dropping*). Drop any paths that hit the drop line.†

**Step 3.** (*Stopping*). Stop if any path hits the stop line. Release the first branch of this path as output.

**Step 4.** (*The best M*). Delete all remaining paths except the best $M$ in terms of distance to the received path.

Figure 6.2-2(a)(b) gives an example of a successful $M$-algorithm search with $M = 3$. The correct path is dashed and dropped paths are circled. A practical $M$-algorithm differs in a few details from this first-symbol version. Most practical versions are decoders of the *sliding block* type that we introduced in Section 4.1. After releasing the first output branch, they do not purge the storage, but continue searching on the remnants of the paths already there until they decode the second branch; the descendants of the first output are, after all, the logical place to look. This process continues indefinitely.†

The first-symbol $M$-algorithm is needed for analysis, but a practical version, based on sliding block principles, is really simpler and we should give a standard version of it. This version drops paths only when they fail to be in the best $M$. Instead of the usual stop line, it releases a branch when the paths reach a certain depth called the *decision depth.* In the distance-depth plane this is a simple vertical line.

†If all paths are dropped, the decoder is considered to make an error.

**The M-Algorithm (Practical Sliding Block Version)**

Repeat **Steps 1–3** for each tree level.

**Step 1.** (*Path extension*). Extend all stored paths to the next depth, creating $b$ new paths from each stored path.

**Step 2.** (*Dropping*). Drop all but the $M$ best paths.

**Step 3.** (*Branch release*). If the paths have reached the decision depth, release as output the first branch of the best path.

Because the search is breadth-first, all paths are the same length and in effect the scheme puts out branches in lock step at a fixed delay behind the forward edge of the search. This is the sliding block principle at work. With its lock step release and its fixed searching per release, the M-algorithm has a property of *output synchronism* that is often an advantage in applications. As for the decision depth, estimates of this are in Sections 6.3 and 6.5.

When releasing output in the sliding block way, the logical branch is the ancestor of the path in storage with the minimum distance. What about paths that do not stem from this branch? Technically, these should be purged if they are present, so that the decoder cannot somehow release them at a later time as output; if it did, some part of the path would have to be in error, since the path would have a jump in it that could not have been produced by the encoder. An extensive purge will cut the path number below $M$ but future searching will soon build it up. In a real implementation this purge is often ignored. The probability is that the offending paths will soon be washed out of storage. Tests have shown that the purge does benefit the overall decoder error rate in good channels, but not in poor ones.

Example 6.2-1: The M-Algorithm with Convolutional Codes.

> Lin [6-5] constructed an $M$-algorithm decoder for the BSC. The event error rate is shown in Fig. 6.2-3 for this decoder and a constraint length 14 rate-1/2 quick-look convolutional code. The energy axis here is expressed in a way that makes comparison easier to the other examples in the chapter. $E_b/N_0$ is the energy-per-data-bit-to-noise-density ratio in a white Gaussian noise channel with a binary antipodal modulator such as BPSK. The modulator has a bit error rate $p$. This is the BSC crossover probability seen by the decoder, but $E_b/N_0$ is plotted in the figure. Recall from Chapter 4 that an error event begins when the state transition sequence in the decoder output splits from that of the encoder, and it ends when the state sequences merge. The overall bit error rate of the decoder is somewhat worse than the figure because the events may contain more than one bit error. As $M$ grows, the event error rate curve steadily improves, until about $M = 90$, when the high-$E_b/N_0$ performance of the Viterbi algorithm is reached. That is, an $M$-algorithm of size 90 has the performance of a Viterbi algorithm of size 8192.

Virtually the same plot as Fig. 6.2-3 results when other rate 1/2 codes are decoded. In fact, the event error curve of the $M$-algorithm depends only on $M$ and not on the code, so long as the code's free distance can support the curve. Beyond $M = 90$ in Fig. 6.2-3 this point has been reached, and a larger $M$, or even the full Viterbi search, will not improve the error curve. Only a better code can do that.

## Catastrophic Path Loss

Like most of the algorithms in this chapter, the M-algorithms described here explore a bounded region of the code tree for each decoded symbol. If the correct path wanders out of the region, it is dropped and an error occurs, as we mentioned in Section 6.1. With some trellis codes it is almost impossible for a bounded-region decoder to acquire the correct path again, once all vestige of it has been dropped. The remaining paths somehow give no direction toward the missing path. A code like this has a property of *catastrophic correct path loss.* Many codes are not so affected and spontaneously reacquire the correct path as they decode subsequent symbols; examples are ISI trellises and most CPM codes. Unfortunately, all the convolutional codes in this book exhibit the property, although convolutional codes without the property have recently been discovered [6-6].

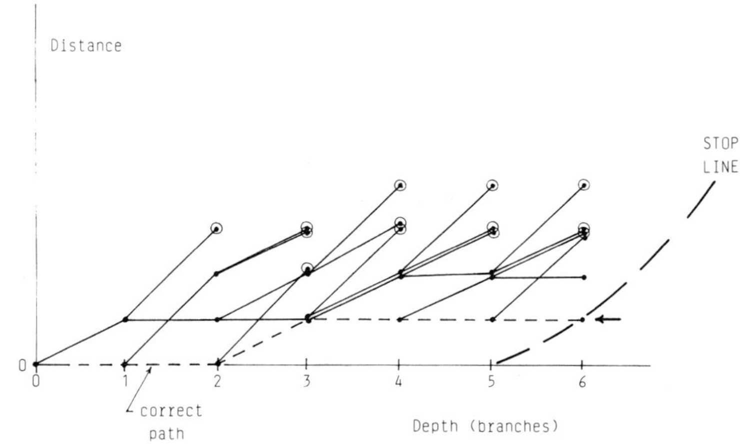


(a)

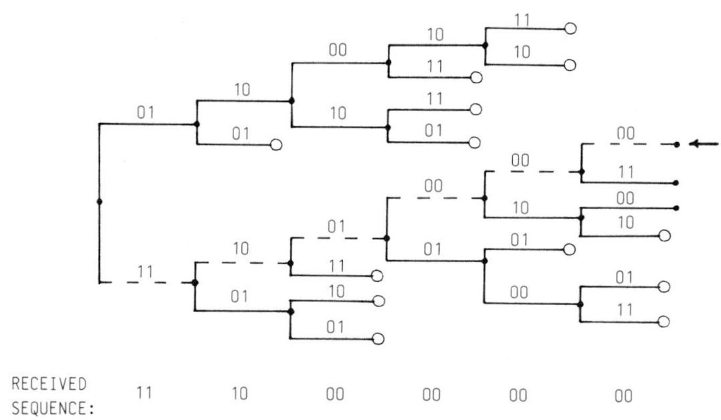


(b)

**Figure 6.2-2.** (a) Example of a successful $M$-algorithm decoding with $M = 3$ and a binary tree; and (b), the rate-1/2 tree that was searched. Three paths (nodes circled) are dropped at each depth; this supplants the usual drop line. The correct path contains a single channel error. Searching stops when the path hits the stop line at depth 6 (arrow).

The combination of a bounded-region decoder and a code with catastrophic path loss means that correct path loss initiates an indefinitely long error event. Whether this is a disadvantage in a practical system depends on its design. Typically, systems carry data in frames, the beginnings of which restart the coding from a known state. Loss of a bit is as bad as loss of an entire frame. Thus, a catastrophic path loss is

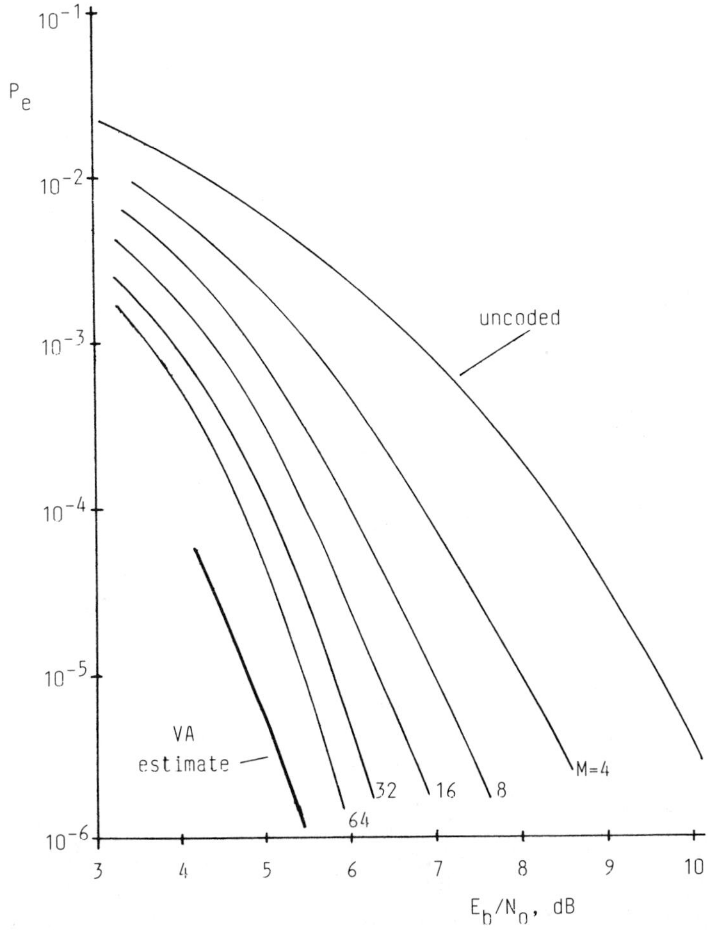


**Figure 6.2-3.** Error event frequency versus AWGN channel $E_b/N_0$ for the $M$-algorithm decoding of a rate-1/2 convolutional code. Quick-look code of memory 13 and free distance 14 with generators 46716 and 20000. $M$ as shown. The Viterbi algorithm estimate is based on the code free distance. Data from ref. [6-5].

no worse than a normal decoding error and recoveries are automatically provided. In other systems, catastrophic path loss is more serious. For these, two part decoder algorithms have been proposed in which one part is a standard decoder like those in this chapter and the other serves solely to recover lost correct paths [6-5], [6-7]. Another solution is given in Section 6.5.

**Example 6.2-2: The M-Algorithm with CPM Codes.**

> Aulin [6-8] reports tests of the quaternary 3RC code with modulation index 1/3. This code has free distance 1.67 and 96 states in its trellis. Some of his symbol error measurements are reproduced in Fig. 6.2-4. They are qualitatively similar to Example 6.2-1: Performance improves as $M$ grows until $M = 7$, after which it is essentially the same as the Viterbi algorithm. As in Example 6.2-1, the overall bit error rate will be a little worse because a symbol error sometimes causes several bit errors. The $M$-algorithm here uses the Euclidean distance measure. Catastrophic correct path loss does not occur.

## Other Breadth-First Schemes

The best known breadth-first decoder is the Viterbi algorithm, which we have discussed in Chapter 4. This algorithm is really an exhaustive decoder because it searches the entire code trellis. In its most common implementation, symbols are released after a fixed decision depth, so that the algorithm has the symbol synch property. Since it searches the entire trellis, it cannot suffer catastrophic path loss. For these reasons and because of its simple recursive structure, the algorithm is popular for codes with small trellises.

One way to use the Viterbi algorithm with a large trellis is to freeze an earlier part of the state history of the decoded path and search exhaustively the trellis part that remains free. This idea has been pursued by Eyuboglu [6-10] for set-partition codes and by Duel-Hallen and Heegard [6-9] for ISI trellises. The latter work may be compared to Seshadri [6-11], who directly decoded an entire ISI tree with the $M$-algorithm.

Interesting breadth-first decoders have been proposed by Foschini [6-12] and by Simmons [6-13]. Foschini's scheme simply explores all paths closer to the received path than a certain distance. It is the optimal search for the $t$-error-correction criterion in Section 6.5. Simmons' scheme retains the paths lying in a band around the received path.

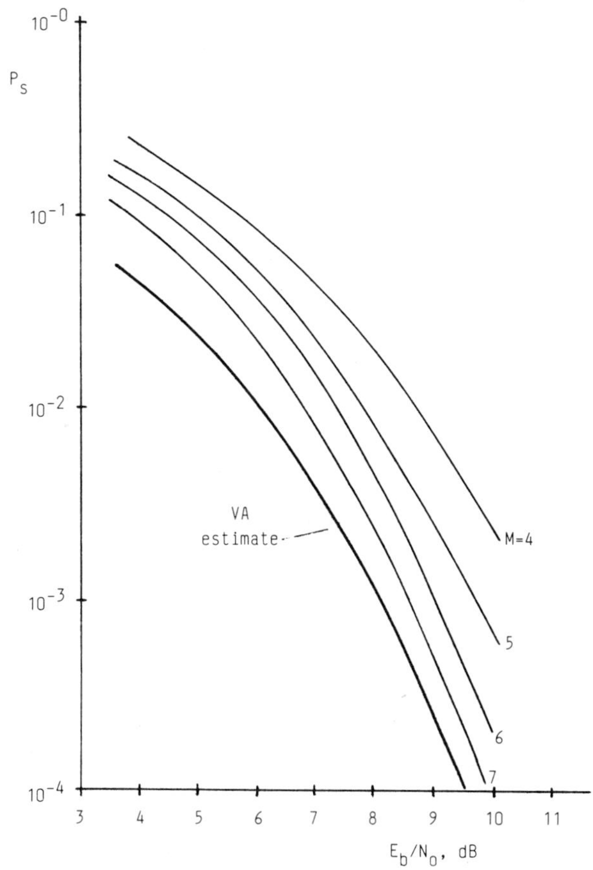


**Figure 6.2-4.** Symbol error frequency versus AWGN channel $E_b/N_0$ for the $M$-algorithm decoding of CPM. Quaternary 3RC code with $h = 1/3$. $M$ as shown. The Viterbi algorithm estimate is based on code free distance. Data from ref. [6-8].

## The Stack Algorithm: A Metric-First Search

The stack algorithm always extends the best path in storage one branch forward. The path metric is the Fano metric. In computer science language the algorithm contains not a stack of paths but a *list*, and it must keep this list ordered at all times in order to know the best path. In terms of paths viewed, the stack algorithm is very efficient but the effort needed to keep the list ordered exacts a toll, as we will see in Section 6.6. The scheme was discovered independently by Zigangirov [6-14] and Jelinek [6-15] and as a consequence is sometimes called the ZJ stack algorithm. A standard first-symbol stack algorithm now follows.

**The Stack Algorithm**

Repeat **Steps 1–2** until a path satisfies the stop criterion.

**Step 1.** (*Path extension*). Extend the best stored path to the next depth, creating a new path; if all $b$ extensions have been made from the original path, it is dropped.

**Step 2.** (*Dropping*). Drop paths that satisfy the drop criterion.

**Step 3.** (*Stopping*). Release as output the first branch of the path that satisfies the stop criterion.

The stack algorithm has not found wide practical use, but it is important because of its simplicity and because it is such an obvious way to search a tree. Figure 6.2-5 shows a stack search of the code in Fig. 6.2-2(b). As in Fig. 6.2-2(a), the axes are depth and distance from the received path, and the search stops successfully. Now, however, path extension is in order of Fano metric; equal-metric contours appear as sloped lines in the figure, with poorer ones to the upper left. Ties for best path may occur at Step 1 and any simple tie breaking scheme will suffice here.

Many variations on the basic algorithm have appeared in the literature. The only path drop criterion in most of these is the simple size of the stack: Once the path number exceeds the size, paths are dropped, worst first. If the correct path is dropped in this way, the decoding fails, an event called *erasure*. Stopping occurs like it does in the practical $M$-algorithm: when a path hits a pre-ordained decision depth for the first time, searching stops and its first branch — or in some versions, the entire path — is released as output. One variant is called the stack-bucket algorithm. This one quantizes the ordering of the stack by placing paths

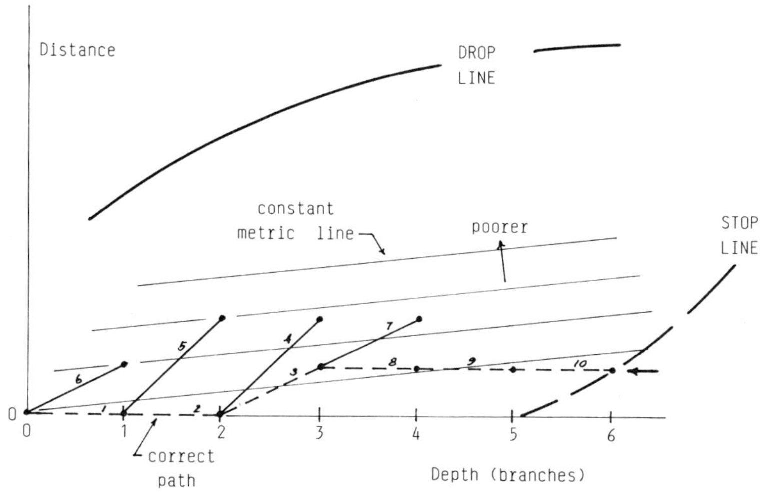


**Figure 6.2-5** Example of a successful stack algorithm search on the same code as in Fig. 6.2-2. Numbers on the branches indicate the order of search. Only 5 paths total are searched, many fewer than Fig. 6.2-2. Searching stops when a path hits the stop line (arrow); no paths hit the drop line in this example. The search is also depth-first with the metric limits shown, if the ties break this way.

of similar metrics together in a "bucket". Since paths are sorted only into the buckets and not perfectly ordered, the search is not purely metric-first, but the sorting effort is significantly reduced without much effect on decoder performance.

The original paper by Jelinek [6-15] reported tests of the stack decoder on the BSC. This and other work has shown that error performance is essentially maximum likelihood, i.e., the same as the Viterbi algorithm, provided that the correct path does not fall through the stack bottom. Put another way, the number of paths above the correct path in metric can never exceed the stack size. This number fluctuates widely. Figure 6.2-6 plots some of Jelinek's data for the worst position of the correct path that was observed in 1000 executions of a first-symbol algorithm. The code rate was 1/2, the decision depth was 1000, and the abscissa in the plot is the BSC crossover $p$. The capacity of the BSC falls to rate 1/2 when $p$ is 0.110. At less than half this probability, the needed stack size already exceeds 9000!

When we analyze the metric-first search in Section 6.4, we will dis-

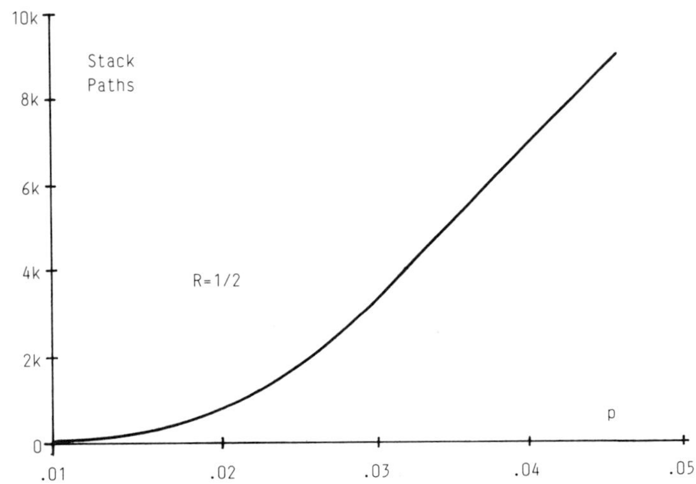


**Figure 6.2-6.** Number of paths in the stack lying above the correct path versus BSC crossover $p$. Long-memory rate-1/2 convolutional code; decision depth 1000. The number shown is the worst case observed in 1000 independent decoding cycles. Data from ref. [6-15].

cover two important facts. The number of paths searched is very small on the average, as one would expect in a metric-first regime. But second, the worst-case path storage is very much larger and is in fact no better than the storage for any other kind of search. In trying to design an efficient stack decoder, we are caught between two unpleasant realities: A short stack will often lose the correct path and a long one will be a chore to reorder.

Aside from its reordering problem, the stack algorithm's properties are similar to the depth-first schemes', which we discuss next.

## Depth-First Searches

The most important depth-first decoder is the Fano algorithm, proposed by him in 1963 [6-1]. Because the algorithm is no longer widely used we will be brief in our treatment. The scheme searches straight forward in the tree until it either reaches a stopping depth or the path's Fano metric hits a metric limit called the *threshold*. The threshold varies in a complex pattern. Basically, the algorithm tightens (lowers in the figures) the threshold if it finds itself visiting a tree node for the first time; if it visits a node for a second or later time and the path hits the threshold, the algorithm backtracks along the path to an earlier node that lies above the threshold and tries a different extension from it; if it finds that all such extensions hit the threshold value, it loosens (raises) the threshold. The idea of these threshold changes is to force the algorithm forward through the tree, while giving it an escape when it enters the wrong part of the tree.

Earlier in the section we discussed depth-first searching against a metric limit that only increased. This is a much simpler procedure. It is important to realize that both these depth-first searches, and the stack algorithm as well, visit approximately the same ISS paths, since all three visit all the ISS paths that have metric greater than the least metric along the correct path. The search in the stack example of Fig. 6.2-5 is in fact depth-first if the ties break right. If one of these schemes has an advantage, we must probe more deeply than simply counting paths.

We have seen already that the stack algorithm must take time to reorder its stack. The Fano algorithm suffers from a computation disability as well, at least in its original version. We can think of the standard Fano search as a mobile processor that crawls along paths in the code tree. When the processor hits a metric threshold, it crawls backward, retracing its steps to an earlier node, and then crawls forward again along a partly different path. During a noisy period in the channel output, the processor will revisit the same nodes over and over as it crawls back and forth, waiting for a sufficient loosening of the threshold. An advantage of this procedure is that the algorithm stores just a single path. In any practical code tree, it can generate the branches fore and aft at will, from the state transition diagram of the code. For this reason it is sometimes said that the Fano algorithm has no storage. The disadvantage of such a scheme is that the algorithm executes a potentially huge number of steps in regenerating paths already visited. The Fano algorithm trades a large storage for a large execution time. (Figure 6.5-3 in Section 6.5 is a plot of this effort).

Tests have shown that the Fano algorithm visits and revisits many more nodes than the stack algorithm does, as the channel worsens [6-15]. Still, the weight of opinion seems to be that the Fano is somewhat preferable to the stack algorithm [6-3], [6-17]. Both should be inferior to a depth-first search that holds on to paths in a storage, but no channel coding tests of such a scheme have been reported.

Among recent texts, references [6-3] and [6-17] give a detailed view of the Fano algorithm. A classic report of tests is Forney and Bower [6-18]. References [6-19] – [6-21] analyze the visits and revisits to nodes. The

Fano and stack algorithms are like the $M$-algorithm in that their error performance depends on the storage and the number of paths searched, and not on the precise code employed, so long as the code is of sufficient strength. Small order performance improvements can be obtained by using optimized codes, such as the Optimized Distance Profile class in Section 4.2, or by modifying the basic search regime. Overall, the Fano algorithm resembles the stack algorithm in the parts of the code tree that it visits.

In source encoding, depth-first searches against a simple metric limit have been proposed by several authors [6-22], [6-23].

## A Mixed-Mode Search: The Generalized Stack Algorithm

Although we have taken some pains to categorize algorithms into types, it should not be assumed that the best algorithms work purely in one mode. An example of a successful algorithm that combines metric- and depth-first searching is the generalized stack algorithm of Chevillat and Costello [6-24], [6-3]. This scheme executes a metric-first search with a relatively small stack. If no path reaches a decision depth, a small number of paths are taken off the stack and used to seed another metric-first search; if the second search is unsuccessful it generates a new search in the same way, and so on, until some path reaches the decision depth. Ways are provided for the search to backtrack from a forward thrust if it turns out poorly. The idea is to force the search forward periodically and to use small, easy-to-order stacks.

Here is an abridged first-symbol version of the algorithm. The initial stack is of size $S_i$, a size smaller than that of the true stack algorithm, but large enough for all but the worst channel noises. Any later stacks that are formed are of size $S'$, where $S' < S_i$. To form a new stack, $T$ paths are taken from a previous stack and form the seed of a new one; $T$ is typically 3-10 and is much smaller than $S'$. The algorithm also contains a best path buffer, in which is stored the best path that has reached the decision depth. The total size of all stacks cannot exceed a limit.

The first step in the algorithm is the standard stack algorithm previously given.

**The Generalized Stack Algorithm**

**Step 1.** Execute a standard stack algorithm until the present stack is full. If the total stacks exceed a limit, release as output the path in the best path buffer and stop. If a path reaches the decision depth, go to (3).

**Step 2.** Make a new stack, starting with $T$ paths of the previous one. Go to (1).

**Step 3.** (*Stopping*). If a path reaches the decision depth and is in the initial stack, release it as output and stop.

**Step 4.** (*Dropping a stack*). Else check if the path is the best yet found at the decision depth, and store it in the best path buffer if it is. In any case, drop the present stack and return to the previous one. Go to Step 1.

**Example 6.2-3: The Generalized Stack Algorithm and an ISI Channel.**

Xiong [6-25] has tested this algorithm with a simulated magnetic recording channel. The intersymbol interference in this channel is defined by the impulse response $h(t) = g(t) - g(t - T)$, where

$$g(t) = [1 + (t/T)^2]^{-1} \quad \text{all } t. \tag{6.2-1}$$

Xiong's receiver is of the whitened matched filter type in Section 4.5. His search algorithm has the following parameters:

| | |
|---|---|
| Initial stack size | $S_i = 3000$ |
| All other stacks | $S' = 100$ |
| Top entries transferred | $T = 3$ |
| Computation limit | 5000 path extensions |
| Decision depth | 266 |

Figure 6.2-7 plots overall bit error probability against a relative SNR scale for this algorithm and for a Viterbi decoder that works on a 512-state model of the ISI. The data represent an average of 2000

data frames and a lower bound based on minimum distance is given for comparison. In the higher SNR region Xiong's algorithm averaged only a little over one path extension per decoded symbol, compared to 512 for the Viterbi algorithm. Xiong's stack algorithm implementation needed 10 times as long to extend a path as did his Viterbi algorithm, but the overall speed advantage of the stack algorithm still varied from 5 to 50, depending on the SNR. A smaller Viterbi decoder may be used, but its error performance will be worse because its ISI model is worse.

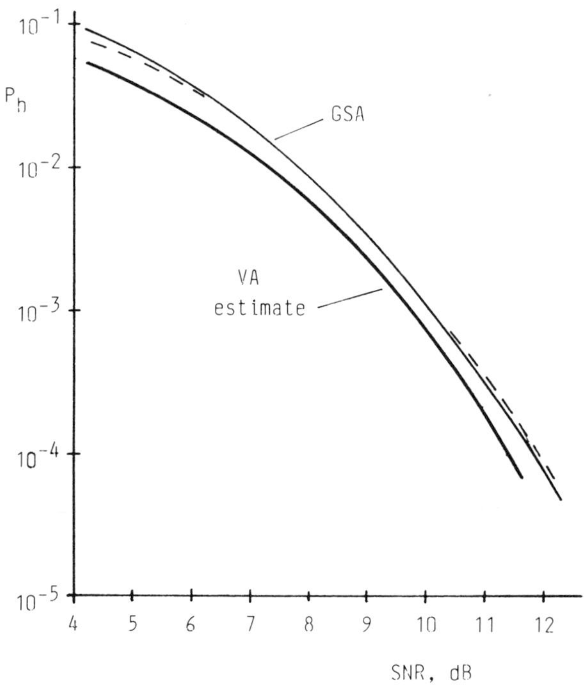


Figure 6.2-7. Bit error frequency versus relative SNR for the generalized stack algorithm with a magnetic recording channel. Dashed curve shows actual 512-state Viterbi algorithm; the VA estimate is based on the ISI free distance. Data from ref. [6-25].

# 6.3 CALCULATION OF THE SEARCH REGION BOUNDARIES

We turn now to the question of how much of the tree a sequential decoder must search in order to achieve a set decoded symbol error probability. We will want an answer for each of the basic search modes in Section 6.2. When can the search stop? When can paths be dropped? These are the subjects of this section. We will take up the *number* of these paths in Section 6.4.

It is worthwhile to summarize the conclusions of both sections. Decoding is possible at finite effort and delay at any code rate below capacity, but these grow enormously fast near capacity. Depth- and metric-first searches are much more efficient than breadth-first ones in terms of average paths searched, but all modes require the same peak storage. Breadth-first searching, on the other hand, has a much shorter coding delay, and of course it is symbol synchronous. Depth-first/metric-first searching employs backtracking while breadth-first does not. In fact, we will see that the important factor in path searching behavior is just this, *whether or not backtracking is allowed.*

All of our analysis is for the BSC and for random tree codes, but ongoing research has verified most of it for convolutional codes and the BSC. Qualitatively similar behavior has been observed with modulation codes and ISI decoders as well.

## The Stop and Drop Lines

As we introduced in Section 6.1, decoding must continue until the output of a first-symbol decoder has error probability $P_e$. This is the stop criterion. One failure mode is that the correct path is dropped, and the drop line is the border of the region out of which the correct path wanders with probability $P_e$ or less. The other failure mode is that an ISS path is the first to hit the stop line, the locus of points in metric and depth that stop the search. The stop line needs to lie far enough out that this happens with probability $P_e$ or less.

So our first task is to derive the stop and drop lines from the required error probability $P_e$. From this will follow the size of the search region and from that the decoding delay. To begin, we partition the event of error according to the whereabouts of the correct path:

$$P_e = P\{\text{Error, and CP hits drop line}\} + P\{\text{Error, and CP hits stop line}\}. \qquad (6.3\text{-}1)$$

We are assuming that if the correct path hits the drop line an error occurs, and so the first term is simply $P\{\text{CP hits drop line}\}$.

In reality, eq. (6.3-1) is two approximate conditions,

$$\log\ P_e = \log P\{\text{CP hits drop line}\} \tag{6.3-2}$$

$$\log\ P_e = \log P\{\text{Error, and CP hits stop line}\} \tag{6.3-3}$$

because $P_e$ is generally very small. We can see this by exploring some cases that satisfy (6.3-1) exactly. We could, for instance, require that both $P\{\text{CP hits drop line}\}$ and $P\{\text{Error, and CP hits stop line}\}$ be equal to $P_e/2$. The best search region is the smallest one, and we could bring one of the two lines closer in by setting one of these terms to $0.9P_e$. But now the other term must be $0.1P_e$, which is a line much farther out. Balanced coefficients are probably best. Anyway, as $P_e$ becomes smaller these leading coefficients make little difference, and in the limit of small $P_e$, (6.3-2) and (6.3-3) hold.

In what follows we will exploit (6.3-2) to compute the drop line and then turn later to the stop line.

We have defined already the random walk (6.1-1) in distance and depth executed by the correct path. In terms of the Fano metric, the steps $\alpha$ and $\beta$ are given on a per bit basis by (6.1-4). Our calculations will be easier if we modify the Fano metric slightly. First, we scale the Fano metric by the factor $1/(\alpha-\beta)$. Second, we make the bias $B$ a free parameter rather than fix it equal to the rate $R$. What results is a path random walk with positive and negative metric steps

$$\frac{\alpha}{(\alpha-\beta)} = \frac{\log_2 2(1-p) - B}{\log_2\left(\dfrac{1-p}{p}\right)} \qquad (\geq 0;\ \text{bits agree; probability } (1-p))$$

$$\frac{\beta}{(\alpha-\beta)} = \frac{\log_2 2p - B}{\log_2\left(\dfrac{1-p}{p}\right)} \qquad (\leq 0;\ \text{bits disagree; probability } p), \tag{6.3-4}$$

where $\alpha$ and $\beta$ are the original Fano steps with bias $B$ instead of $R$. We will call (6.3-4) a *scaled* metric to distinguish it from the Fano metric.

The theory of random walks provides the means to compute the drop line. In that theory, the CP metric total is a particle, or in some writings a "drunkard", that walks along a one-dimensional sidewalk. So inebriated is this particle that its steps are purely random according to (6.3-4). Suppose a police station lies at one end of the sidewalk at position $-\zeta$ and the drunk starts at 0. According to random walk theory, the probability that the drunk's position $z$ ever hits the police station is bounded by

$$P\{z < -\zeta\} < 2^{h\zeta}, \ \zeta > 0 \tag{6.3-5}$$

where $h$ is the negative root of

$$1 = \sum_i P\{\text{Step size } i\} 2^{h(\text{step size } i)}.$$

A classic derivation of this is in Feller [6-26], Chap. 14. For steps distributed as (6.3-4), the $h$ equation becomes

$$1 = (1-p)2^{h\alpha/(\alpha-\beta)} + p2^{h\beta/(\alpha-\beta)}. \tag{6.3-6}$$

To this fact we will add the drop line constraint which says that the event probability in (6.3-5) must at least satisfy

$$2^{h\zeta} = P_e \tag{6.3-7}$$

Figure 6.3-1 is a plot of CP metric $z$ versus path depth in bits, $n$, which shows the region of the event $\{z \geq -\zeta\}$; the figure replots the same region in terms of distance $k$ versus depth. The conversion between the two is particularly simple because of the metric scaling:

$$z = \frac{n\alpha}{(\alpha-\beta)} - k \tag{6.3-8a}$$

$$k = \frac{n\alpha}{(\alpha-\beta)} - z \tag{6.3-8b}$$

In distance vs. depth the region is bounded at the top by a line with slope

$$s = \frac{\alpha}{(\alpha-\beta)} = \frac{\log 2(1-p) - B}{\log\left(\dfrac{1-p}{p}\right)} \tag{6.3-9}$$

whose distance-axis intercept is $\zeta$. The probability that the correct path lies between this line and the depth axis is at least $1 - P_e$. Note that good paths fall above the others in the metric plot, but below the others in the distance plot.

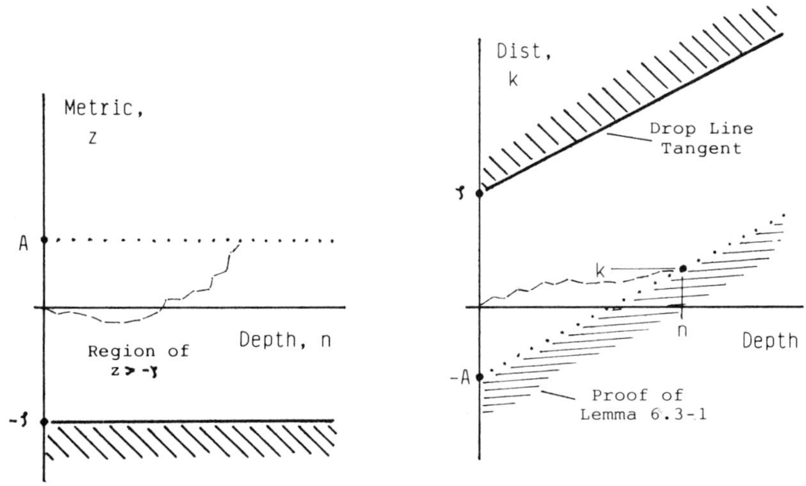


**Figure 6.3-1.** Two plots of the correct path random walk: As scaled metric versus depth and as distance versus depth. The unshaded region in either case is the event $\{z \geq \zeta\}$.

Every choice of the bias $B$ generates a different sloped line, and since (6.3-5) is a tight bound, each is approximately a *tangent* to the true drop line contour. Three such tangents are illustrated in Fig. 6.3-2(a). With enough tangents, we can draw the whole drop line.

We would like to write down an explicit procedure to generate all the tangent lines, and to do so it will be convenient to write (6.3-6) in the equivalent form

$$1 + \frac{1}{h'} \log \left[ (1-p)^{1+h'} + p^{1+h'} \right] = B \tag{6.3-10}$$

with

$$h' = \frac{h}{\log\left(\dfrac{1-p}{p}\right)}.$$

Then for each $B$ the procedure is as follows.

i) Choose $B$. Compute $h$ from (6.3-10).
ii) Compute the tangent's slope from (6.3-9).
iii) Compute the tangent's distance-axis intercept $\zeta$ from (6.3-7).

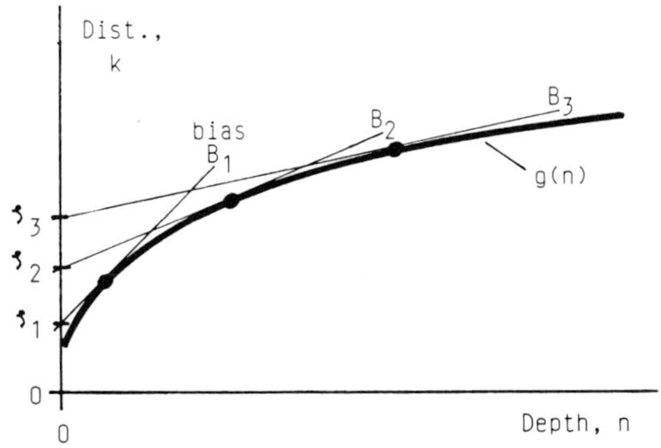


(a)

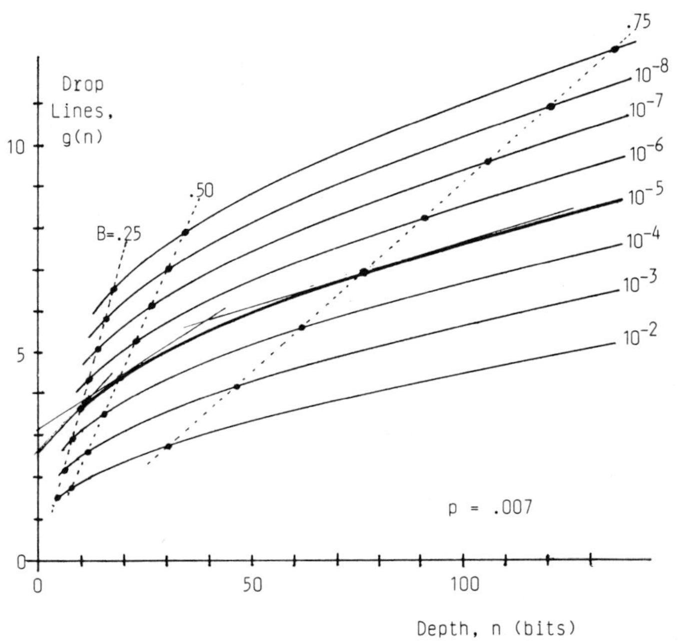


(b)

**Figure 6.3-2.** (a) Illustration of the drop line construction, showing three of the tangents to the line; (b) The actual drop lines for crossover $p = 0.007$ and $P_e = 10^{-2}$ to $10^{-9}$. The points shown are the $B$ values 0.25, 0.50 and 0.75; the spacing of these for each $B$ is linear in $\log P_e$.

Figure 6.3-2(b) plots the locus of all the tangency points for the BSC with crossover $p = .007$ and a selection of decoder error probabilities. (The points were actually produced by a more subtle procedure [6-29] that directly produces the points). It can be shown that points on the drop lines for successive $\log P_e$ are spaced out linearly in the distance–depth plane, for each choice of the bias $B$. This is illustrated graphically in Fig. 6.3-2(b) for three biases. The linearity follows from step (iii).

We have used random walk properties to locate this approximate drop line. With some effort one can find the exact line from the binomial distribution of the correct path distance. Since the distances of paths are integer-valued, the exact line should progress in integer jumps. An exact line is shown in Fig. 6.3-4 and it closely follows our approximation.

## The Stop Line: Breadth-First Searching

To compute how long a path must be pursued, we begin with (6.3-3). Decoding stops when any path in the breadth-first search hits the stop line. We will be conservative and assume that if the correct path distance lies at or above the minimum ISS distance, an error occurs. Another path in the correct subset may lie below the ISS paths, but we are again too conservative to count this as a victory.

The following simplified argument shows approximately where the stop line must lie. We will first define the location in terms of a probabilistic event and then find the asymptotic form of its probability.

Partition the stop line into $Q$ small regions such that $P\{\text{Error, CP hits stop line}\}$ is divided into $Q$ equal parts. An illustration of this appears in Fig. 6.3-3. $Q$ is any integer but it is fixed a priori for all $P_e$. For the $i$-th of these regions it must be that

$$\frac{P_e}{Q} = P\{\text{Error, CP hits in region } i\}. \tag{6.3-11}$$

We can estimate the right-hand side in terms of a point $(n_i, k_i)$ near the center of the region. The probability is close to

$$P\{\text{Min ISS distance } \leq k_i \text{ at depth } n_i, \\ \text{and CP distance near } k_i \text{ at } n_i\}. \tag{6.3-12}$$

(Note ties and near-ties count as errors). Since events in different subsets are independent, (6.3-12) becomes

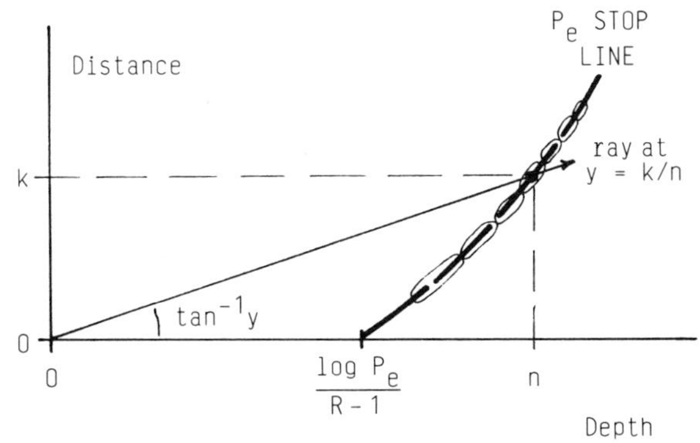


**Figure 6.3-3.** Illustration of a partition of the stop line error event and of the ratio parameter $y$ in eq. (6.3-17) and Lemmas 6.3-1 and 6.3-2. $y$ is a path's ratio of distance to the correct path $k$ to path length $n$.

$$P\{\text{Min ISS} \leq k_i \text{ at } n_i\}P\{\text{CP near } k_i \text{ at } n_i\}. \qquad (6.3\text{-}13)$$

Equating this to $P_e/Q$ and taking logs gives

$$\log P_e - \log Q = \log P\{\text{Min ISS } \leq k_i \text{ at } n_i\} + \log P\{\text{CP near } k_i \text{ at } n_i\}. \qquad (6.3\text{-}14)$$

We give the asymptotic form of these probabilities as two lemmas, which are proven in Appendix C. The first states an asymptotic property of random trees, while the second is an asymptotic property of the binomial distribution. Both express $k$ and $n$ in terms of their *ratio*, which is denoted $y$.

**Lemma 6.3-1.** For a random tree of rate $R$ and depth $n$ code word bits, the minimum distance satisfies

$$\frac{1}{n}\log P\{\text{Min ISS} \leq k \text{ at } n\} \to h_B(y) + R - 1, \qquad y = \frac{k}{n}, \qquad (6.3\text{-}15)$$

as $k$ and $n$ tend to infinity. Here $h_B(\ \ )$ is the binary entropy function and the logs are base-2.

**Lemma 6.3-2.** Over a BSC with crossover $p$, the correct path distance satisfies for $k/n > p$

$$\frac{1}{n}\log P\{\text{CP near } k \text{ at } n\} \to h_B(y) + y\log p + (1-y)\log(1-p), \qquad y = \frac{k}{n} \tag{6.3-16}$$

as $k$ and $n$ tend to infinity.

Although these formulas are exact only at large distance and depth, they do provide a good approximate equation for the stop line. We can replace the first term in (6.3-14) directly with (6.3-15). In replacing the second term, there are two cases: When the distance/depth ratio $k/n$ exceeds $p$ (6.3-16) applies, but otherwise $k$ is very likely to exceed $np$, and we instead substitute simply 1. The result is the approximate breadth-first stop line formula

$$\log_2 P_e \approx \begin{cases} n\Big[R - 1 + 2h_B(y) + y\log_2\Big(\frac{p}{(1-p)}\Big) + \log_2(1-p)\Big], & \text{if } y > p \\ n\big[R - 1 + h_B(y)\big], & \text{if } 0 \le y \le p, \end{cases} \tag{6.3-17}$$

where $y = k/n$. The symbol $\approx$ means the ratio of the two sides tends to one as $n$ grows. This holds no matter how large $Q$ is.

Eq. (6.3-17) relates $P_e$ to the distance/depth ratio and the stop depth. A ratio represents a ray from the origin in the distance - depth plot, and for each such ratio, the formula shows how deep the search must go to ensure the stop criterion. Figure 6.3-3 shows this geometry. At this point, we need to recall the logic of the stop line design: Each position $(n, k)$ should be as close in as possible so that the search stops as soon as it can, consistent with error probability $P_e$. Consequently, each term like (6.3-11) in a partition should be as large as it can be, but not larger. Little can be gained by apportioning the error more to some stop line regions than to others, since the line behaves like a chain that is as strong as its weakest link.

The stop line begins at depth $n = (\log P_e)/(R-1)$, which is the solution of (6.3-17) at ray $y = 0$. It ends when the line hits the drop line, that is, when the CP term in (6.3-14) equals $\log P_e$ and the ISS term is therefore 0. From (6.3-14), this occurs at $y = k/n = h_B^{-1}(1-R)$. It means that the ISS is almost certain to contain a path at depth $n$ lying closer than $k$ to the correct path, so that the maintenance of $P_e$ in (6.3-14) must come entirely from the CP term.

**Example 6.3-1: Breadth-First Stop and Drop Lines for the BSC.**

These lines were computed in [6-29] and appear in Figs. 6.3-4 to 6.3-6. Figure 6.3-4 shows how the lines depend on the criterion $P_e$ when the rate is 1/2 and the BSC crossover is .01; the growth of the search region is approximately uniform in $\log P_e$. In Fig. 6.3-5 the crossover grows while $R = 1/2$ and $P_e = 10^{-5}$. Now the region grows rapidly. In Fig. 6.3-6 the rate grows while $P_e$ and the crossover are fixed. The search region grows very rapidly with rate.

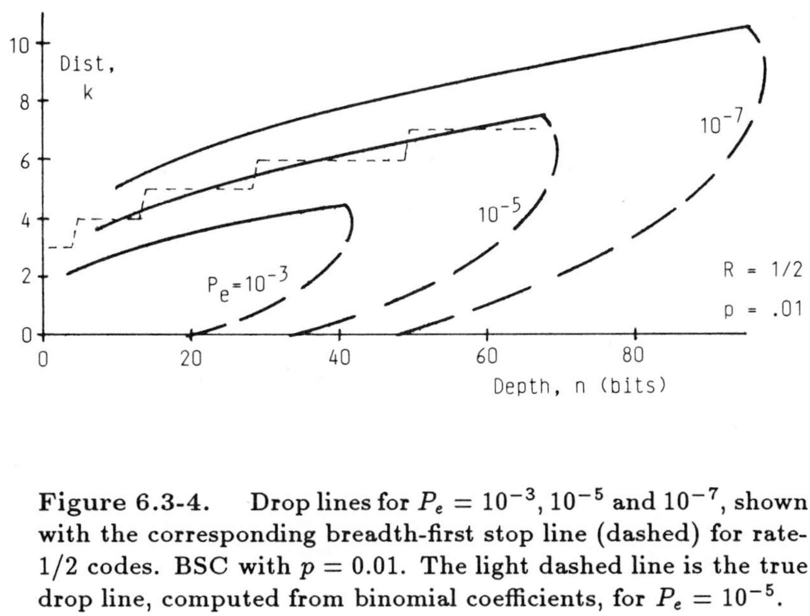


Figure 6.3-4. Drop lines for $P_e = 10^{-3}, 10^{-5}$ and $10^{-7}$, shown with the corresponding breadth-first stop line (dashed) for rate-1/2 codes. BSC with $p = 0.01$. The light dashed line is the true drop line, computed from binomial coefficients, for $P_e = 10^{-5}$.

The greatest depth reached by the stop line is both the decoding delay and the maximum length of paths in storage. Most of the time the decoder will stop at a distance/depth ratio near $p$ along the line, which is generally at a shorter depth, but we must allow for the worst case in a design. To do otherwise will create failures that raise the decoding error rate above $P_e$. A close analysis of (6.3-17) [6-29] shows that the greatest search depth at small $P_e$ satisfies

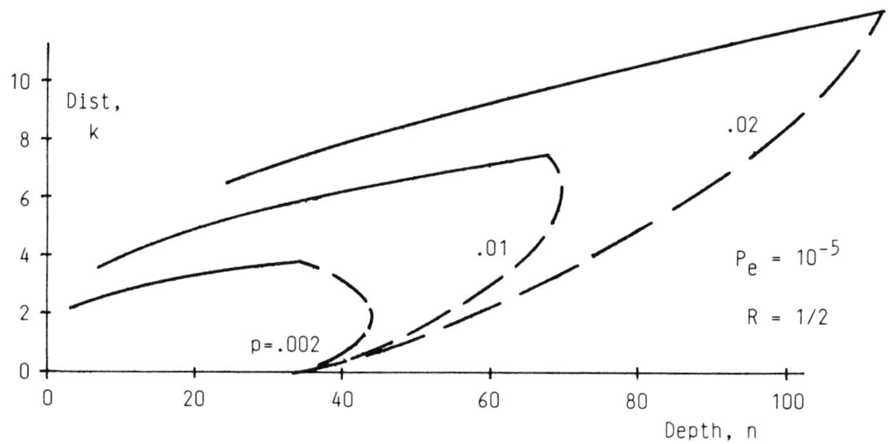


**Figure 6.3-5. Breadth-first stop (dashed) and drop lines for $P_e = 10^{-5}$ and rate 1/2. The BSC crossover $p$ is 0.002, 0.01 and 0.02.**

$$n = \begin{cases} \dfrac{\log_2 P_e}{R - 1 + 2h_B(y_m) + y_m \log_2\left(\dfrac{p}{1-p}\right) + \log_2(1-p)}, \\ \quad \text{for } 0 \le R < 1 - h_B(y_m); \\ \\ \dfrac{\log_2 P_e}{1 - R + h_B^{-1}(1-R)\log_2\left(\dfrac{p}{1-p}\right) + \log_2(1-p)}, \\ \quad \text{for } 1 - h_B(y_m) \le R < C, \end{cases} \tag{6.3-18}$$

where $y_m = 1/(1+\sqrt{(1-p)/p})$ and $C$ is the BSC capacity. An example of these two formulas appears on Fig. 6.3-6: The greatest depths at the lower rates 1/3 and 1/2 come from the first formula, while the rate 2/3 depth comes from the second.

These depth estimates are for a random tree code of average strength. It can be shown that the length of a random block code that achieves $P_e$ follows an identical relationship. (This is done by computing the *reliability exponent* for such codes [6-27], [6-3]). We see that the "strength" of tree codes, as measured by the length of code needed, is no worse than that of block codes.

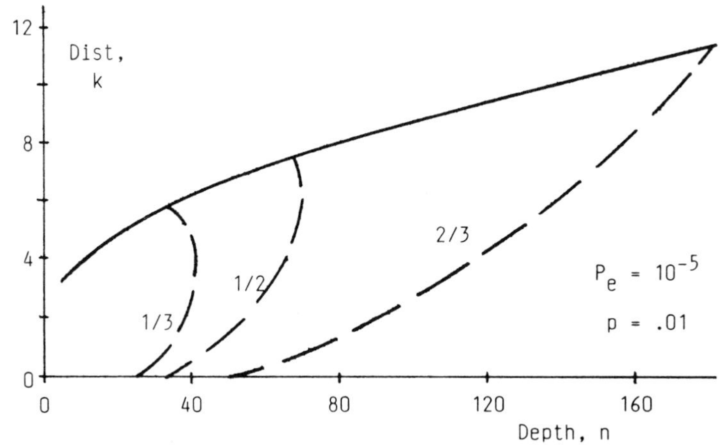


**Figure 6.3-6.** Breadth-first stop (dashed) and drop lines for $P_e = 10^{-5}$ and BSC crossover 0.01. Code rates are 1/3, 1/2 and 2/3.

## The Stop Line: Backtracking Searches

We have seen in Section 6.2 that backtracking tree searches, whether metric- or depth-first, search with reference to a metric limit. They search all paths below the limit, increment the limit and search again, until some path hits the stop line. For this kind of search it turns out that the decoder needs a new stop line for each limit. As the limit loosens (upward in Fig. 6.2-1), there will be a succession of stop lines, each inside the previous one. The tighter metric limits need stop lines further out because the search is less sure of the correct path – the path may be stuck temporarily at the metric limit – and it must observe correct and incorrect paths longer to be sure which is which.

To calculate the stop line in this new circumstance, we begin as before with (6.3-3) and proceed to (6.3-14). This time the term $P\{\text{CP near } k_i \text{ at } n_i\}$ is more complicated. Because the correct path may have hit the metric limit at an earlier depth, it may not appear at the stop depth, even if it is the least distant path there. There are now two causes of a stop-line decoder error, the correct path may be frozen at the metric limit or it may not be the least distant at the stop line. From the union bound of probability, we can overbound our term as

$$P\{\text{CP near } k_i \text{ at } n_i \text{ (with metric limit)}\} \leq P\{\text{CP hits metric limit}\} + P\{\text{CP near } k_i \text{ at } n_i \text{ (breadth first)}\}. \qquad (6.3\text{-}19)$$

Here the term labeled "breadth first" is identical to the one in (6.3-14) and Lemma 6.3-2.

One or the other of the terms in (6.3-19) dominates the value of $\log P\{\text{CP near } k_i \text{ at } n_i \text{ (with metric limit)}\}$ when $P_e$ is small. The second one is estimated by Lemma 6.3-2. The first is bounded by (6.3-5), in which $-\zeta$ is the value of the metric limit. The following procedure gives the stop line for metric limit $-\zeta$.

i) *The Lower Segment.* Initially, the second term in (6.3-19) dominates, and the stop line is the breadth-first one of (6.3-17). In this part of the stop line, distances are small and the correct path is unlikely to have hit the metric limit. The lower segment grows as the metric limit loosens (becomes more negative).
ii) *The Upper Segment.* Now the first term in (6.3-19) dominates, with the result that the stop line satisfies

$$\log_2 P_e = n\big[h_B(y) + R - 1\big] + h\zeta, \qquad y = \frac{k}{n}. \qquad (6.3\text{-}20)$$

As the correct path wanders higher, it is more likely to hit the metric limit than survive until the stop line and cause an error there.

Figure 6.3-7 revisits Example 6.3-1 and shows some backtracking stop lines for the BSC at crossover .01, rate 1/2, and $P_e = 10^{-5}$. The Fano metric limits to the search appear as lighter lines and there is a different stop line for each limit. Several general facts about backtracking stop lines are clear in this figure. The successive lines peel off of the breadth-first stop line. The decoder design must allow for longer paths, about twice as long at rate-1/2 (this diminishes somewhat at higher rates). Finally, the greatest depth of search occurs at the initial metric limit, 0; further analysis [6-29] shows that this is

$$n = \frac{\log_2 P_e}{h_B(y_0) + R - 1}, \qquad (6.3\text{-}21)$$

where

$$y_0 = \frac{R - 1 - \log_2(1-p)}{\log_2\left(\dfrac{p}{1-p}\right)}.$$

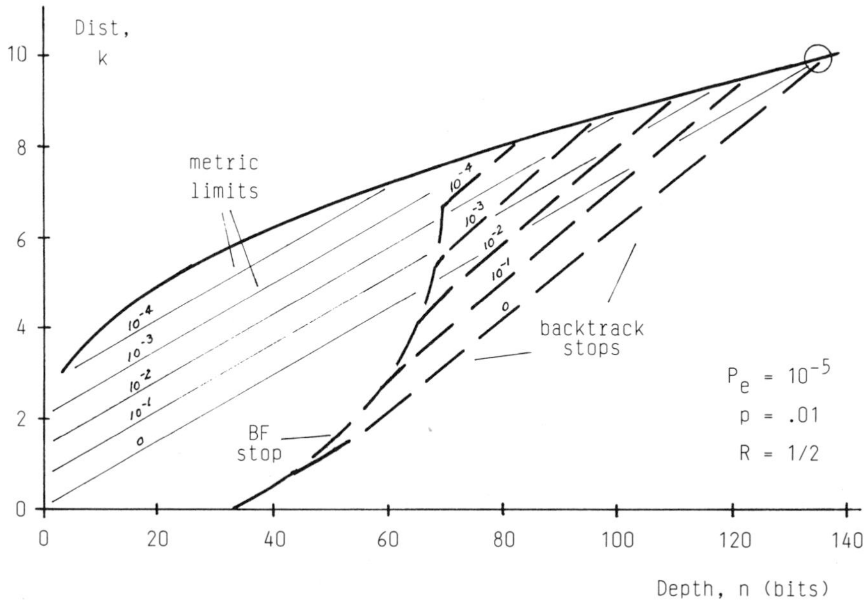


**Figure 6.3-7.** Stop lines for a backtracking search (dashed) for $P_e = 10^{-5}$, rate-1/2 and crossover 0.01. A stop line exists for each metric limit; metric limits appear as lighter lines. The drop line is the same as the breadth-first one. The greatest search depth is circled.

# 6.4 NUMBER OF PATHS SEARCHED

It remains to count the number of paths that an algorithm views within its search region. To a first approximation, algorithms fall into two categories, breadth-first and backtracking. Each has its own estimate. As we justified in Section 6.2, the metric-first and depth-first modes, in the present state of the art, search essentially the same number of incorrect subset paths. We will assume that these backtracking searches employ the Fano metric (6.1-4). We will find that breadth-first searches view many more paths on the average than do backtracking searches as the channel worsens or the rate grows. But both have the same peak number.

As in the previous section, the encoder here uses a random code, and the path estimates are expectations over both the choice of the code and the outcome of the channel. We count only ISS paths, because in the long run only this many distinct paths are viewed per output symbol. A good estimate of the total in storage at one time is this number times $b$, the tree's branching factor. The section concludes with some calculations of the ISS paths in BSC decoding.

The original calculations of path numbers appeared in the 1960's in the work of Jacobs, Berlekamp, Savage, Gallager, Jelinek and Forney (see refs. [6-19] – [6-21], [6-27], and [6-35]). Although these calculations were for a different kind of decoder than we now use, some of the results reappear here and in Section 6.5. Other results here are new. In addition, we use a simpler analytical method based on difference equations. These equations were first used in sequential decoding by Zigangirov [6-14].

### Breadth-First Searches

The idea of the breadth-first calculation is to estimate the number of paths that lie below a drop line tangent. This overbounds the number that lie below the drop line itself. Every tangent provides its own bound and what we seek is the least of these bounds. It may seem that a mere tangent provides a weak bound to the paths under the actual drop line, but the path numbers in fact concentrate intensely in just a small part of the search region, especially when $P_e$ is small, and we need only choose the tangent that most tightly encloses the concentration. This we do by optimizing over the tangents.

The calculation takes place in the scaled metric vs. depth plane of Fig. 6.3-1. The equivalent activities in distance vs. depth also appear in the figure. Consider a horizontal line in the metric vs. depth plane that lies at $z = -\zeta$. This line maps to a tangent in the distance - depth plane whose slope is (6.3-9); the bias $B$ in that formula is the solution of (6.3-7), (6.3-6) and (6.3-4) for the assumed $\zeta$.

In the metric-depth plane, define $EN_\zeta(z)$ to be the expected number of ISS paths that hit the $-\zeta$ line in front of a node with metric $z$, before all hit metric $A$. Figure 6.4-1 explains this definition. The descendants counted are a single branch out of the node plus all the descendants in front of the branch that hit the line. Path extension stops when paths hit either metric $-\zeta$ or $A$. $N_\zeta(z)$ is four in the figure.

We can solve easily for $EN_\zeta(z)$ using a difference equation method that was pioneered by Zigangirov [6-14]. $EN_\zeta(z)$ satisfies

$$EN_\zeta(z) = b \sum_\mu P(\mu) EN_\zeta(z + \mu) \tag{6.4-1}$$

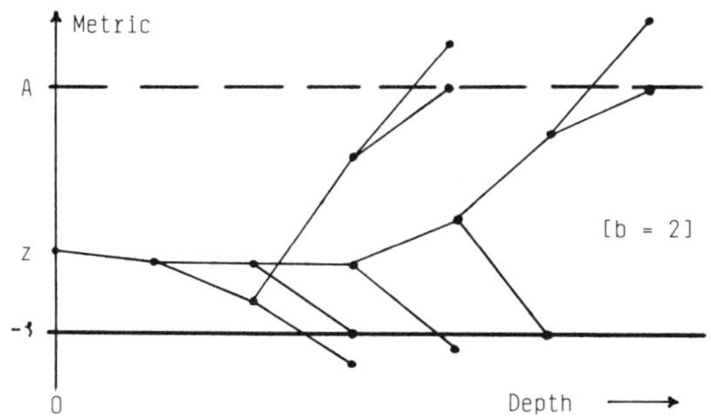


**Figure 6.4-1.** Example of path counting in definition of $EN_\zeta(z)$. Four paths hit $-\zeta$ before all have hit either $-\zeta$ or $A$. Note that only a single branch stems from the initial node.

Here $\mu$, the metric increment on the branch in front of the original node, takes the values $as, as-1, as-2, ..., as-a$, in which $s$ is $\alpha/(\alpha-\beta)$, the value of the slope (6.3-9). The metric increment takes these values at random because the code bits are random, and its probabilities are the binomial distribution

$$P(\mu) = \binom{a}{j}\left(\frac{1}{2}\right)^a, \qquad \text{for } \mu = as - j. \tag{6.4-2}$$

Figure 6.4-2 shows why (6.4-1) holds. The paths that will hit $-\zeta$ in front of the original node are the same as those that will hit in front of the $b$ branches out of the next node. Because of the random code, the path numbers distribute uniformly in front of each of these $b$. Each of the expectations is $EN_\zeta(z+\mu)$ because metric $\mu$ has been added to the original node metric $z$. If the next node has metric $-\zeta$ or less, it has hit the $-\zeta$ line, and we need to count one hit. If the node has metric $A$ or more, there will be no further extension, and $N_\zeta$ is zero. We can summarize these thoughts in the boundary conditions

$$EN_\zeta(z) = \begin{cases} \frac{1}{b}, & \text{if } -\zeta + as - a \le z \le -\zeta \\ 0, & \text{if } A \le z \le A + as. \end{cases} \tag{6.4-3}$$

The boundary $A > 0$ here is artificial and we will eliminate it soon by making $A$ arbitrarily large. In that case there is really only one barrier to path extension, the metric horizontal at $-\zeta$.

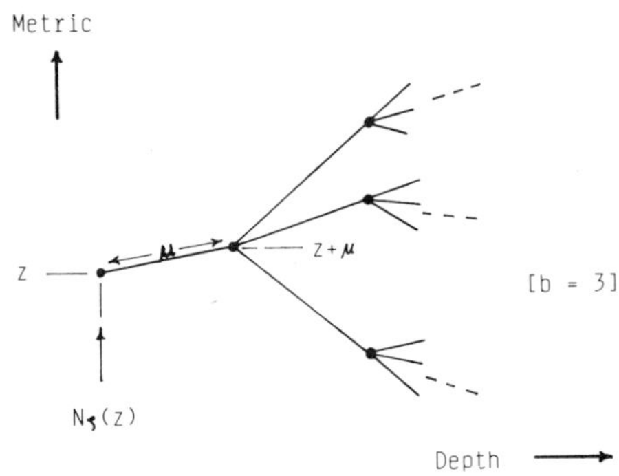


**Figure 6.4-2.** Illustration of the path number difference equation (6.4-1). $\mu$ is the metric increment on the first branch.

The solution of a linear difference equation like (6.4-1) follows standard steps. First, we find exponential solutions to (6.4-1) of the form $K\sigma^z$. $\sigma$ must satisfy the *characteristic equation*

$$1 + \frac{1}{\sigma} = 2^{1-R}\sigma^{-s}, \tag{6.4-4}$$

an equation that results from substituting $EN_\zeta(z) = K\sigma^z$ into (6.4-1). Eq. (6.4-4) has two real roots when $0 < s < h_B^{-1}(1-R)$. Second, we find a combination of these exponential solutions $K_1(\sigma_1)^z + K_2(\sigma_2)^z$ that overbounds $EN_\zeta(z)$ in its boundary regions. It can be shown that a solution with this property overbounds $EN_\zeta(z)$ throughout its domain; that is, a solution to (6.4-1) that overbounds (6.4-3) also overbounds the solution to (6.4-1) that satisfies (6.4-3). Figure 6.4-3 shows such a bound compared to the true $EN_\zeta(z)$.

The last step is to find the exponential solutions that comprise the tightest bound. This will be the combination that is tightest in the boundary regions. Close examination shows that only the left boundary in Fig. 6.4-3 matters and that the tightest solution is

$$\sigma^z\left(\frac{\sigma^{\zeta-as+a}}{b}\right), \tag{6.4-5}$$

where $\sigma$ is the smallest real solution to (6.4-4).

Now we have the desired object, a bound to the number of paths that hit a metric horizontal in front of a node. We turn to the problem at

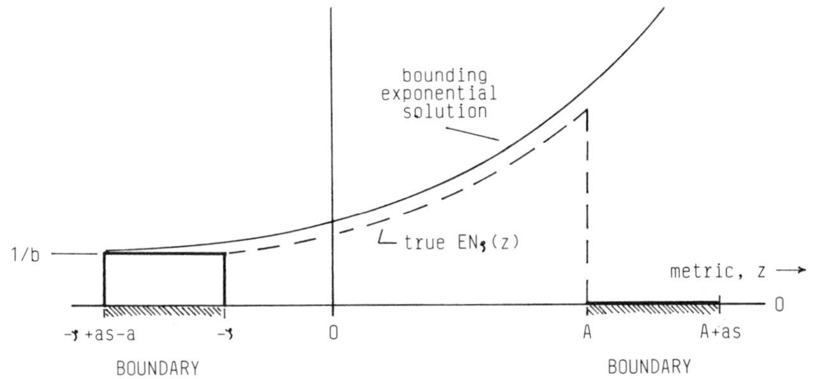


**Figure 6.4-3.** Two solutions to difference equation (6.4-1). The true solution satisfies the boundary values exactly. The other solution overbounds the boundaries, and consequently the true solution.

hand: The descendants in an ISS of the root node that hit a drop line tangent. $\zeta$ defines the tangent and $EN_\zeta(0)$ is the expected number in one subset. If we ignore constant factors, the estimate of the expectation is

$$EN_\zeta(0) \approx \sigma^\zeta \tag{6.4-6}$$

per subset. The symbol $\approx$ means that the ratio of the logs of the two sides tends to one as $\zeta$ grows.

From the drop line derivation in Section 6.3 we have the constraint that $\zeta = (\log P_e)/h$ [eq. (6.3-7)]. In terms of the decoder error probability, then, breadth-first searches explore about

$$\mathcal{E}[\text{ISS paths}] \approx \sigma^{(\log P_e)/h} \tag{6.4-7}$$

per subset. Every tangent provides its own bound (6.4-7) and we want the smallest one; put another way, we want the largest value of $\sigma^{1/h}$. Some calculus shows [6-29] that the optimum occurs when the bias $B = R$ and

$$\sigma = \left[\frac{1-p}{p}\right]^{1+h'}; \tag{6.4-8}$$

the optimum itself is

$$[\sigma^{1/h}]_{opt} = 2\left[\frac{1-p}{p}\right]^{1/h}, \tag{6.4-9}$$

where $h$ satisfies (6.3-6) with $B = R$.

## Backtracking Searches

When the search may backtrack, the search region varies, as we discussed in Section 6.1. An estimate of path numbers is now truly an average, in the sense that the actual number in a particular search varies widely. Whether in the metric- or depth-first mode, the search explores paths below a metric limit in the distance - depth plane, set at $\zeta$ in the right-hand picture of Fig. 6.3-1. This limit is precisely the tangent referred to in the breadth-first analysis. While the limit is at the present position, eq. (6.4-6) gives the average ISS paths searched if the Fano metric has bias $B$. As the limit loosens, the ISS path estimate becomes (6.4-6) at the looser positions. Our approach now will be to average (6.4-6) over the position of the loosest limit that occurs during the search.

By (6.3-5) the correct path hits the limit set at $\zeta$ (i.e., falls to metric $-\zeta$) with probability less than $2^{h\zeta}$, and about $\sigma^{\zeta}$ ISS paths lie below the limit. So the overall ISS paths are bounded by

$$\mathcal{E}\ [\text{ISS paths}] \lessapprox \int 2^{h\zeta}\sigma^{\zeta}\, d\zeta, \tag{6.4-10}$$

where the integral ranges from 0 up to $(\log P_e)/h$, the loosest metric limit allowed. Here we have replaced the probability that the limit is near $\zeta$ with a bound, (6.3-5).

Carrying out the integration in (6.4-10) gives

$$\mathcal{E}[\text{ISS paths}] \le \frac{-1}{h\left(\ln 2 + \dfrac{\ln\sigma}{h}\right)}\left[1 - e^{\left(\ln 2 + \frac{\ln\sigma}{h}\right)(\log P_e)}\right]. \tag{6.4-11}$$

As in the breadth-first calculation, $h$ and $\sigma$ here depend on the rate and the Fano bias through (6.3-6), (6.3-9) and (6.4-4).

Eq. (6.4-11) is simpler than it looks. Depending on the code rate and the BSC crossover it reduces to one of two forms. The transition between the forms occurs when

$$\ln 2 + \frac{\ln\sigma}{h} = 0. \tag{6.4-12}$$

When the left side of (6.4-12) is positive, (6.4-11) is small and is close to $-1/(h \ln 2 + \ln \sigma)$. Otherwise, (6.4-11) grows rapidly with $\log_2 P_e$ and has the form

$$\left(\frac{1}{\ln \sigma}\right) P_e [\sigma^{1/h}]^{\log P_e}. \tag{6.4-13}$$

In this last expression we see, again, the quantity $\sigma^{1/h}$ that appeared in the breadth-first analysis and we know how to minimize it: Set the search bias equal the rate. For the backtracking case, apparently, when the search gets tough we should have $B = R$.

When $B = R$, it can be shown that (6.4-12) reduces to the equation $R = R_0(p)$, where $R_0(p)$, called the *computational cut-off rate*, is defined by

$$R_0(p) = 1 - \log_2(1 + 2\sqrt{p(1-p)}). \tag{6.4-14}$$

When the code rate falls below this function of the crossover, the average path numbers are small; otherwise they grow rapidly in the form (6.4-13).

We can summarize as follows: Backtracking searches satisfy

$$\mathcal{E}[\text{ISS paths}] \approx \begin{cases} \dfrac{-1}{(h \ \ln 2 + \ln \sigma)}, & \text{when } 0 \leq R < R_0(p) \\ P_e \sigma^{(\log P_e)/h}, & \text{when } R_0(p) \leq R < C. \end{cases} \tag{6.4-15}$$

Here

$$\sigma = \left[\frac{1-p}{p}\right]^{1+h'}, \quad h' = \frac{h}{\log_2\left(\dfrac{1-p}{p}\right)}, \tag{6.4-8}$$

$$\sigma^{1/h} = 2\left[\frac{1-p}{p}\right]^{1/h} \tag{6.4-9}$$

and $h'$ solves

$$1 + \frac{1}{h'} \log_2 \left[(1-p)^{1+h'} + p^{1+h'}\right] = R.$$

There is a wealth of information here about the metric- and depth-first modes. At low rates the number of paths searched *does not depend on the decoder error probability.* Calculation shows that it is less than unity and does not depend very much on the code rate, as long as $R$ is less than $R_0(p)$. At high rates the number depends strongly on $P_e$ and its expectation is *less than breadth-first searching by a factor* $P_e$. Even so, all search modes have the same peak storage, as we observed in Section 6.2. As the rate nears capacity path numbers grow dramatically, because $h \rightarrow 0$.

The cut-off rate, $R_0$, has some other interesting properties that are not apparent here, and we will return to these in Section 6.5.

## Numerical Comparison of Search Modes

Figure 6.4-4 compares the number of paths searched per decoder output symbol for backtracking and breadth-first searches and for the $t$-criterion decoder that we will discuss in Section 6.5. The figure plots expected path numbers against crossover probability for the three rates 1/3, 1/2 and 2/3. The formulas used are the random coding overbounds (6.4-11), (6.4-7) and (6.5-1). The cut-off rate transition appears as the $p$ at which $R_0(p)$ equals these rates. All the decoders achieve the same error probability, $P_e = 10^{-5}$, the same one for the search regions in Figs. 6.3-5 to 6.3-7.

The superiority of backtracking is obvious from Fig. 6.4-4. Below cut-off rate, the expected paths are less than unity, and the expected search always falls far below the breadth-first number. All of these numbers, however, are expectations and they hide the fact that backtracking searches vary widely from output to output, while breadth-first and $t$-criterion searches vary hardly at all. An expectation below unity in a backtracking search, for instance, means that most of the time the incorrect subsets are not searched at all but sometimes extensive searches occur.

Figure 6.4-5 plots the backtracking expectation against rate, this time for a set of fixed crossovers. The breadth-first searches are larger by $1/P_e$, roughly speaking, and $P_e$ is again $10^{-5}$.

The estimates in this section all apply to random codes, but there is little evidence that practical convolutional codes behave differently. We will do more comparing to practice in Section 6.5 when we have built up more analytical evidence. In the meantime, we can revisit Example 6.2-1, in which the breadth-first M-algorithm decoded a convolutional code. There was no drop line in this example – the decoder simply drops all paths in excess of M – but the breadth-first estimate (6.4-7) at least roughly predicts the size of a decoder that achieves $P_e$. From Fig. 6.2-3, the decoder achieves $P_e = 10^{-5}$ with about 5 paths at $E_b/N_o = 7.8$ dB, about 10 paths at 6.7 dB, and about 20 paths at 6.3 dB; these SNR's correspond to $p = .007$, .015, and .02 in the BSC model of the example's

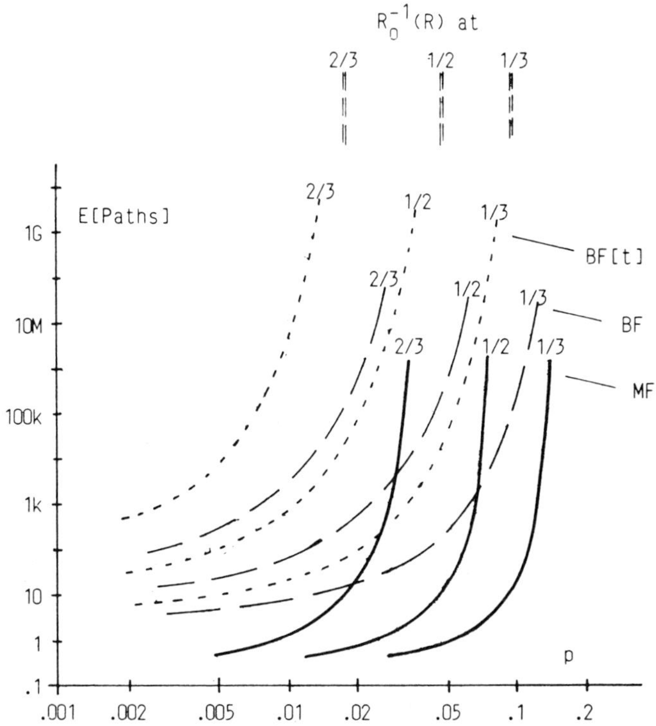


**Figure 6.4-4** Expected paths searched per incorrect subset versus BSC crossover $p$, for rates 1/3, 1/2 and 2/3. The search modes are backtracking [MF, solid line], breadth-first [BF, long dashes] and breadth-first by the $t$-criterion [BF($t$), short dashes]. The $t$-criterion is discussed in Section 6.5. Note log scales. $P_e = 10^{-5}$.

Gaussian channel. For poorer channels the path numbers rapidly grow. Evaluation of (6.4-7) yields perhaps 10 times these numbers, but we can attribute this to the overbounds that led to (6.4-7). As the channel worsens relative to the rate, the logs of the two measurements will grow closer. Our aim in this section has been to display the essential character of the search modes, and in this regard (6.4-7) and (6.4-15) are accurate.

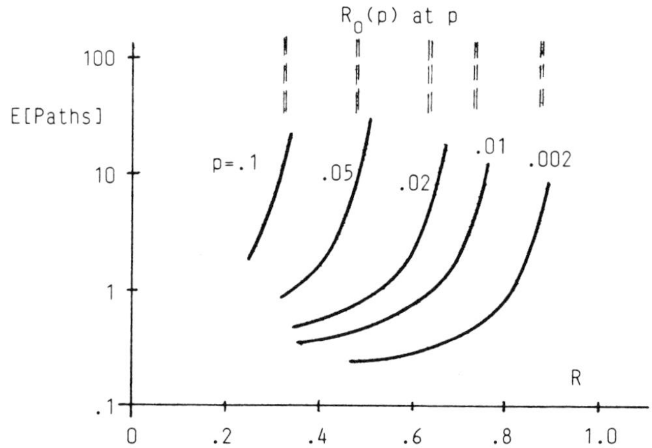


Figure 6.4-5 Expected paths searched per incorrect subset versus code rate $R$, for several BSC crossovers. Backtracking searches. $P_e = 10^{-5}$.

# 6.5 OTHER ANALYSES OF SEQUENTIAL DECODING

We will explore two other kinds of sequential decoding. The first corrects a specified number of channel errors, instead of maintaining a decoder error probability. The error mechanism in the channel need not be defined. The second kind works again with a defined probabilistic channel like the BSC, and decodes perfectly, without error. These decoders sometimes demand unreasonable resources and they completely fail in a band of rates below capacity.

## t-Error Correcting Sequential Decoders

What if nothing is known about the error mechanism in a channel? Certainly, we cannot compute a decoder error probability and so the results of Sections 6.3 and 6.4 are of little use. But it is still possible to design a decoder that corrects a set *number* of errors. Figure 6.5-1 shows the search region of a decoder that is guaranteed to correct every combination of $t$ or fewer errors occurring within a decision depth $L_D$. Since the vertical axis is distance to the received path, the correct path must lie in the search region if it has $t$ or fewer errors. If no ISS path lies this close, then correct subset paths alone eventually lie in the search region, no matter how the search proceeds. In the figure, the correct path has fewer than $t$ errors and it alone reaches the decision depth.

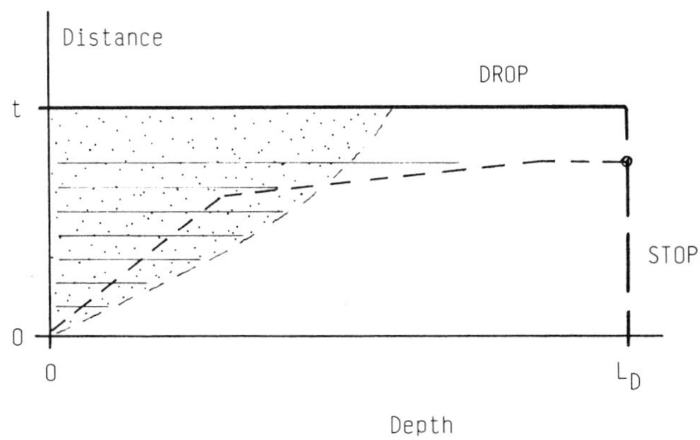


**Figure 6.5-1.** A $t$-criterion code search; incorrect subset stippled and correct path dashed. A breadth-first mode must search the entire area below the drop line. The horizontals depict a layered backtracking search. Compare to Fig. 6.2-1.

To analyze these decoders further, suppose the channel is a binary one with Hamming distance, but we do not know its error mechanism. The code is a random one. We have two aims: (i) Estimate the paths searched by a first-symbol decoder per subset, and (ii) Estimate $L_D$, the decision depth at which the search can stop. The decoders must satisfy a new stop criterion, the *t-criterion*, that up to $t$ errors are to be corrected.

A breadth-first search cannot backtrack and therefore must search the entire region in Fig. 6.5-1, if it is to satisfy this $t$-criterion. A backtracking search can explore the more likely parts first and return to other parts as needed. Although we know little about the channel, it is a good guess that fewer errors are more probable than more errors. A backtracking decoder might thus search in layers like the horizontal lines in the figure, and stop as soon as some path hits the decision depth.

Since the ISS paths in the search region grow rapidly with $t$, layered backtracking will be much more efficient on the average than a breadth-first search, although its peak storage will be the same.

The number of ISS paths $N_t$ in a random code that hit the $t$ line in Fig. 6.5-1 has been analyzed in ref. [6-28]. As $t$ grows, it satisfies

$$EN_t \approx \left(2^{1-R} - 1\right)^{-t}; \tag{6.5-1}$$

as usual, $\approx$ here means that the ratio of the logs of the two sides converges to 1 as $t$ grows. As with the path counting in Section 6.4, (6.5-1) follows from a simple difference equation. It is

$$EN_t(y) = b \sum_d P(d) EN_t(y + d), \qquad 0 < y < t, \tag{6.5-2}$$

which is the same formula as (6.4-1), but this time $d$ is Hamming distance, not metric, and $EN_t(y)$ is the expected number of paths to hit distance $t$ in front of a node of distance $y$. Because of the random code, the distance $d$ on a branch takes values $0, 1, \ldots, a$, with the binomial probabilities $\binom{a}{d} 2^{-a}$. The boundary conditions are $EN_t(y) = 1/b$ on the right boundary and 0 on the left. Following steps like those after (6.4-1) gives (6.5-1).

Table 6.5-1 lists the storage growth rates (6.5-1) for several code rates. Particularly at high rates, storage grows rapidly with $t$. This shows the advantage of layered backtracking. A closer look at one rate, $R = 1/2$, appears in Fig. 6.5-2. The storage estimate (6.5-1) is compared there to the actual requirement of the best $d_{\text{free}}$ and ODP codes in Section 4.1. These concrete convolutional codes lie in a tight range that closely follows the trend of (6.5-1).

It remains to estimate the decision depth. This will be the maximum length of paths in storage as well as the coding delay. There are two ways to perform the estimate. For binary random tree codes [6-28] (and for convolutional codes with randomly selected generators[6-30]) it is known that the Hamming distance between the received path and the closest ISS path grows at the rate of $h_B^{-1}(1 - R)$ Hamming units per code word bit. To guarantee correction of all $t$-error patterns, we require distance $2t + 1$, and so the decision depth in tree levels is

$$L_D \approx \frac{2t}{a h_B^{-1}(1 - R)} \quad \text{levels.} \tag{6.5-3}$$

This is $L_D a$ code word bits. $L_D$ is listed in Table 6.5-1 in this asymptotic form.

**Table 6.5-1.** THEORETICAL DECISION DEPTH AND STORAGE SIZE FOR $t$-ERROR CORRECTING CODES OF RATE $R$. THE CODE RATE IS $(\log_2 b)/a$. SOME MEASURED TRENDS FOR BEST $d_{\text{free}}$ AND ODP CODES ARE SHOWN. * CODE MEMORY MUST BE SUFFICIENT TO GIVE FREE DISTANCE ABOVE $2t$; DECISION DEPTH LIES ABOVE THE TREND AT SMALL $t$.

| Rate, R | (a,b) | Decision Depth, $L_D(2t)$, levels eq. (6.5-3) | Measured trend, levels* | Storage growth $(2^{1-R} - 1)^{-t}$ |
|---|---|---|---|---|
| 1/4 | 4,2 | 2.33t | 2.3t - 6 | $(1.47)^t$ |
| 1/3 | 3,2 | 3.83t | 3.8t - 10 | $(1.70)^t$ |
| 1/2 | 2,2 | 9.09t | 9.1t - 21 | $(2.41)^t$ |
| 2/3 | 3,4 | 10.8t | 11t - 16 | $(3.85)^t$ |
| 3/4 | 4,8 | 12.0t | 12t - 14 | $(5.89)^t$ |
| 7/8 | 8,128 | 14.6t | —— | $(11.05)^t$ |

For concrete codes we can be more exact. For a given convolutional code the required decision depth for a code with $d_{\text{free}}$ was discussed in Section 4.2. The concept there needs only a little generalization now. Instead of measuring the code depth when all ISS paths exceed $d_{\text{free}}$, we should measure the depth when all exceed $2t$, since the code is to correct only $t$ errors. Such depths have been tabulated for most codes [6-31].

**Example 6.5-1: $t$-Criterion Decoding of Rate 1/2 Convolutional Codes.**

> As an example of path numbers and lengths for concrete codes, Table 6.5-2 lists these for the best $d_{\text{free}}$ codes of rate 1/2 and a breadth-first search. By reading down columns one can see that there is little variation among codes asked to correct $t$ errors; *only the size of the correction matters*, not the code. Numbers for other codes appear in ref. [6-28].

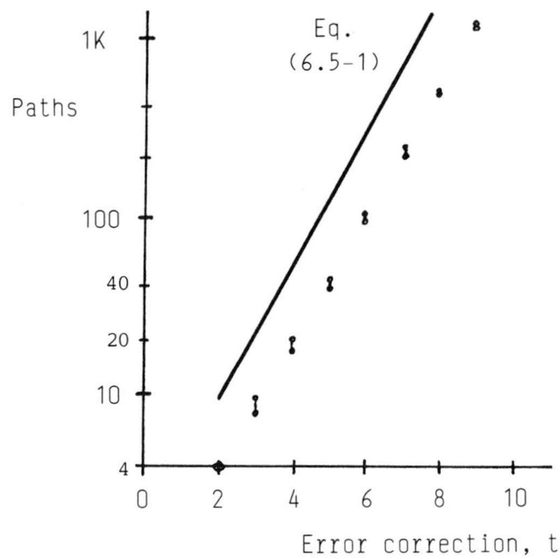


**Figure 6.5-2.** Estimates to the $M$-algorithm storage versus error correction $t$ at rate 1/2. The bound (6.5-1) is compared to the actual worst-case requirements of the best $d_{\text{free}}$ and the ODP codes.

## Comparisons to Other Decoders

If we have probabilistic knowledge of the channel, we can associate an error probability with a $t$-error correcting decoder, or put another way, we can choose a $t$ to meet a specification on $P_e$. This gives a $P_e$-criterion decoder. Its stop line will be similar to the breadth-first one in Section 6.3, but its drop line will be the horizontal in Fig. 6.5-1. The number of ISS paths this decoder searches is compared to the $P_e$-criterion decoders in Fig. 6.4-4. It is much worse. This demonstrates the great value of channel probabilities in the decoder design.

An even less subtle decoder than the $t$-error one is the Viterbi algorithm. The size of the Viterbi decoder that corrects all $t$-error combinations is the state space size of the smallest code with $d_{\text{free}} > 2t$. We can estimate this size from known codes and from tight bounds on free distance that exist; at rates 1/3 and 1/2, for instance, the state space grows as $(2.52)^t$ and $4^t$, respectively. On the other hand, Table 6.5-1 shows that breadth-first $t$-criterion decoders need to grow no faster than $(1.702)^t$ and $(2.414)^t$. This shows that Viterbi decoders are larger in size by $(1.48)^t$ and $(1.66)^t$. An actual comparison for rate 1/2 appears in Table

**Table 6.5-2** Decision Depth [in Brackets] and Storage Size Versus Error Correction $t$ for the Rate 1/2 Best $d_{\text{free}}$ Codes. $d_f$ Is the Code Free Distance. Storage Size Shown Is for the Worst Case of $t$ Errors. The Depth and Size Follow Eqs. (6.5-3) and (6.5-1).

| m | Gens. | $d_f$ | V. Alg. States | Storage Size[Depth] at t = 2 | 3 | 4 | 5 | 6 | 7 | 8 | 9 |
|---|---|---|---|---|---|---|---|---|---|---|---|
| 2 | 5 = $g^1$ | 5 | 4 | 4 | | | | | | | |
| | 7 = $g^2$ | | | [6] | | | | | | | |
| 3 | 64 | 6 | 8 | 5 | | | | | | | |
| | 74 | | | [7] | | | | | | | |
| 4 | 46 | 7 | 16 | 4 | 9 | | | | | | |
| | 72 | | | [9] | [13] | | | | | | |
| 5 | 65 | 8 | 32 | 4 | 8 | | | | | | |
| | 57 | | | [9] | [14] | | | | | | |
| 6 | 554 | 10 | 64 | 4 | 8 | 20 | | | | | |
| | 744 | | | [8] | [16] | [20] | | | | | |
| 7 | 712 | 10 | 128 | 4 | 9 | 21 | | | | | |
| | 476 | | | [8] | [16] | [22] | | | | | |
| 8 | 561 | 12 | 256 | 4 | 8 | 18 | 42 | | | | |
| | 753 | | | [6] | [12] | [22] | [31] | | | | |
| 9 | 4734 | 12 | 512 | 4 | 10 | 18 | 43 | | | | |
| | 6624 | | | [7] | [18] | [23] | [31] | | | | |
| 10 | 4672 | 14 | 1024 | 4 | 8 | 19 | 42 | 93 | | | |
| | 7542 | | | [6] | [15] | [21] | [30] | [36] | | | |
| 11 | 4335 | 15 | 2048 | 5 | 10 | 20 | 46 | 103 | 236 | | |
| | 5723 | | | [7] | [15] | [29] | [35] | [42] | [46] | | |
| 12 | 42554 | 16 | 4096 | | 9 | 21 | 44 | 99 | 224 | | |
| | 77304 | | | | [13] | [22] | [32] | [40] | [47] | | |
| 13 | 43572 | 16 | 8192 | | 9 | 21 | 46 | 105 | 239 | | |
| | 56246 | | | | [19] | [24] | [32] | [40] | [46] | | |
| 14 | 56721 | 18 | 16384 | | 9 | 19 | 40 | 92 | 215 | 494 | |
| | 61713 | | | | [11] | [19] | [25] | [33] | [43] | [52] | |
| 15 | 447254 | 19 | 32768 | | | 19 | 41 | 93 | 220 | 500 | 1172 |
| | 627324 | | | | | [19] | [26] | [35] | [43] | [50] | [59] |
| 16 | 716502 | 20 | 65536 | | | 18 | 42 | 93 | 214 | 491 | 1144 |
| | 514576 | | | | | [20] | [26] | [35] | [42] | [52] | [58] |

6.5-2, since the codes there are the smallest that exist at each $d_{\text{free}}$. For example, the smallest Viterbi decoder that corrects 8 errors has 16384 states and views 32768 paths, but the table gives a code that retains only 491 paths for 8 errors.

Another way to compare $P_e$- and $t$-criterion decoders and the Viterbi algorithm appears in Table 6.5-3. Suppose that a decoder design is "useful" if its average paths searched lies in the range 10 – 100 per decoded symbol. The table assumes that the channel is known to be a

BSC with crossover $p$ and it estimates path numbers to be the bounds in Section 6.4 divided by 10. It then gives the range of $p$ over which the different decoder designs are useful. All the decoders achieve an error rate near $10^{-5}$. For very good channels the Viterbi algorithm is acceptable and the more sophisticated schemes are simply a waste of software talent. As the channel grows worse, a $P_e$-criterion design is required, but breadth-first searching is enough. Eventually, decoders stay within the useful range only by adding backtracking to their design.

**Table 6.5-3.** APPROXIMATE USEFUL RANGE OF CHANNEL $p$ FOR BSC DECODERS THAT ACHIEVE ERROR PROBABILITY $10^{-5}$. A "USEFUL" DECODER IS ONE THAT SEARCHES AN AVERAGE OF 10 – 100 PATHS PER DECODED SYMBOL. PATH NUMBERS ARE TAKEN AS THE RANDOM CODING ESTIMATES OF SECTION 6.4 DIVIDED BY 10.

Range of crossover p in BSC

| | Rate 1/3 | Rate 1/2 | Rate 2/3 |
|---|---|---|---|
| Viterbi Alg. or Breadth-first / t-criterion | 0 - .03 | 0 - .015 | 0 - .002 |
| Breadth-first / $P_e$ -criterion | .03 - .06 | .015 - .03 | .002 - .008 |
| Backtracking | .06 - .12 | .03 - .07 | .008 - .03 |
| p when C(p) = R | .174 | .110 | .061 |

## With-Probability-One Decoders

Another criterion for stopping a decoder is that it never stops, at least in theory. The search overcomes any run of errors and decodes the first symbol correctly with probability one. Is this a fantasy? In one sense, it is not: Decoders can be designed whose ISS paths have finite expectation even though they have zero probability of mistaking the first symbol. Yet it is only common sense that a decoder of finite size will fail sooner or later. We will call the design philosophy here – that the decoder aims at error-free decoding even though it will occasionally fail – the *with-probability-one* stop criterion.

It will turn out that these decoders have finite expected ISS paths so long as the code rate falls below $R_0(p)$, the computational cut-off rate defined in (6.4-14). Although $R_0$ there was defined for the BSC, the coupling between $R_0$ and finite path numbers holds true for a wide range of channels [6-3], [6-27], [6-32]. The actual path number to decode a symbol varies widely, however, and exceeds any limit with some probability. Such an excursion beyond the size of a decoder is called an *erasure*. A decoding error is the usual result. The behavior here and the philosophy of design should be compared to that in the previous sections. There it is assumed that decoding errors occur and the design controls their frequency. The search region is fixed and although the path numbers vary in some decoders, they have a fixed limit. In a probability-one decoder, the design fixes no limits. Error occurs because the decoder does not quite fulfill the design.

With the tools in Section 6.4, we can estimate the ISS path numbers for a probability-one decoder. These have a *Pareto distribution*

$$Pr\{N \geq \eta\} = K_0 \eta^{-\rho}, \qquad 0 < \rho < \infty \tag{6.5-4}$$

where the parameter $\rho$ is the Pareto exponent. As distributions go, this is an unpleasant one because it says that the probability of large path numbers decays more slowly than exponentially. The $i$-th moment of a Pareto variable,

$$\mathcal{E}[N^i] = \frac{K_0 \rho}{(i-\rho)}, \qquad \rho > i, \tag{6.5-5}$$

exists only if $\rho > i$. In Appendix C, we show the Pareto character of the ISS path numbers as given by (6.5-4).

As for $\mathcal{E}N$ itself, it is (C-6), which is

$$\mathcal{E}N \leq \frac{-1}{h(\ln\ 2 + (\ln \sigma)/h)} \tag{6.5-6}$$

when $\ln\ 2 + (\ln \sigma)/h > 0$, and infinity otherwise. Formula (6.5-6) can be written as

$$\mathcal{E}N \leq \left( \frac{-1}{\ln\left(\dfrac{1-p}{p}\right)} \right) \left(\frac{1+\rho}{1-\rho}\right) \tag{6.5-7}$$

in terms of the Pareto exponent, or

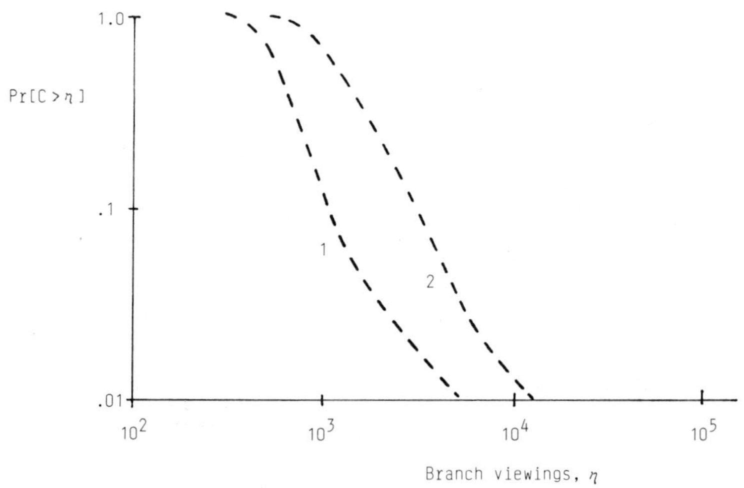


**Figure 6.5-3.** Measured distribution function of the number of branch viewings in a version of the Fano algorithm; this distribution is similar to that of the ISS paths visited in other backtracking searches. Log-log plot. The decision depth function and the related column distance function of Chapter 4 are better for code 1 than for code 2. The slopes in the plot are the Pareto exponents. Code rate 1/2, BSC with $p = 0.033$, 1000 runs, 256-bit deep search, code memory 20. (From Chevillat and Costello [6-36], ©IEEE 1976, used by permission).

$$\mathcal{E}N \le \frac{-1}{\ln\left(\dfrac{1-p}{p}\right) + (2\ln 2)h} \tag{6.5-8}$$

in terms of $h$.

It is clear from (6.5-7) that $\mathcal{E}N$ diverges when the Pareto exponent hits one. We have already seen in Section 6.4 that the transition in $\rho$ occurs when the cut-off rate $R_0(p)$ of the channel equals the code rate. Put another way, probability-one decoding is possible in finite average effort until the channel crossover $p$ solves $R_0(p) = R$; for larger $p$, perfect decoding requires infinite work in the long term. This is hardly unjust. On the contrary, it is remarkable that perfect decoding is possible at all in finite work.

The cut-off rate appears in still more places in coding theory. It can be shown, for instance, that for many channels the error probability with random block codes is bounded by $2^{n(R-R_0)}$ at low code rates.

The Pareto character of path numbers holds for concrete codes as well as random ones. Figure 6.5-3 shows the measured computation distribution for a convolutional code, plotted on log-log scales. Eventually, the distribution assumes a straight line, the slope of which is the effective Pareto exponent.

The Pareto distribution also related to $P_e$-criterion decoding, through the argument leading to (6.5-4). For a backtracking $P_e$ decoder, (6.5-4) is the distribution of path numbers searched in an incorrect subset. We already know that the number is near (6.4-7) in the worst case, and that the worst case occurs with probability close to $P_e$.

## 6.6 ALGORITHM STEPS

Until now our study of sequential decoders has focussed on the numbers and lengths of paths. Early tree coding researchers noticed that some algorithms executed slowly even though they searched a small number of paths. A particular example was the stack algorithm. Common sense dictated that no search could be faster than a metric-first one, yet noise bursts led to long delays and poor overall speed. Why was this? Eventually it was understood that one must not only count path numbers, but also the software steps required per path.

We turn now to estimates of these steps. As in the earlier sections, we will analyze first-symbol decoders; that is, we will estimate the algorithm steps to decode the first tree branch. Sliding block decoders will not differ by much. An algorithm that needs about $\kappa N$ steps to explore $N$ paths, $\kappa$ a constant, is a *linear* algorithm. We cannot hope for better. Among linear algorithms, some will have a larger $\kappa$ than others, but as the channel worsens and $N$ grows, our first priority in algorithm design is to avoid one with more than linear growth.

A discussion of steps inevitably focuses on the noisy periods in a message transmission. In a practical channel the error rate before coding is typically 1% or less. It is helpful to imagine the channel noise as falling into two types, quiet periods during which the decoder directly finds the correct path and unlucky periods during which the decoder searches frantically. Some algorithms, like the Viterbi algorithm, cannot reduce effort during quite periods, but others can limit searching virtually to the correct path alone. During noise bursts these latter decoders search intensively, and it is now that efficiency, whether of steps or path numbers, is important. Over the long term, the efficiency of an algorithm is a mixture of these two modes, with the noisy mode coming to the fore as the channel worsens.

## Breadth-First Decoders

*The Viterbi Algorithm.* This widely used breadth-first decoder was featured throughout Chapter 4. It is clearly a linear scheme. It extends paths by updating the path associated with each state in the code trellis, an act that consists of independently computing a metric and a survivor at each state. These are simple jobs. After $L$ trellis levels, the algorithm has viewed about $LbS$ paths in $\kappa LbS$ steps, where $S$ is the number of states in the trellis and $\kappa$ is the update effort. $N$ paths thus consume about $\kappa N$ steps. Compared to other algorithms, the $\kappa$ factor here is small, but $N$ is large for the Viterbi algorithm, so much so that the total step count is often larger than for other schemes.

*Optimal $P_e$- or t-Criterion Decoders.* Here again, a breadth-first search of a fixed search region (see Fig. 6.2-1 or 6.5-1) is clearly linear: The algorithm simply picks paths from storage, extends them, checks the drop and stop conditions, and puts them back in storage. Here it is easier to think about a first-symbol decoder, for which the search region is truly fixed, but a sliding-block version will vary the region only slightly, and then in response to the output symbols, not the paths in contention; the important point is that the region shape is independent of the contending paths.

*The M-Algorithm.* The behavior of the $M$-algorithm is more interesting and it introduces a new philosophy, biasing the design toward quiet channel performance. The $M$-algorithm does not search paths within a region, but retains instead a fixed number at each depth, the best $M$. Practical $M$-algorithms have resembled a naive but effective scheme which is roughly as follows:

Repeat for each of the $M$ paths in storage at depth $\ell$:

i) Extend to the $b$ paths at depth $\ell+1$.

ii) Compare each to a temporary list of the best $M$ paths so far at depth $\ell+1$; if a new path is better than the worst of the $M$, drop that one and enter the new one.

iii) Else drop the new paths.

The number of steps here is about $bM^2$ per tree depth. Step (ii) constitutes $bM$ comparisons (once an initial stage is past) and the scheme repeats $M$ times per depth. By depth $L$, about $L(b-1)M$ paths have been viewed in $LbM^2$ steps. Replacing $L(b-1)M$ with $N$, we get a step count of $\kappa bN^2/L(b-1)^2$ to view $N$ paths, where $\kappa$ measures the effort of an extension and comparison. As the channel worsens, $N$ is of much higher order than $L$, and we can approximate the step count as $\kappa N^2$.

There are good reasons to use a scheme like this. In practice $M$ is small enough that $LbM^2$ steps per depth is not onerous. Second, the scheme can be modified to run fast in a quiet channel by comparing new paths to a single variable that contains the worst path kept so far. This is a linear procedure in a channel that stays quiet, and so in a good, practical channel the procedure is barely quadratic. The most important advantage of the scheme is its simplicity. But the $M$-algorithm is actually linear. To see this the key is to cast the algorithm as follows.

i) Form all $bM$ path extensions to depth $\ell+1$.
ii) Obtain the $M$-th poorest path in this list.
iii) Compare all $bM$ to this one, and keep only those better.

It can be shown that step (ii) is of order $M$ (see Knuth [6-4], p. 216), and the other steps are order $M$, so the whole procedure is order $M$. In $L$ levels, $L(b-1)M$ paths are explored in $\kappa LbM$ steps, which is an effort of $\kappa Nb/(b-1)$ to explore $N$ paths. Here $\kappa$ stems primarily from step (ii) and is relatively large.

## Metric-First Decoders: The Stack Algorithm

The effort of ordering paths by metric makes the stack algorithm a nonlinear algorithm. As we did with the $M$-algorithm, we look first at a naive implementation, which is the way the stack algorithm is usually constructed, and then at an asymptotically optimal scheme.

The naive approach to a metric-first search is to iterate through an unordered path list at each extension and find the highest-metric path. The work of doing this at the peak storage dominates the entire effort. Also, we must assume the best path is randomly placed in the list, because this is the case during the confusion of a noise burst. The extension of $N$ paths by this method requires an order $N^2$ comparisons.

An alternative is to search from a strictly ordered list. Each newly extended path is placed in order in the list, thus keeping it ordered. Here again, about $N^2$ steps are required to search $N$ paths. While these naive algorithms may work when $N$ is small, we saw in Section 6.5 (Table 6.5-3) that backtracking decoders find their best application when $N$ ranges widely. In practice, the quadratic order of steps has proved intolerable.

One way to reduce the stack search steps is Jelinek's bucket algorithm [6-15], [6-3], [6-33]. In this scheme, the ordered list of paths is broken into ordered "buckets" of paths. A new path is sorted into the right bucket, but not ordered further. This partial ordering will reduce the order of the steps, provided that the number of buckets grows more slowly than the path number $N$. Tests show that the bucket algorithm reduces algorithm steps while not much affecting the decoder error rate, but little is known about the asymptotic properties of bucket schemes and, in any case, they are not truly metric-first decoders.

It has been shown that true metric-first decoders require at least an order $N \log N$ steps as $N$ grows large, and two implementations with this order have been found [6-34]. One is based on list merging and the other on AVL trees, a method of organizing the path storage. As methods of handling stored data, these two techniques are well known in computer science and may be found in Knuth [6-4] or in other texts. As ways to implement a decoder, they are optimal but clumsy.

Turning now specifically to $P_e$-criterion decoders, we can compare the steps in optimal breadth-first and metric-first searching. Breadth-first schemes achieve order $N$ steps and metric-first schemes achieve $N \log N$, while eqs. (6.4-7) and (6.4-15), respectively, give $N$. We can assume that the searching in all subsets totals about $bN$. Carrying out the calculation shows that at practical rates metric-first decoders indeed are better in average step count, but near channel capacity *metric-first decoding is worse than breadth-first.* For an order $N^2$ stack search, the transition occurs somewhat above cut-off rate.

## Depth-First Decoders

If our goal is to minimize steps, depth-first decoders are of great interest because they are linear algorithms and at the same time they search about the same path number $N$ as the metric-first schemes. It is easy to see that the depth-first steps are of order $N$. During decoding, paths are extended by simply taking them from the store of tentative paths, extending them until they hit a limit, and putting them back in the store. Roughly, there will be one metric calculation and comparison per branch in the search region. It is easy to show that the number of distinct paths in the search region roughly equals the number of branches in it, and so the steps in the search are of order $N$.

Since no metric-first search can be better in steps than $N \log N$, we conclude that in terms of steps *the depth-first mode is indisputably best as $N$ grows.* What does this say about practical decoders for a noisy channel? We can reduce this section to two principles. If symbol synchronism or lack of search variability is important, the decoder should approximate a breadth-first mode. Otherwise, a depth-design is inherently the strongest one.

## 6.7 SOURCE CODING REVISITED

Since leaving source coding in Chapter 2, our discussions have focussed on channel coding. We now return to source coding to discuss the application of the tree searching algorithms already described in the context of channel decoding. We point out the similarities and differences that exist between the two disciplines.

In both source coding and channel decoding, the aim of the algorithms is to search the code tree in some manner according to a suitable search criterion. In both cases they select a tree path that best matches a given sequence of symbols – source symbols in the case of source coding and received symbols in the case of channel decoding.

The concept of correct path that exists in channel decoding does not exist in source encoding. What the source encoder tries to find is a "good" tree path, from among the many paths the source encoder has searched, that approximates the given source symbol sequence to within a given distortion. Figure 6.7-1 illustrates the idea. Two tree paths have the least Hamming distance of 2 to the given source sequence among all the paths. Assume that the breadth first $M$-algorithm, with $M = 2$ for example, is employed to search the tree, and that the search must find a path with per-source-bit distortion of less than or equal to 0.4. The algorithm will find the two paths shown in Figure 6.7-1; one of them is a path with distance two, which is acceptable. Assume now a depth-first algorithm is used instead. In searching depth-first, the algorithm chooses branches out of a node in order from bottom to top. This algorithm will also select a distance-2 path, but it is different from the one selected by the breadth-first algorithm, as can be seen from the figure. This path is also acceptable since the path searched satisfies the per-source-bit criterion. In fact, if the breadth-first algorithm is allowed to retain three or more paths at each level, it will also find the path selected by the depth-first algorithm. At the end of the search, it will release one of the two Hamming-distance-2 paths that it has searched. Depending on the criterion used to resolve ties, the path released may or may not be the same path released earlier when $M$ was equal to 2.

From the preceding discussion, we see that the algorithms of Section 6.2 apply to source encoding as well. They may need to be modified

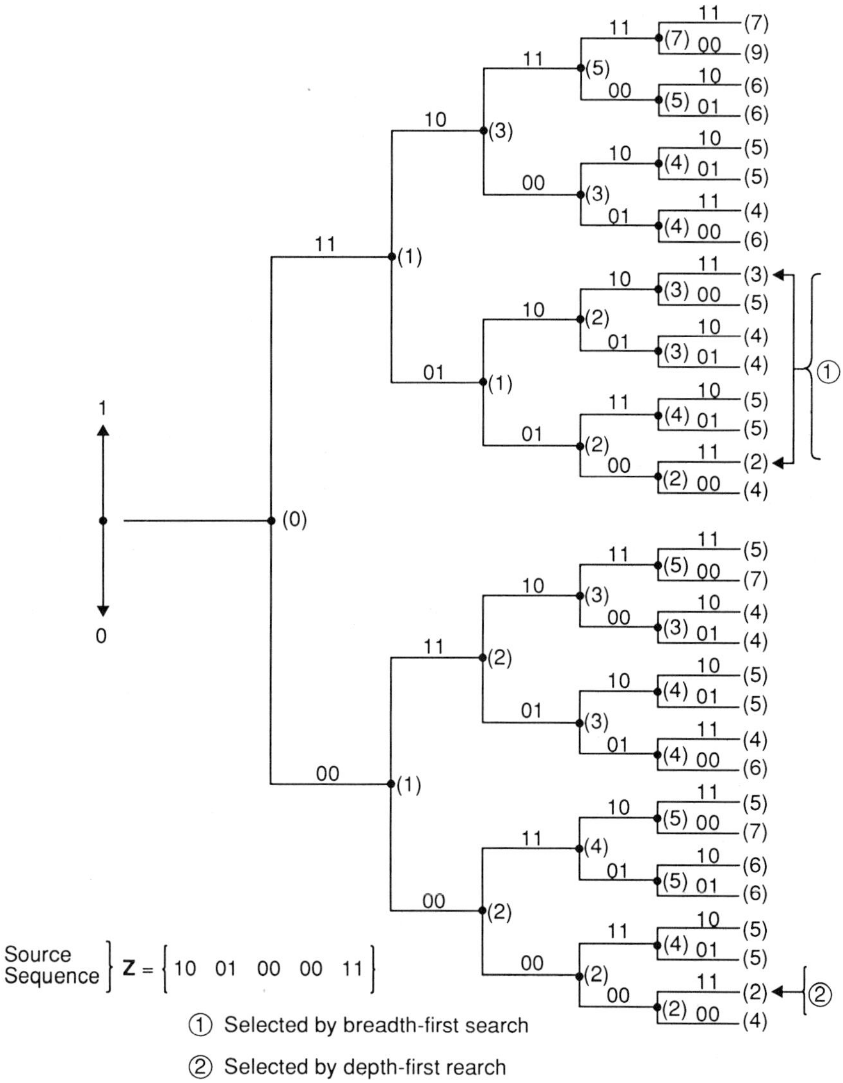


**Figure 6.7-1.** Figure illustrating the breadth- and depth-first algorithms. The breadth-first algorithm used is the $M$-algorithm, with $M = 2$. The depth-first algorithm searches the branches out of any node in order from bottom to top and selects the first path that satisfies the per-source-bit distortion criterion, which is assumed to be 0.4.

somewhat to make them practical and instrumentable. Many instrumentable versions of these algorithms are described in the references; see for example [6-15], [6-16], [6-24], [6-33], [6-34], and [6-41].

Given a code tree, we can search it. But how should one generate a code? We know from the rate-distortion coding theorems (Theorems 2.3-1 and 2.3-2) that, for discrete memoryless sources, memoryless Gaussian and autoregressive sources, random block codes exist that achieve the rate-distortion bound. Unfortunately, as we have now learned, these theorems only point to the existence of such codes and do not construct them. Moreover, it is certainly desirable to impose structures on codes in order to make them practical. One such structure is the tree structure. Do tree codes exist achieving the rate-distortion bound? If they do, how do we construct such codes? The first question was answered in the affirmative by Jelinek [6-37] when he proved the following theorem.

**Theorem 6.7-1:** For a bounded per-letter distortion measure and a discrete memoryless source, and given any $\varepsilon > 0$, and $D > 0$, there exists a tree code of sufficiently large depth with rate $R(D) + \varepsilon$ and average distortion less than $D + \varepsilon$.

*Proof*: See [6-38].

As we evaluate the performance of sequential source encoders, we will see answers to the question of code construction.

## Memoryless Binary Source with PE Distortion

For an equiprobable memoryless binary source, Ex. 2.3-1 provides a clue to code design; simply populate the tree branches randomly with zeros and ones, with $q(0) = q(1) = 0.5$. Reference [6-16] provides simulation results for rate-1/2 codes. Three classes of codes are evaluated, random rate-1/2 tree codes, convolutional codes, and random complementary codes,† and Table 6.7-1 reproduces some of the results. The table reveals the interesting fact that carefully chosen convolutional codes perform close to the rate-distortion bound for large $M$; convolutional and random complementary codes achieve similar distortion performance. As $M$ grows, all codes perform closer to the $D$ that solves $R(D) = 1/2$, which is 0.110.

†These are random tree codes in which the source symbols on one of the two branches out of a node are chosen randomly and those on the other branch are the complements of these.

**Table 6.7-1:** THE $M$-ALGORITHM SIMULATION RESULTS WHEN USED WITH THREE CLASSES OF CODES, CONVOLUTIONAL CODES, RANDOM TREE CODES, AND COMPLEMENTARY RANDOM TREE CODES. ALSO LISTED ARE THE GENERATORS (IN OCTAL) FOR THE BEST RATE-1/2 CONVOLUTIONAL CODES AND THEIR DISTORTION PERFORMANCE. THE PARAMETER $L$ REFERS TO THE SEARCH DEPTH OR THE LENGTH OF ANY PATH SEARCHED IN TERMS OF THE NUMBER OF BRANCHES. NOTE THAT $R^{-1}(1/2) = 0.110$. FROM [6-16], ©IEEE 1971, USED BY PERMISSION.

| M | L | Best rate -1/2 Convolutional Code $g_1, g_2$ | Distortion | Convolutional Ensemble Average | Complementary Ensemble Average | Random Ensemble Average |
|---|---|---|---|---|---|---|
| 10 | 25 | 1,1532 | 0.141 | 0.149 | 0.148 | 0.166 |
| 25 | 25 | 1,1667 | 0.139 | 0.1445 | 0.144 | 0.156 |
| 50 | 50 | 127742106, 164264505 | 0.1295 | 0.1335 | 0.134 | 0.1405 |

## Gaussian Memoryless Source with Squared-Error Distortion

Let us now examine the process of generating tree codes for Gaussian independent sources with squared-error distortion. Recall from (2.3-19) that the optimum reproducer word density, given the source output $z$ is

$$q(w|z) = \frac{1}{\sqrt{2\pi\alpha D}}\, e^{-\frac{(w-\alpha z)^2}{2\alpha D}}, \quad \alpha = 1 - \frac{D}{\sigma^2}, \tag{2.3-19}$$

where $Z$ is Gaussian with zero mean and variance $\sigma^2$. Observe that $q(w|z)$ is the Gausssian density function with mean $\alpha z$ and variance $\alpha D$; that is $\mathcal{E}[W|Z=z] = \alpha z$ and $\mathcal{E}[W^2|Z=z] = \alpha D + (\alpha z)^2$. We can average $\mathcal{E}[W|Z=z]$ and $\mathcal{E}[W^2|Z=z]$ over $Z$ to get

$$\mathcal{E}[W] = 0,$$

and

$$\mathcal{E}[W^2] = \sigma^2 - D.$$

$W$ is therefore Gaussian with zero mean and variance $\sigma^2 - D$. The ensemble of tree codes obtained by populating the tree branches with numbers chosen from this distribution achieves the rate-distortion bound for independent Gaussian sources, which is

$$R(D) = \frac{1}{2}\log\frac{\sigma^2}{D}, \qquad D \le \sigma^2. \tag{2.3-20}$$

For example, consider a Gaussian source with zero mean and unit variance. A rate-1 tree code is obtained by populating the tree branches with numbers from the Gaussian distribution with zero mean and variance $1 - D = 0.75$, since from Fig. 2.3-2, $D$ is 0.25 when $R(D)$ is 1.

Reference [6-39] reports simulation of these tree codes for the independent Gaussian source with zero mean and unit variance using the Viterbi and the stack algorithm; signal-to-distortion ratios about 1 dB below the rate-distortion limit are obtained. Figure 6.7-2 depicts the results of an $M$-algorithm search of the rate-1 tree code populated with zero-mean and variance 0.75 Gaussian variates [6-40]. Also shown in this figure are the results of searching random complementary tree codes, in which one branch is a Gaussian and the other is the same variate with opposite sign. For $M = 10$, the complementary tree codes achieve a signal-to-noise ratio 1 dB below the rate distortion limit. Compare this with the rate-1 Max quantizer which achieves 4.4 dB, about 1.6 dB below the rate-distortion limit. Intuitively speaking, the randomly generated code tree can be thought of as providing a set of sequences whose elements are not limited to a small set of values; the Max quantizer, on the other hand, provides a set of sequences whose elements are limited to one of two values only. Consequently, the random tree code provides a better match to the random source sequence than does the tree of the Max quantizer.

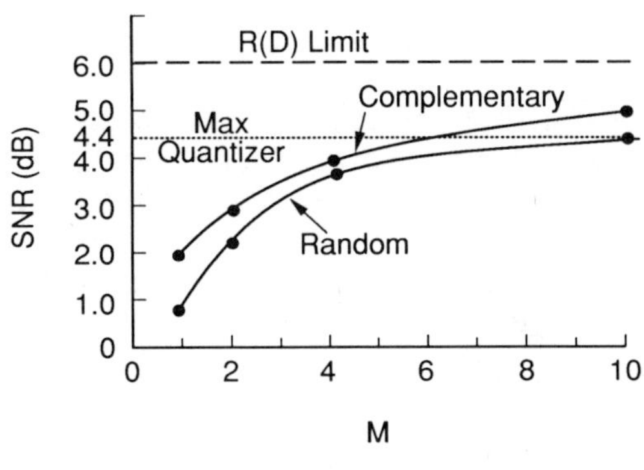


**Figure 6.7-2.** $M$-algorithm search of the rate-1 random and complementary random tree codes populated with zero-mean and variance 0.75 Gaussian variates, adapted from [6-40].

**Table 6.7-2:** AUTOCORRELATIONS, PREDICTOR COEFFICIENTS, AND $B$-FILTER COEFFICIENTS FOR McDONALD AND BUTTERWORTH SOURCE MODELS; $\sigma = 1$. FROM [6-41], ©IEEE 1975, REPRODUCED BY PERMISSION.

| Auto Correlation | Coefficients of A Filter | Coefficients of B Filter D = 0.1 | D = 0.05 | D = 0.01 |
|---|---|---|---|---|
| Butterworth | | | | |
| r (1) = 0.790 | $a_1$ = 1.231 | $b_0$ = 0.894 | 0.990 | |
| r (2) = 0.441 | $a_2$ = -0.625 | $b_1$ = 0.096 | 0.0085 | . . . . |
| r (3) = 0.170 | $a_3$ = 0.119 | $b_2$ = -0.0164 | -0.0015 | |
| McDonald | | | | |
| 0.864 | 1.748 | | 0.87 | 0.976 |
| 0.557 | -1.222 | . . . . | 0.15 | 0.0255 |
| 0.227 | 0.301 | | -0.030 | -0.0054 |

## Correlated Gaussian and Speech Sources

We have seen in Section 2.4 that the class of optimal random tree codes achieving the rate distortion bound can be generated by passing independent Gaussian variates with zero mean and variance $\sigma^2$ through the filter

$$\frac{B(z)}{1 - A(z)},$$

where $B(z)$ satisfies (2.4-11), which is

$$|B(z)|^2 = \sigma^2 - D|A(z)|^2;$$

A(z) is the $z$-transform of the coefficients $a_i$ that occur in (2.3-21) governing the definition of the autoregressive source, and $\sigma^2$ is the variance of the input excitation $u_j$ (see Fig. 2.3-3). Table 6.7-2 from Anderson and Bodie [6-41] lists the $B$ filter coefficients for two different $A$ filters, which have proven useful in the encoding of speech. Spectrum plots of the $B$ filters reveal that they are essentially low pass filters and that the severity of the filters increases with increasing $D$.

We made the observation in Section 2.4 that the equation governing the DPCM code generation process (2.4-7) and that governing the autoregressive source generation (2.3-21) are identical. This implies that the class of optimum tree codes for autoregressive sources should in fact be useful for encoding practical sources such as speech. The trick is to first fit an autoregressive model for speech based on its autocorrelation coefficients by computing $\mathbf{a}$ as in Section 2.4. The filter $B(z)$ can then be designed by solving (2.4-11). The $B$ filters in Table 6.7-2 are obtained this way.

We have suggested another class of waveform coder in Section 2.4, the smoothed DPCM class of codes (2.4-10), which are similar in structure to the rate-distortion derived codes, with $F(z)$ replacing $B(z)$. $F(z)$ was a low pass filter with zeros at one-half the sampling frequency of speech, to suppress the jitter due to quantization at this frequency. Figure 6.7-3 shows the $M$-algorithm search using both the unsmoothed and the smoothed DPCM codes. As the figure illustrates, the smoothed class of codes is capable of yielding a better performance compared to its unsmoothed counter part; its performance matches that of the class of rate-distortion derived codes.

Table 6.7-3 compares $M$-algorithm coding of actual speech to the DPCM performance and the rate-distortion limits for the autoregressive source matched to speech. The table lists the $D_0$ defined in Theorem 2.3-2 for the AR models. The Max quantizer mean-squared error for the Gaussian source with zero mean and unit variance (given by Table 2.2-2) for 2 and 3 bits are, respectively, 0.118 and 0.0345. For these quantizer errors, the table lists the DPCM's SNR performance for the AR sources as estimated by (2.4-6). The next two columns in the table list $R(D)$ in dB SNR for 2 and 3 bits. The DPCM and M-algorithm results for speech are listed next. Either the smoothed DPCM code or the rate-distortion code is used, whichever performed better. Use of tree coding results in 4–7 dB SNR performance improvement over DPCM.

Another code class for speech [6-42] is based on (2.4-9):

$$\widehat{z}_j = \sum_{i=0}^{\infty} c_i \widehat{u}_{j-i}, \tag{2.4-9}$$

where $\widehat{z}_j$ are the reproducer values, $\widehat{u}_j$ are a fixed set of quantizer outputs, and $c_i$ are the filter coefficients. Reference [6-40] suggests that for a 4-level quantizer corresponding to a rate-2 tree, the quantizer reproducer levels be in the ratio $-1, -0.23, 0.23, 1$. Usually the reproducer values $\widehat{z}_j$ are normalized such that the average energy in the set of reproducer values matches the speech source. Reference [6-42] recommends the filter

$$c_0 = 0.7704,\ \ c_1 = 1.5254,\ \ c_2 = 1.6332,\ \ c_3 = 1.2054,\ \ c_5 = 0.0492.$$

Figure 6.7-4 illustrates the resulting tree code, suitably normalized. (A normalization constant $\approx 7.6$ is used.) The code lends itself to ease of implementation as a look-up table, since there are just 1024 distinct reproducer words. The reproducer word on any branch is determined by the current and the preceding four branches. Thus the most recent set of five path map symbols can be used as the address to retrieve the code word from a 1024-word look-up table.

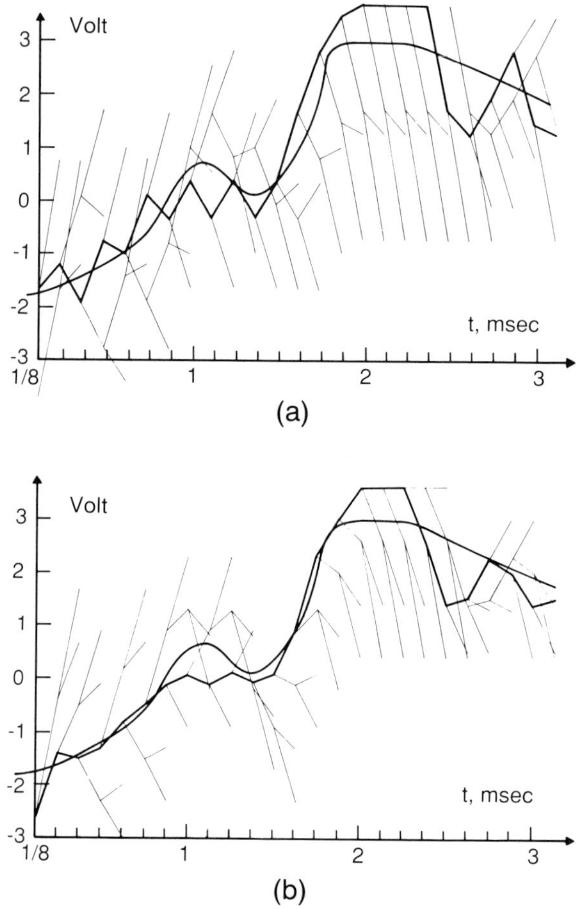


**Figure 6.7-3.** $M$-algorithm ($M = 2$) encoding of an analog waveform. Heavy lines are waveform and final encoding, light lines are tree paths dropped. Butterworth predictor with sampling frequency 8 kHz and (a) no smoothing and (b) smoothing. From [6-41], ©IEEE 1975, reproduced by permission.

The stack algorithm of Section 6.2 has also been employed to encode speech [6-43]. In Section 6.4, we have noticed that the metric-first search performs better than breadth-first in terms of expected search size. This conclusion is intuitively reasonable, since when conditions are favorable, the metric-first search does very little work. Stack algorithm speech coding achieves most of the gain over DPCM with fewer than 2 node

**Table 6.7-3:** COMPARISON OF $M$-ALGORITHM AND DPCM PERFORMANCE WHEN USED WITH MATCHING AUTOREGRESSIVE SPEECH MODELS; ALL SNR FIGURES ARE IN DB. FROM [6-41], ©IEEE 1975, USED BY PERMISSION.

| | | MARS Model | | | | Actual Speech | | | |
|---|---|---|---|---|---|---|---|---|---|
| | | LMS Pred/ Quant.* | | R(D) Limit | | LMS Pred/ Quant. | | Multi-Path Search (M=8) | |
| McDonald | | 2bit | 3bit | Δ(2) | Δ(3) | 1bit | 2bit | 1bit | 2bit |
| 3 tap | $D_0$ = 0.055 | 19.3 | 24.7 | <22.0 | 28.1dB | 7.9 | 14.0 | >12.5 | >21dB |
| 2 tap | 0.094 | 18.9 | 24.2 | 21.6 | 27.7 | • • • | | • • • | |
| 1 tap | 0.288 | 15.3 | 20.6 | 18.0 | 24.1 | • • • | | • • • | |
| Butterworth | | | | | | | | | |
| 3 tap | 0.113 | 14.8 | 20.1 | 17.5 | 23.6 | 6.3 | 16.1 | >12 | >20 |

* Approximate
< Indicates Overbound
> Indicates Underbound

visits per source symbol encoded†. (Observe that the number of node visits must be at least 1 and this is the case with DPCM.) The $M$-algorithm achieves comparable performance with 4 or more node visits per source symbol encoded. For further discussions on speech encoding by the stack algorithm and the resulting performance curves, the reader is referred to [6-43]. More discussion also appears there on how paths should be arranged in storage for achieving computational optimality in metric-first searches.

The stack algorithm's average work may be reduced by releasing more than one symbol every time the algorithm releases symbols. Intuitively, one can see that the $M$-algorithm's average search will not differ significantly since it must always search $bM$ paths at each tree level; on the other hand, the stack algorithm will never have to explore the paths dropped off that originate out of nodes on the released branches. This conclusion is verified in the simulation results of [6-44]. In terms of SNR performance, however, the single-symbol-release technique outperforms the multiple-symbol-release technique. While [6-44] demonstrates the validity of this conclusion for speech, ref. [6-45] does it for rate-1/2

†The count of nodes visited is not the same as the count of paths retained, which was discussed in preceding sections, but the two are closely related.

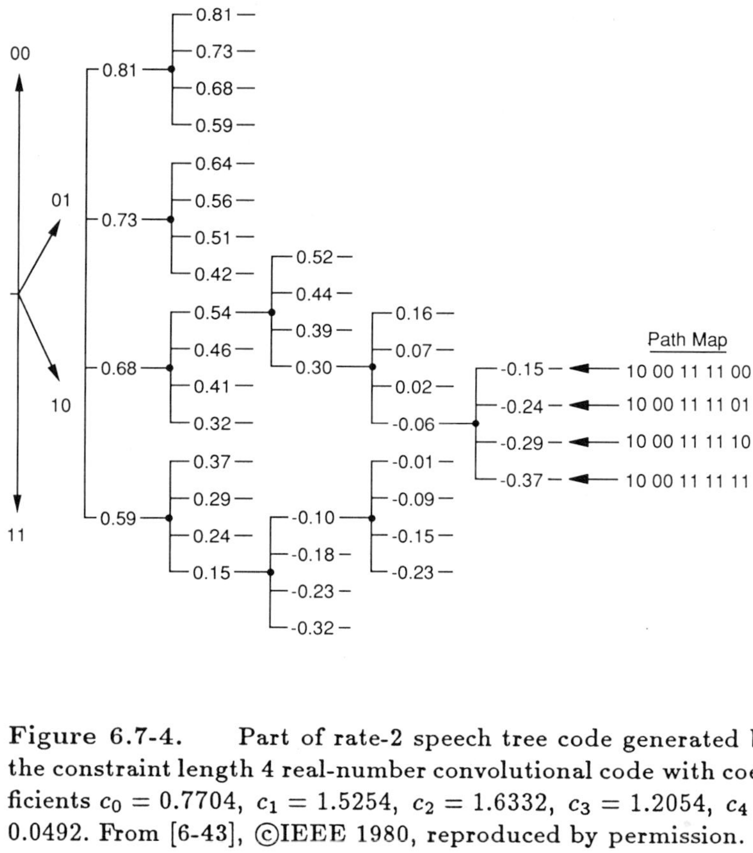


**Figure 6.7-4.** Part of rate-2 speech tree code generated by the constraint length 4 real-number convolutional code with coefficients $c_0 = 0.7704$, $c_1 = 1.5254$, $c_2 = 1.6332$, $c_3 = 1.2054$, $c_4 = 0.0492$. From [6-43], ©IEEE 1980, reproduced by permission.

Gaussian memoryless sources. One can show that the branch metric in a Gaussian code tree is exponentially distributed (see Problem 6-16). Equipped with this knowledge, we can generate a *metric tree* whose branches are assigned variates that are exponentially distributed and then search the tree using a search algorithm. This is the approach used in [6-45] to determine the performance of tree searching.

We leave it for the reader to explore aspects of tree searching such as adaptivity [6-46], [6-47] or frequency-weighted distortion measures [6-48]. Tree searching has been applied as well to other types of sources such as line drawings [6-49] or images [6-50].

# PROBLEMS

6-1 Extend (6.1-4) to the case of a discrete memoryless channel with two inputs and many outputs.

6-2 Consider the rate-1/2 convolutional code in Fig. 4.2-5. Find the smallest number of paths a breadth-first decoder can keep and still correct a single channel error in the first trellis branch. (To do this, assume the error is in one of the first two positions and attempt to decode the first trellis branch; then look for the worst position in storage occupied by the correct path).

6-3 a) Construct a $2 \times 2$ table of the Fano metric branch increments that can occur in the decoding of a rate-1/2 code like that in Fig. 4.2-5. Use your table to fill in the cumulative metric at each node in the trellis of Fig. 4.2-5. Assume that the channel crossover probability is $p = .02$ and the Fano bias equals the code rate.
b) A metric-first decoder like the stack algorithm is to be used on this trellis. Chart its progress by numbering the order in which the nodes are reached in the trellis. Do this for three cases: No errors, an initial error, two initial errors. Consider only the decoding of the initial trellis branch. Assign a suitable decision depth (see, for instance, Table 6.5-2).

6-4 Repeat the decoding in Problem 6-3(b), but this time with the $M$-algorithm. Use $M = 2$ and try all three cases. No matter how ties are broken, does the algorithm always succeed in the single error case? Can it fail in the double error case?

6-5 Repeat Problem 6-3 for the rate-1/3 code in Figs. 4.1-1 and 4.2-3. Consider the same three error cases. What effect does the lower rate have on the decision depth? What effect does it have on the metric-first search?

6-6 Repeat Problem 6-3(b), this time with a depth-first search. Limit the search with a decision depth barrier and a metric limit that is relaxed in steps of $-1$. If you have done Problem 6-3, compare the progress of metric- and depth-first searching.

6-7 Compute the stop and drop lines for a breadth-first search at rate 1/2, $P_e = 10^{-5}$, and $p = .0005$. Plot them in the distance - depth plane. The case here corresponds to a very good channel, so your region should be small.

6-8 For the case in Problem 6-7, find the maximum required search depth in bits. Repeat for $P_e = 10^{-10}$. These depths are the approximate range for practical decoding at rate 1/2 in a good channel.

6-9 Show that the maximum depth of a breadth-first search, the $n$ in eq. (6.3-18), satisfies

$$2^{n\left(R-R_0(p)\right)}$$

at low code rates, where $R$ is the code rate and $R_0(p)$ is the cut-off rate (6.4-14). This shows that breadth-first searching achieves the Gallager exponent for block codes, at least for low rates.

6-10 Plot on the same graph the capacity and the cut-off rate of a BSC versus the crossover $p$. Characterize how these compare to each other.

6-11 Show that the search rate of growth $\sigma$ satisfies (6.4-8) when the Fano bias $B$ equals the code rate $R$. (*Hint*: Substitute (6.4-8) into (6.3-6) and rearrange the result to produce (6.4-4)).

6-12 Show that the derivative of $\sigma^{1/h}$ with respect to the Fano bias $B$ is zero when $B = R$. Except for a few details, this shows that setting the bias to $R$ minimizes a backtracking search. (*Hint*: Once you have an expression for the derivative, show that it amounts to (6.4-8), which you know to be true when $B = R$ by Problem 6-11).

6-13 Show that the rate at which $\ln 2 + (\ln \sigma)/h = 0$ in eq. (6.5-6) is the cut-off rate (6.4-14). Remember that the bias equals the rate here.

6-14 Consider decoding a BSC at rate 1/2. What crossover probabilities $p$ correspond to Pareto exponents $\rho = 1$ and 2 in the distribution (6.5-4)? Assume that $K_o = 1$ and plot $P\{N \geq \eta\}$ vs. $N$ for the two cases. Make this a log-log plot.

6-15 Search the tree of Fig. 6.7-4 using the $M$- and the stack algorithm. Use $M = 2$. Assume that either algorithm can retain paths at most three branches long. (This means that whenever either algorithm tries to extend a path it has retained that is more than three branches long, it must release the eldest symbol of the best path in storage, advance the root node to the end node of the released branch, and purge all tree paths not originating from the new root node. These steps of the algorithm form what has been termed the *ambiguity check*). For the $M$-algorithm use the squared-error distortion measure. For the stack algorithm use $D^* - (z - \hat{z})^2$ as the branch metric, with $D^* = 0.01$. Assume that the stack can only retain 8 paths. Use the source sequence $\{0.5, 0.65, 0.3, -0.2, -0.6\}$. Are the encodings different in the two cases? What is the performance improvement over DPCM?

6-16 Assume both the source and reproducer words are zero-mean, unit-variance Gaussian distributions. Assume that the reproducer letters are placed two per branch in a rate-1/2 tree. Define the branch metric as

$$(z_1 - \hat{z}_1)^2 + (z_2 - \hat{z}_2)^2,$$

where $z_1$ and $z_2$ are source words and $\widehat{z}_1$ and $\widehat{z}_2$ are reproducer words. Show that the branch metric is exponentially distributed.

# REFERENCES

[6-1] R. M. Fano, "A heuristic discussion of probabilistic decoding," *IEEE Trans. Information Theory*, vol. IT-9, pp. 64–74, April 1963.

[6-2] J. L. Massey, "Variable length codes and the Fano metric," *IEEE Trans. Information Theory*, vol. IT-18, pp. 196–198, Jan. 1972.

[6-3] S. Lin and D. J. Costello, Jr., *Error Control Coding: Fundamentals and Applications*, Prentice-Hall, Englewood Cliffs, N.J., 1983.

[6-4] D. E. Knuth, *The Art of Computer Programming Vol. 3, Searching and Sorting*, Addison-Wesley, Reading, Mass., 1973.

[6-5] C.-F. Lin "A truncated Viterbi algorithm approach to trellis codes," Ph.D. dissertation, Dept. Elec., Computer and Systems Eng., Rensselaer Poly. Inst., Troy, N.Y., Sept. 1986.

[6-6] T. Aulin, Chalmers Technical Univ., Gothenberg, Sweden, and R. Johannesson, Tech. Univ. of Lund, Lund, Sweden, personal communication, 1987.

[6-7] U. Roethlisberger, "A path recovery scheme for the truncated Viterbi algorithm," M. Sc. Thesis, Dept. Elec., Computer and Systems Eng., Rensselaer Poly. Inst., Troy, N.Y., April 1988.

[6-8] T. Aulin, "Study of a new trellis decoding algorithm and its applications," Final Report, ESTEC Contract 6039/84/NL/DG, European Space Agency, Noordwijk, Holland, Dec. 1985.

[6-9] A. Duel-Hallen and C. Heegard, "Delayed decision-feedback sequence estimation," *IEEE Trans. Communications*, vol. COM-37, pp. 428–436, May 1989.

[6-10] M. V. Eyuboglu and S. Qureshi, "Reduced-state sequence estimation for coded modulation on interference channels," *IEEE J. Sel. Areas Communs.*, vol, SAC-7, pp. 989–995, Aug. 1989.

[6-11] N. Seshadri and J. B. Anderson, "Decoding of severely filtered modulation codes using the (M,L) algorithm," *ibid.*, pp. 1006–1016, Aug. 1989.

[6-12] G. J. Foschini, "A reduced state variant of maximum likelihood sequence detection attaining optimum performance for high signal-to-noise ratios," *IEEE Trans. Information Theory*, vol. IT-23, pp. 605–609, Sept. 1977.

[6-13] S. J. Simmons, "Breadth-first trellis decoding with adaptive effort," *IEEE Trans. Communs.*, vol. 38, pp. 3-12 , Jan. 1990.

[6-14] K. Zigangirov, "Some sequential decoding procedures," *Probl. Peredachi Inf.*, vol. 2, pp. 13–25, 1966.

[6-15] F. Jelinek, "A fast sequential decoding algorithm using a stack," *IBM J. Res. and Dev.*, vol. 13, pp. 675–685, Nov. 1969.

[6-16] F. Jelinek and J. B. Anderson, "Instrumentable tree encoding of information sources," *IEEE Trans. Information Theory*, vol. IT-17, pp. 118–119, Jan. 1971.

[6-17] G. C. Clark, Jr. and J. Bibb Cain, *Error-Correction Coding for Digital Communications*, Plenum, New York, 1981.

[6-18] G. D. Forney, Jr. and E. K. Bower, "A high-speed sequential decoder: Prototype design and test," *IEEE Trans. Commun. Tech.*, vol. COM-19, pp. 821–835, Oct. 1971.

[6-19] J. E. Savage, "Sequential decoding - the computation problem," *Bell Syst. Tech. J.*, vol. 45, pp. 149–175, Jan. 1966.

[6-20] I. M. Jacobs and E. R. Berlekamp, "A lower bound to the distribution of computation for sequential decoding," *IEEE Trans. Information Theory*, vol. IT–13, pp. 167–174, April 1967.

[6-21] F. Jelinek, "An upper bound on moments of sequential decoding effort," *ibid.*, vol. IT-15, pp. 140–149, Jan. 1969.

[6-22] R. G. Gallager, "Tree encoding for symmetric sources with a distortion measure," *ibid.*, vol. IT-20, pp. 65–76, Jan. 1974.

[6-23] J. B. Anderson and F. Jelinek, "A 2-cycle algorithm for source coding with a fidelity criterion," *ibid.*, vol. IT-19, pp. 77–92, Jan. 1973.

[6-24] P. R. Chevillat and D. J. Costello, Jr., "A multiple stack algorithm for erasure-free decoding of convolutional codes," *IEEE Trans. Communications*, COM-25, pp. 1460–1470, Dec. 1977.

[6-25] F. Xiong, "Sequential sequence estimation for channels with intersymbol interference," Ph.D. Thesis, Dept. Elec. Eng., Univ. of Manitoba, Winnipeg, Canada, Aug. 1989.

[6-26] W. Feller, *An Introduction to Probability Theory and Its Applications*, vol. I, 3rd Ed., Wiley, New York, 1968.

[6-27] R. G. Gallager, *Information Theory and Reliable Communication*, Wiley, New York, 1968.

[6-28] J. B. Anderson, "Limited search trellis decoding of convolutional codes," *IEEE Trans. Information Theory*, vol. IT-35, pp. 944–955, Sept. 1989.

[6-29] J. B. Anderson, "Sequential decoding based on an error criterion," *IEEE Trans. Information Theory*, in submission, 1990.

[6-30] R. Johannesson and K. Zigangirov, "Distances and distance bounds for convolutional codes - An overview," in *Lecture Notes in Control and Information Sciences, vol. 128: Topics in Coding Theory*, Springer, Berlin, 1989.

[6-31] J. B. Anderson and K. Balachandran, "Decision depths of convolutional codes," *IEEE Trans. Information Theory*, vol. IT-35, pp. 455–459, March 1989.

[6-32] A. J. Viterbi and J. K. Omura, *Principles of Digital Communication and Coding*, McGraw-Hill, New York, 1979.

[6-33] J. B. Anderson and S. Mohan, "Sequential coding algorithms: A survey and cost analysis," *IEEE Trans. Communications*, vol. COM-32, pp. 169–176, Feb. 1984.

[6-34] S. Mohan and J. B. Anderson, "Computationally optimal metric-first code tree search algorithms," *ibid.*, pp. 710–717, June 1984.

[6-35] G. D. Forney, Jr., "Convolutional codes III. Sequential decoding," *Inf. Control*, vol. 25, pp. 267–297, July 1974.

[6-36] P. R. Chevillat and D. J. Costello, Jr., "Distance and computation in sequential decoding," *IEEE Trans. Communications*, vol. COM-24, pp. 440–447, April 1976.

[6-37] F. Jelinek, "Tree encoding of memoryless time-discrete sources with a fidelity criterion " *IEEE Trans. Information Theory*, vol. IT-15, pp. 584–590, Sept. 1969.

[6-38] C.R. Davis and M.E. Hellman, "On tree coding with a fidelity criterion, " *IEEE Trans. Information Theory*, vol. IT-21, pp. 373–378, July 1975.

[6-39] R.J. Dick, T. Berger, and F. Jelinek, "Tree encoding of Gaussian sources, " *IEEE Trans. Information Theory*, vol. IT-20, pp. 332–336, May 1974.

[6-40] J.B. Bodie, "Tree encoding of analog data sources," M.Eng. Thesis, Dept. Elec. Eng., McMaster University, Hamilton, Canada, April 1974.

[6-41] J.B. Anderson and J.B. Bodie, "Tree encoding of speech," *IEEE Trans. Information Theory*, vol. IT-21, pp. 379–387, July 1975.

[6-42] J.B. Anderson and C.-W. Law, "Real-number convolutional codes for speech-like quasi-stationary sources," *IEEE Trans. Information Theory*, vol. IT-23, pp. 778–782, nov. 1977.

[6-43] S. Mohan and J.B. Anderson, "Speech encoding by a stack algorithm," *IEEE Trans. Communications*, vol. COM-28, pp. 825–829, June 1980.

[6-44] S. Mohan, C.-M. Lin, and D. Kryskowski, "Stack algorithm speech encoding with fixed and variable symbol release rules," *IEEE Trans. Communications*, vol. COM-39, pp. 1015–1018, Sept. 1985.

[6-45] W.-W. Chang and J.D. Gibson, "Path map symbol release rules and the exponential metric tree," *IEEE Trans. Information Theory*, vol. IT-36, pp. 1134–1139, Sept. 1990.

[6-46] H. Chan and J.B. Anderson, "Adaptivity and tree searching in DPCM," *IEEE Trans. Communications*, vol. COM-30, pp. 1254-1259, May 1982.

[6-47] N.S. Jayant and S.A. Christensen, "Tree encoding of speech using the $(M, L)$-algorithm and adaptive quantization, " *IEEE Trans. Communications*, vol. COM-26, pp. 1376–1379, Sept. 1978.

[6-48] S. G. Wilson and S. Hussain, "Adaptive tree encoding of speech at 8000 bits/s with a frequency-weighted fidelity criterion," *IEEE Trans. Communications*, vol. COM-27, pp. 165–170, Jan. 1979.

[6-49] M. Vembar and S. Mohan, "Tree encoding of line drawings," *IEEE Trans. Acoustics Speech and Signal Proc.*, vol. ASSP-36, pp. 1542–1549, Sept. 1988.

[6-50] J.W. Modestino, V. Bhaskaran, and J.B. Anderson, "Tree encoding of images in the presence of channel errors," *IEEE Trans. Information Theory*, vol. IT-27, pp. 677–697, Nov. 1981.

# CHAPTER 7

# Architectures for Coding Algorithms

We began our book by observing in Chapter 1 that the field of coded communications is a synergism among three evolving disciplines: communications theory, coding and signal processing algorithms, and special-purpose processors and architectures suitable for implementing coding algorithms. So far in this book, we have treated coded communications algorithms in detail and communications theory just enough to facilitate discussion of algorithms. This chapter introduces the third discipline, architectures for coding algorithms. This subject is interdisciplinary in nature, involving areas such as VLSI processor arrays and parallel algorithms for sorting and searching, besides coded communications theory and algorithms. Consequently, the subject is of recent origin and the results are limited. Nevertheless, it is an important and growing field and we seek to give the reader a flavor for what the field is all about. The literature of VLSI design, array processors, and parallel algorithms is vast (see, for example, [7-1] – [7-5]) and we do not attempt a general discussion. Results from these fields are discussed and applied as needed to illustrate some specific architectures for coding that have appeared in recent years.

Section 7.1 introduces parallel computer architecture and describes some parallel algorithms that we will need later. Section 7.2 describes two architectures for the $M$-algorithm [7-6], [7-7]; one is based on parallel sorting and the other on selection. Section 7.3 discusses several architectures for Viterbi decoding [7-8] – [7-10], perfect shuffle linear processor arrays, a hypercube structure and a two-dimensional processor array. We have excluded decoding structures for algebraic block codes; some of these appear in Chapter 3.

# 7.1 PARALLEL ARCHITECTURES AND ALGORITHMS

There is a always a demand for higher data rates and faster processing. One way to achieve this is to use faster devices and the other is to process data in parallel. We take the parallel processing approach here. The emergence of sophisticated coding methods coupled with high data rates makes parallel processing a necessity.

## Parallel Architectures

Parallel architectures for computers and signal processors can be mainly classified into two important classes: *single instruction stream / multiple data stream* (SIMD) computers and *multiple instruction stream/ multiple data stream* (MIMD) computers.

A SIMD machine (Fig. 7.1-1) consists of a number of processors, with a subset of them active at any given moment. All the processors that are active work in lock-step synchronism under the direction of a central control. At any given time, all the active processors execute the same instruction broadcast to them by the control unit. They may, however, operate on different sets of data, hence the name single instruction stream/multiple data stream. Each processor may have its own local storage for program and data. Processors may share their results by either writing to, and retrieving from, a shared memory or by communicating with each other via an interconnection network. A number of interconnection networks have been proposed, some examples being *linear*, *loop*, *mesh*, *hypercube*, *tree*, and *perfect shuffle*, as shown in Fig. 7.1-2. Illiac IV, NASA's massively parallel processor and the connection machine serve as examples of SIMD computers. One of the $M$-algorithm architectures and the three Viterbi algorithm architectures discussed here are based on SIMD principles.

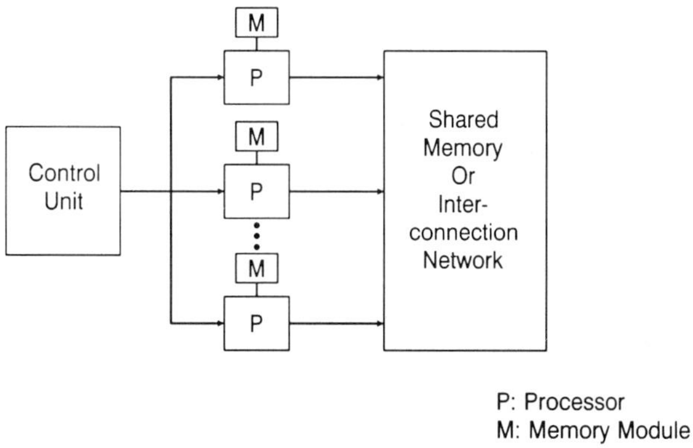


Figure 7.1-1. A single instruction stream/multiple data stream (SIMD) computer.

A MIMD machine (Fig. 7.1-3) consists of a number of processors, each acting independently under its own control unit, and executing and processing its own program and data. As in the SIMD model, the processors may communicate with each other via a shared memory or an interconnection network. The parallel stack algorithm architecture proposed in [7-11] is based on the MIMD model of computing.

## Parallel Sorting Algorithms

Three sorting schemes are discussed now that are useful in coding algorithms. We will describe Batcher odd-even merge sorting, odd-even transposition sorting, and sorting based on the perfect shuffle. We do not give a formal proof for any of the schemes, but instead simply describe and illustrate them with examples.

*Batcher Odd-Even Merge Sorting*

Batcher's work [7-12] was the first result on parallel sorting to appear in the literature. The sorting network is built out of a number of 2-input 2-output devices called comparators. A comparator puts out the smaller of the two inputs onto the upper output terminal labeled L and the larger one onto the lower terminal labeled H, as in Fig. 7.1-4. The figure also shows a merging network that merges two sequences $\{\alpha_1, \alpha_2\}$ and $\{\beta_1, \beta_2\}$, with $\alpha_1, \geq \alpha_2$ and $\beta_1 \geq \beta_2$. This sorting network is called a 2 x 2 merge sorting network.

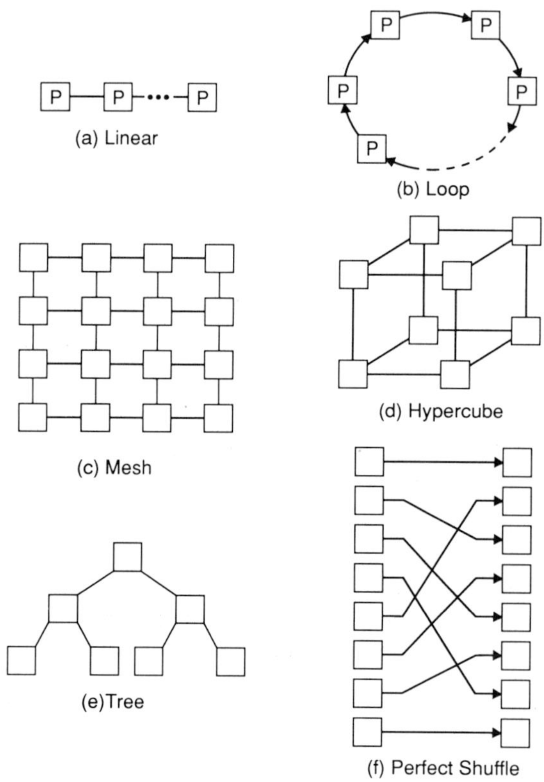


**Figure 7.1-2.** Examples of some interconnection networks.

The Batcher network is easily generalized by defining the recursive structure shown in Fig. 7.1-5. Consider designing an $n \times n$ merging network that merges two sequences $\{\alpha_1, \alpha_2, \ldots, \alpha_n\}$ and $\{\beta_1, \beta_2, \ldots, \beta_n\}$ such that $\alpha_1 \geq \alpha_2 \geq \cdots \geq \alpha_n$ and $\beta_1 \geq \beta_2 \geq \cdots \geq \beta_n$, with $n = 2^m, m$, being an integer. Two $n/2 \times n/2$ merging networks are employed in Fig. 7.1-5. One of them merges the odd-numbered elements $\{\alpha_1, \alpha_3, \ldots \alpha_{n-1}\}$ with the odd-numbered elements $\{\beta_1, \beta_3, \ldots, \beta_{n-1}\}$ to produce the merged sequence $\{\gamma_1, \gamma_2, \ldots, \gamma_n\}$. The other $n/2 \times n/2$ merging network simultaneously merges the even-numbered elements $\{\alpha_2, \alpha_4, \ldots, \alpha_n\}$ with $\{\beta_2, \beta_4, \ldots, \beta_n\}$ to produce $\{\delta_1, \delta_2, \ldots, \delta_n\}$. The final output sequence $\{\epsilon_1, \epsilon_2, \ldots, \epsilon_{2n}\}$ is given by

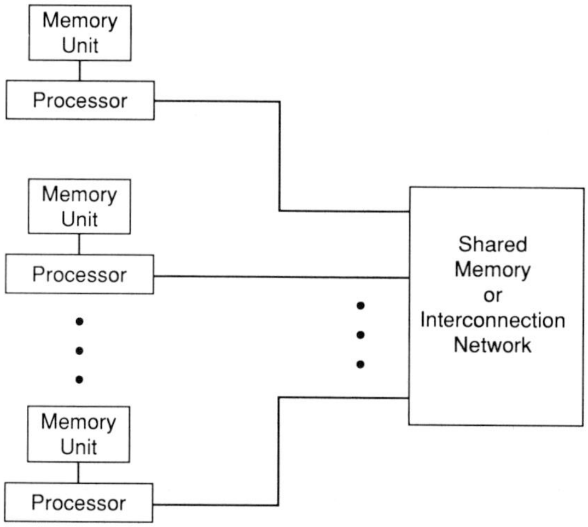


**Figure 7.1-3.** A multiple instruction stream/multiple data stream (MIMD) computer.

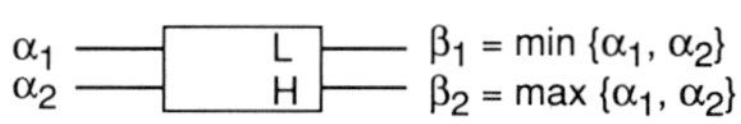


(a)

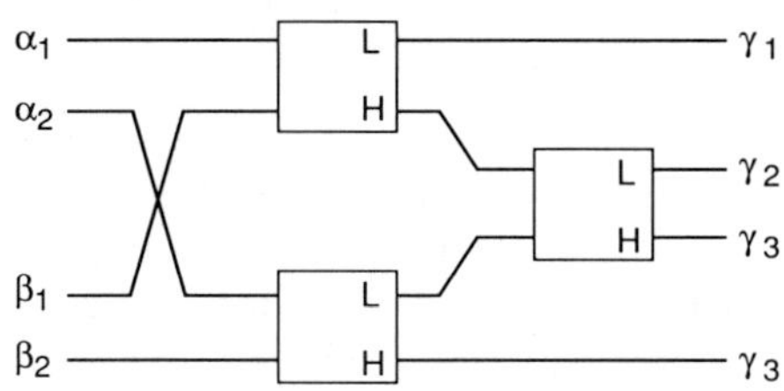


(b)

**Figure 7.1-4.** a) A 2-input 2-output comparator; b) A $2 \times 2$ merge network.

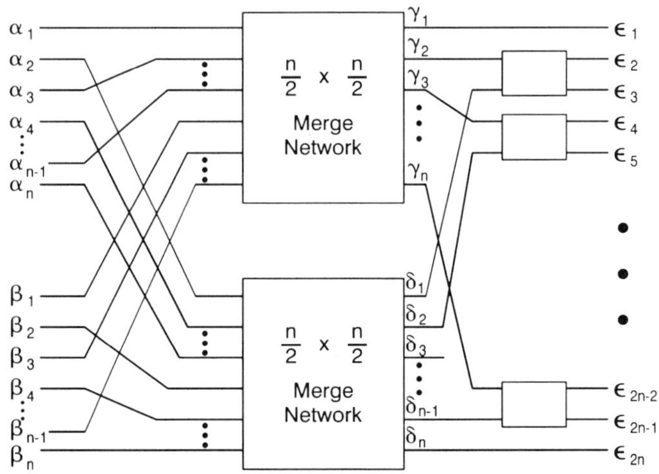


**Figure 7.1-5.** Recursive design of an $n \times n$ Batcher merging network.

$$\begin{aligned} \epsilon_1 &= \gamma_1, \\ \epsilon_{2n} &= \delta_n, \\ \epsilon_{2\ell} &= \min\{\gamma_{\ell+1}, \delta_\ell\}, \qquad & \ell = 1, 2, \ldots, n-1, \\ \epsilon_{2\ell+1} &= \max\{\gamma_{\ell+1}, \delta_\ell\}, \qquad & \ell = 1, 2, \ldots, n-1. \end{aligned}$$

A $1 \times 1$ merging network is simply a single comparator. A $2 \times 2$ merging network appears in Fig. 7.1-4. Using the $2 \times 2$ merging network and the preceding recursive design principle, we can formulate the $4 \times 4$ merging network shown in Fig. 7.1-6. This figure also illustrates how the 8-input Batcher sorter can be designed. During the first stage of merging the eight input numbers, considered as 8 sequences of one number each, are merged using four comparators ($1 \times 1$ merge networks), to produce four sorted sequences, each consisting of two numbers. During the second stage of merging, two $2 \times 2$ mergers are employed to produce two sequences, each consisting of four numbers. The final stage merges the two sequences to produce a single sorted sequence of eight numbers. The figure illustrates how the input sequence {7, 3, 5, 6, 2, 4, 1, 0} flows through the network and appears at the output sorted. The design can be extended recursively to sort 16 numbers, then 32, and so on.

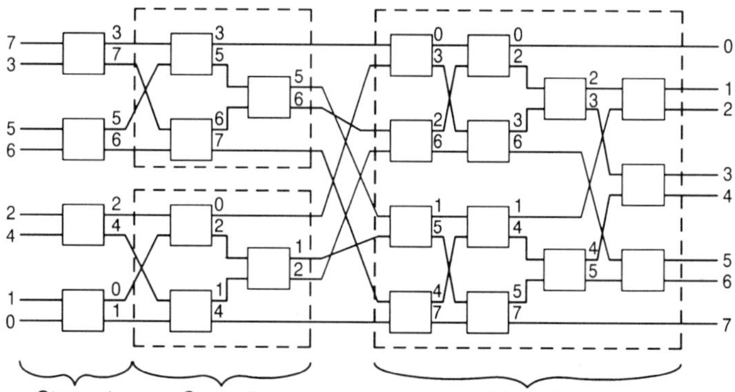


**Figure 7.1-6.** An 8-element sorting network based on Batcher odd-even merging.

*Odd-Even Transposition Sorting*

Unlike Batcher odd-even merge sorting, which requires a network of comparators, the odd-even transposition sorting scheme requires only a linear array of processors. Let the $n$ processors be denoted $P_1, P_2, \ldots, P_n$. Denote by $\beta_i$ the element that $P_i$ has in its storage. Initially, each processor $P_i$ is loaded with $\alpha_i$, where $\{\alpha_1, \alpha_2, \ldots, \alpha_n\}$ is the set of $n$ numbers to be sorted. Hence, initially $\beta_i = \alpha_i$. The algorithm is carried out in $n$ parallel steps. During odd-numbered steps, each odd-numbered processor $P_i$ fetches $\beta_{i+1}$ from $P_{i+1}$, compares $\beta_i$ and $\beta_{i+1}$, and if $\beta_i$ is greater than $\beta_{i+1}$ then $P_i$ and $P_{i+1}$ exchange the elements they have; that is, $P_i$ will have $\beta_{i+1}$ and $P_{i+1}$ will have $\beta_i$. If $\beta_i \leq \beta_{i+1}$, $P_i$ and $P_{i+1}$ retain what they already have. During even-numbered steps each even-numbered processor $P_i$ compares $\beta_i$ with $\beta_{i+1}$ and exchanges it for $\beta_{i+1}$ if $\beta_i > \beta_{i+1}$. Fig. 7.1-7 illustrates the algorithm using the same sequence as in Fig. 7.1-6.

*Sorting Using the Perfect Shuffle*

Figure 7.1-2(f) shows the perfect-shuffle interconnection network. A perfect shuffle is one that produces the "ideal" interlacing of a deck of cards. That is, the deck of cards is split into two halves and merged so that any two neighboring cards from one of the halves is separated by a card from the other. For example, the sequence $\{0,1,2,3,4,5,6,7\}$ will, after a perfect shuffle, result in the sequence $\{0,4,1,5,2,6,3,7\}$. An interesting property of the perfect shuffle is that after $\log_2 n$ shuffles, the

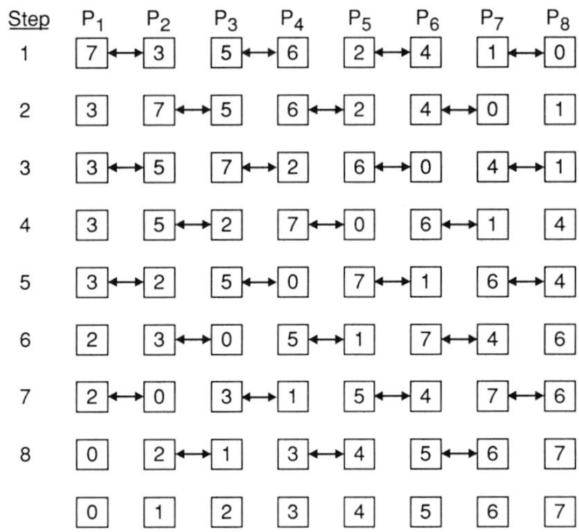


**Figure 7.1-7.** Example illustrating the odd-even transposition algorithm. The processors connected by a double arrow ($\leftrightarrow$) communicate with each other to determine if they should exchange data between themselves.

original sequence of $n$ elements is restored.

Before we use the perfect-shuffle interconnection pattern to perform sorting, we need to define the concept of a *bitonic* sequence. A sequence $\{\alpha_1, \alpha_2, \ldots, \alpha_n\}$ is said to be bitonic if either of the following two conditions are satisfied:

a) There exists some $\ell, 1 \leq \ell \leq n$ such that $\alpha_1 \leq \alpha_2 \leq \cdots \leq \alpha_\ell$ and $\alpha_\ell \geq \alpha_{\ell+1} \geq \cdots \geq \alpha_n$; that is, the elements of the sequence first increase monotonically and then decrease monotonically;

b) There exists some cyclic rotation of the original sequence so that the cyclically rotated sequence satisfies (a).

For example, the sequences $\{0, 1, 2, 3, 7, 6, 5, 4\}$ and $\{3, 7, 6, 5, 4, 0, 1, 2\}$ are both bitonic but $\{0, 1, 2, 4, 3, 7, 5, 6\}$ is not. A single comparator produces a bitonic sequence of length 2. Fig. 7.1-8 shows a set of two comparators that can produce a bitonic sequence of length 4. Also shown in the figure is a recursive structure to construct a bitonic sorter to sort $n$ items using two bitonic sorters that sort $n/2$ items each. The principle is similar to that of the Batcher merging network in Fig. 7.1-5.

Figure 7.1-9 shows in 8-item bitonic sorter based on the perfect shuf-

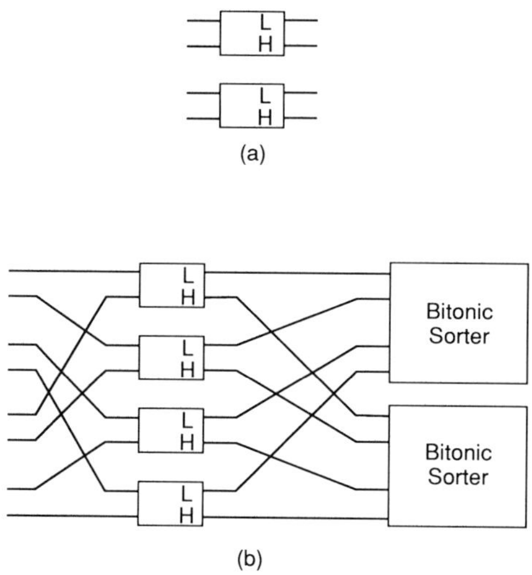


Figure 7.1-8. a) A structure that produces bitonic sequences of length 4; b) A recursive structure for a bitonic sorter.

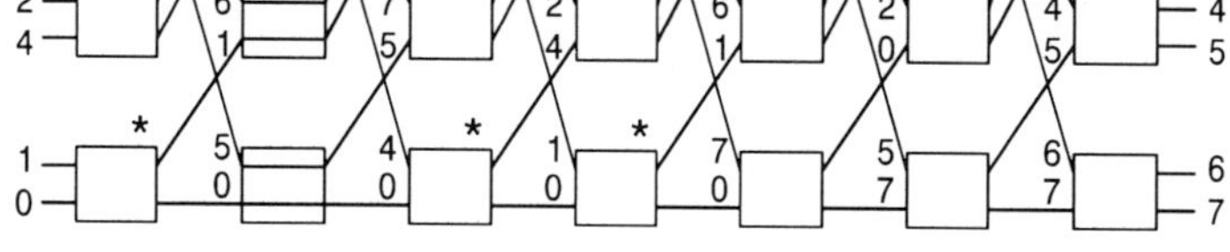

Figure 7.1-9. A perfect-shuffle based bitonic sorter. The normal convention for comparators is that the upper output is the minimum and the lower output is the maximum of the two input numbers; those comparators marked with * put out the larger of the two input numbers onto the upper terminal and the smaller one onto the lower; the comparators with a horizontal line connecting the input to the output terminals transmit the input to the output unaltered.

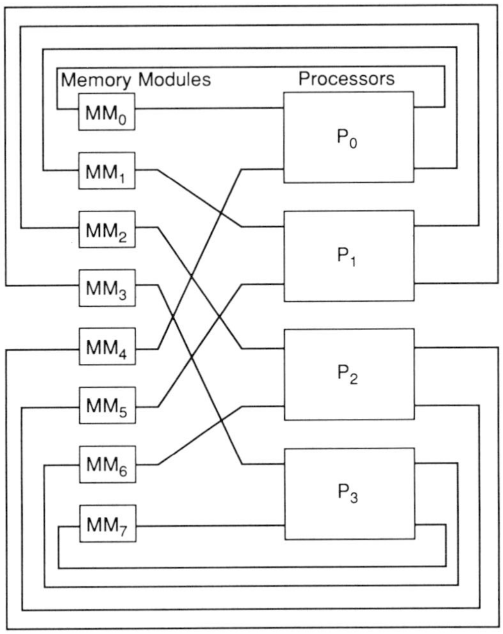


**Figure 7.1-10.** The perfect-shuffle computer to implement the bitonic sorting network of Fig. 7.1-9.

fle. Details are omitted (see Stone [7-13] or Akl [7-4]). The first set of comparators produces two bitonic sequences, each containing four of the eight input elements. These two bitonic sequences are then merged by the next set of three comparison stages to produce a single bitonic sequence containing eight elements. This sequence is then sorted by the remaining three stages. The figure shows how the same input sequence as in the previous two schemes is transformed as it flows through the sorter.

Observe that the interconnection pattern between two adjacent sets of comparators is a perfect shuffle. In fact, it is also the trellis of a rate-$1/a$ convolutional coder with constraint length 3, a useful fact in designing a Viterbi decoder (see Section 7.3). Since only one set of comparators is active at any given instant, it is not necessary to use 28 processors as in Fig. 7.1-9. In fact, four processors will suffice as shown in Fig. 7.1-10. The output of each processor is recirculated to the input using the perfect-shuffle interconnection network. The processors need to implement the three functional capabilities required of them in

Fig. 7.1-9: (1) Processors act as comparators and put out the minimum of two input elements onto the top output terminal and the maximum onto the bottom, or (2) processors reverse these output positions, or (3) processors pass the inputs to the output unaltered. Besides being able to implement the bitonic sorting network of Fig. 7.1-9, Fig. 7.1-10 can also implement the Viterbi algorithm and the fast Fourier transform.

## 7.2 ARCHITECTURES FOR THE M-ALGORITHM

The practical sliding block version of the $M$-algorithm discussed in Section 6.2 inspires the architectures described here. In Chapter 6, we took the view that sequential coding algorithms work with a tree instead of a trellis. When algorithms work with a trellis, they should discard all paths except the best that enter a common state in the trellis. Our architecture [7-6] that we describe next adopts this viewpoint. Parenthetically, we note that even if state-duplicating paths are not rejected at the end of every trellis level, they will be rejected later at the output symbol release, when all those paths are purged whose earliest symbol differs from the released symbol. This approach, however, lets some state-duplicating paths remain temporarily in storage, thereby eating away at the total retained, effectively reducing the $M$. Tests show that duplications degrade the algorithm's performance a little on good channels.

### A SIMD Architecture for the M-Algorithm

Figure 7.2-1 portrays the sliding block version of the $M$-algorithm in block diagram form. As we will see shortly, this figure lends itself to parallel processor implementation using the Batcher sorting network discussed earlier. The first block implements Step 1 of the algorithm; the next three blocks implement Step 2, selecting the best $M$ paths; the last block implements Step 3. We assume $b = 2$ to keep the discussions simple. (Rate-$1/a$ convolutional coders generate trellises with $b = 2$). The architecture is applicable to more general codes.

The block labeled path extension and metric computation employs $M$ processors to extend the $M$ input paths $(a_1, a_2, \ldots, a_M)$ to $bM$ new paths $(b_{11}, b_{12}, \ldots, b_{M1}, b_{M2})$. The $bM$ new paths are then sorted on their states by the state sorter. The state comparator views the set of state-sorted paths $(c_{11}, c_{12}, \ldots, c_{M1}, c_{M2})$ and eliminates all paths but the best that

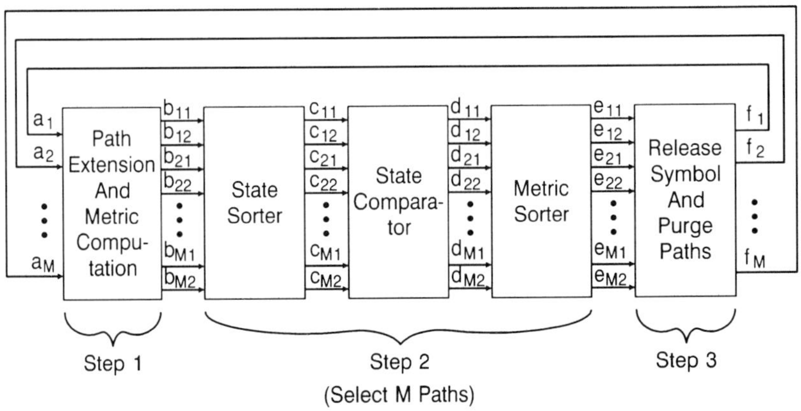


Figure 7.2-1. A block diagram for the $M$-algorithm: Towards the realization of a parallel processors architecture.

enter a common state. This is done by comparing the neighboring paths for duplicate states and assigning a very large metric to the worst of the paths having the same state. These will be eliminated subsequently by the metric-sorter that sorts the $bM$ paths based on their metric. The final block releases the earliest symbol of the best path ($e_{11}$) and purges all those paths differing in their eldest symbol from the released symbol. The output paths ($f_1, f_2, \ldots, f_M$) from this block are then fed back as input to the first block for path extension.

Either state sorting or metric sorting is accomplished by the Batcher sorting network such as the one in Fig. 7.1-6, with the input sequence being the set of states or metrics of paths. However, it is convenient to modify the network as shown in Fig. 7.2-2. This version employs the same number of processors at each step of computation in the sorting network. This facilitates implementation of the sorting network using the SIMD architecture shown in Fig. 7.2-3 that employs only $M$ processors. The processors perform the functions required of them at each sorting step, acting either as comparators or switches, and the interconnection network routes the output of the processors appropriately back to their input. Referring to Fig. 7.2-2 and viewing the interconnections as permutations, we see that the following three permutations are required:

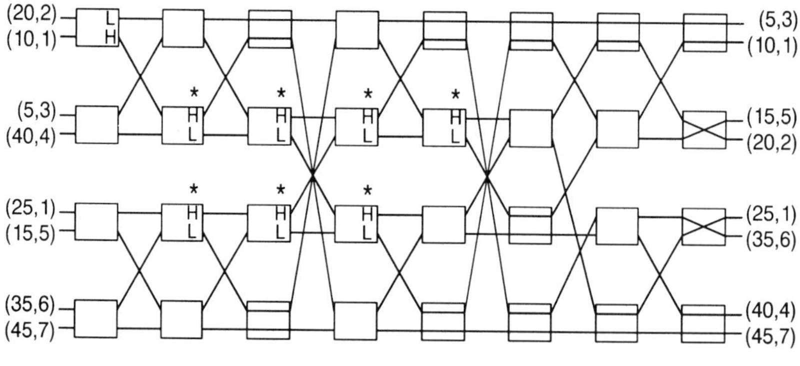


**Figure 7.2-2.** An eight-input reorganized Batcher odd-even merge-sort network with exactly four processors at each stage of computation; the input to the network is the set of metric-state tuples associated with the eight (in general $2M$) input paths. The network pictured carries out metric sorting; it performs other stages of step 2 in Fig. 7.2-1 similarly.

$$\pi_1 = \begin{pmatrix} 1 & 2 & 3 & 4 & 5 & 6 & 7 & 8 \\ 1 & 3 & 2 & 4 & 5 & 7 & 6 & 8 \end{pmatrix},$$

$$\pi_2 = \begin{pmatrix} 1 & 2 & 3 & 4 & 5 & 6 & 7 & 8 \\ 1 & 7 & 3 & 5 & 4 & 6 & 2 & 8 \end{pmatrix},$$

$$\pi_3 = \begin{pmatrix} 1 & 2 & 3 & 4 & 5 & 6 & 7 & 8 \\ 1 & 3 & 2 & 7 & 4 & 6 & 5 & 8 \end{pmatrix}.$$

A variety of interconnection networks are surveyed in [7-14]. One such network, the rearrangeable network, is capable of achieving all possible input-output connections. For a $2M$-input $2M$-output interconnection network, there are $(2M)!$ input-output connections. Specialized interconnection networks may be designed for those networks that require much smaller than $(2M)!$ permutations. For example, the network in Fig. 7.2-4 realizes all the three permutations $\pi_1, \pi_2$ and $\pi_3$. The switches in the figure have two inputs and two outputs and are capable of being in one of two states, the through state or the cross state. The network achieves input-output connections given by $\pi_1$ when switches 1, 7, and 8 are in the cross state and all the others are in the through state. Input-output connections given by $\pi_2$ and $\pi_3$ are similarly achieved using other combinations of switch states.

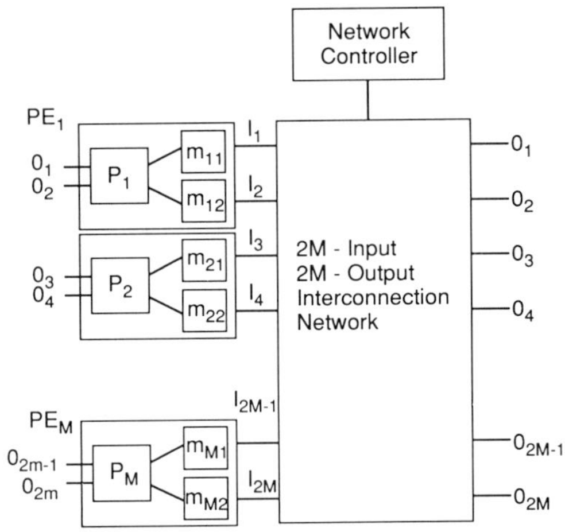


**Figure 7.2-3.** A SIMD machine architecture for the $M$-algorithm employing $M$ processing elements; each processing element $(\mathrm{PE})_i, i = 1, 2, \ldots, M$, consists of a processor $\mathrm{P}_i$ and two (in general $b$) memory elements, $\mathrm{m}_{i1}$ and $\mathrm{m}_{i2}, i = 1, 2, \ldots, M$.

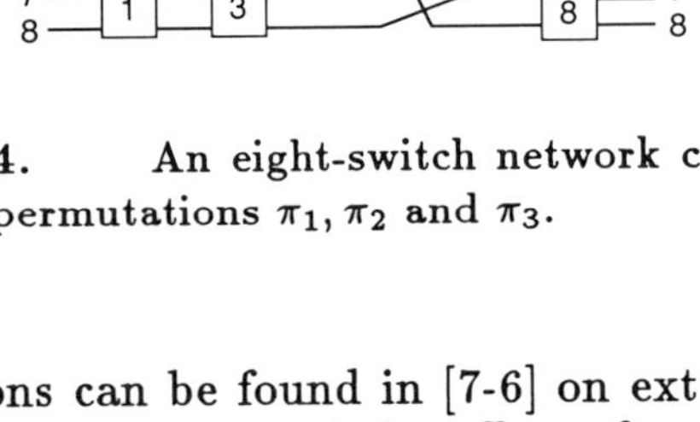


**Figure 7.2-4.** An eight-switch network capable of implementing the permutations $\pi_1, \pi_2$ and $\pi_3$.

Further discussions can be found in [7-6] on extending the architecture to less than $M$ processors and the effect of varying the number of processors on the number of sorting steps. Our discussion can easily be extended to other sorting schemes such as the odd-even transposition or perfect shuffle. In closing, we point out that Batcher sorting network finds use beyond the sorting and tree searching discussed here; see, for example, [7-15] for its use in the design of switches for broadband networks.

## A Selection-Based Architecture for the M-Algorithm

We discuss only the principle behind the selection process and refer the reader to [7-7] for further details concerning logic circuit design, not only for selecting $M$ paths out of $bM$ but also for extracting the best path, extending the paths in storage, releasing the output symbol, and purging paths. This architecture employs parallelism in the path selection process only; other activities of the algorithm such as path extension and best path extraction are carried out sequentially.

Consider selecting the $M$ smallest numbers out of a set $\mathcal{X}$ containing $bM$ numbers. Let each number be represented by a $\ell$-bit binary number. By examining the most significant bit, we can partition the set of numbers into two sets $\mathcal{X}_0$ and $\mathcal{X}_1$; $\mathcal{X}_0$ contains those numbers among $bM$ with the most significant bit equal to 0 and $\mathcal{X}_1$ all the remaining numbers. Three cases arise now.

- If $\mathcal{X}$ contains exactly $M$ numbers, the algorithm has selected the $M$ smallest numbers.
- If $\mathcal{X}$ contains less than $M$ numbers, then all of these must belong to the set of $M$ numbers desired, $\mathcal{X}_M$. In this case, we look upon $\mathcal{X}_1$ as the new reduced set $\mathcal{X}$ containing $bM - |\mathcal{X}_M|$ numbers, where $|\mathcal{X}_M|$ denotes the size of $\mathcal{X}_M$. We need to select now only $M - |\mathcal{X}_M|$ numbers out of $bM - |\mathcal{X}_M|$.
- If $\mathcal{X}_0$ contains more than $M$ numbers, then none of the numbers from $\mathcal{X}_1$ will be in $\mathcal{X}_M$ and these are therefore rejected and dropped from further consideration. Now, $\mathcal{X}_0$ serves as the new reduced set $\mathcal{X}$ containing $bM - |\mathcal{X}_R|$ numbers, where $|\mathcal{X}_R|$ is the size of the rejected set. We need to select now $M$ numbers out of $bM - |\mathcal{X}_R|$.

In either of the last two cases the algorithm is repeated, but this time the reduced set $\mathcal{X}$ is partitioned into two subsets based on the next less significant bit than the one just looked at. The algorithm terminates either when $\mathcal{X}_M$ has $M$ numbers or all the $\ell$ bits have been viewed. $\mathcal{X}_M$ then contains the desired numbers. The example below illustrates the selection process.

### Example 7.2-1: Selection of $M$ Numbers Out of $bM$.

Let $M = 4, b = 2$ and $\mathcal{X} = \{1101, 0010, 0100, 0111, 1010, 0101, 1001, 0001\}$. We use the selection algorithm above to select 4 out of 8 numbers.

Table 7.2-1 illustrates the selection process. At the end of iteration 1, $\mathcal{X}_1$ has three numbers that are dropped from further consideration. This is indicated by adding them to $\mathcal{X}_R$. $\mathcal{X}_0$ with five numbers forms the new reduced set $\mathcal{X}$ for iteration 2. We now need to select 4 out of 5 numbers. At the end of iteration 2, $\mathcal{X}$ has two numbers which are among the four numbers desired. These are added to $\mathcal{X}_M$. The set $\mathcal{X}_1$ with three numbers becomes set $\mathcal{X}$ out of which we need to select two. The algorithm selects one of these three during iteration 3, and one more out of the two that remain during the last iteration.

**Table 7.2-1.** SELECTION 4 OUT OF 8 NUMBERS; A CHECK MARK ($\surd$) UNDER A SET INDICATES THAT THE NUMBER CHECKED BELONGS TO THAT SET.

| $\mathcal{X}$ | Iterations | | | | | | | | | | | | | | | |
|---|---|---|---|---|---|---|---|---|---|---|---|---|---|---|---|---|
| | 1 | | | | 2 | | | | 3 | | | | 4 | | | |
| | Select 4 out of 8 | | | | Select 4 out of 5 | | | | Select 2 out of 3 | | | | Select 1 out of 2 | | | |
| | $\mathcal{X}_0$ | $\mathcal{X}_1$ | $\mathcal{X}_R$ | $\mathcal{X}_M$ | $\mathcal{X}_0$ | $\mathcal{X}_1$ | $\mathcal{X}_R$ | $\mathcal{X}_M$ | $\mathcal{X}_0$ | $\mathcal{X}_1$ | $\mathcal{X}_R$ | $\mathcal{X}_M$ | $\mathcal{X}_0$ | $\mathcal{X}_1$ | $\mathcal{X}_R$ | $\mathcal{X}_M$ |
| 1101 | | ✓ | ✓ | | | | ✓ | | | | ✓ | | | | ✓ | |
| 0010 | ✓ | | | | ✓ | | | ✓ | | | | ✓ | | | | ✓ |
| 0100 | ✓ | | | | | ✓ | | | ✓ | | | ✓ | | | | ✓ |
| 0110 | ✓ | | | | | ✓ | | | | ✓ | | | ✓ | | | ✓ |
| 1010 | | ✓ | ✓ | | | | ✓ | | | | ✓ | | | | ✓ | |
| 0111 | ✓ | | | | | ✓ | | | | ✓ | | | | ✓ | ✓ | |
| 1001 | | ✓ | ✓ | | | | ✓ | | | | ✓ | | | | ✓ | |
| 0001 | ✓ | | | | ✓ | | | ✓ | | | | ✓ | | | | ✓ |
| $\lvert\mathcal{X}_0\rvert$ | >4 | | | | >4 | | | | >2 | | | | = 1 | | | |
| $\lvert\mathcal{X}_M\rvert$ | | | | 0 | | | | 2 | | | | 3 | | | | 4 |

## 7.3 ARCHITECTURES FOR THE VITERBI ALGORITHM

### The Perfect-Shuffle-Based Viterbi Decoder

We have pointed out in Sec. 7.1 that the perfect-shuffle computer of Fig. 7.1-10 can, in fact, be used to implement the Viterbi algorithm. To see how, assume that at any stage of the algorithm, memory modules $MM_\ell$ and $MM_{\ell+1}$ in Fig. 7.1-10, $\ell = 0, 2, 4, 6$, have stored the cumulative metric $\mu(\ell/2)$† and the path map $x(\ell/2)$ of the best path in the trellis from the root node to state $\ell/2$. During the next stage of trellis computation, processor $P_0$ retrieves metrics and the best path to state 0 and state 2 from $MM_0$ and $MM_4$, respectively. It then computes the branch metrics $\mu_{00}$ and $\mu_{20}$ (see Fig. 7.3-1), adds them respectively to the metric from $MM_0$ and $MM_4$, selects the best out of these two metrics, and stores the best metric and updated path back again into $MM_0$ and $MM_1$. These are identical to the computations as required by the trellis section in Fig. 7.3-1. Simultaneously, other processors carry out similar computations to complete a stage of trellis computation carried out by the Viterbi algorithm.

Architectures based on perfect shuffle and VLSI systolic arrays are discussed in Gulak, et al. [7-10], [7-16]. Next, we discuss two linear systolic processor array architectures. In terms of VLSI layout, a linear processor array has the advantage of requiring much smaller wire area compared to the perfect shuffle.

### Linear Systolic Processor Array Architectures

Figure 7.3-2, adapted from [7-10], explains how a linear systolic array‡ of processors can route and exchange data among themselves to achieve the perfect shuffle interconnection. The architecture consists of twice the number of processors (in general $b$ times, where $b$ is the number of trellis branches) compared to the perfect shuffle. Each pair of processors $P_\ell$ stores the cumulative metric and the best path to state $\ell$. In the

†The notation $\mu(S)$ refers to the metric of the path from the root node of the trellis to some state in the trellis stage of interest. We have omitted the index corresponding to the trellis stage being considered. We also use $\mu_{S_1S_2}$ to denote the metric of the branch connecting states $S_1$ and $S_2$.

‡A systolic array consists of a set of processors that work rhythmically, computing, processing, and passing along data through the array.

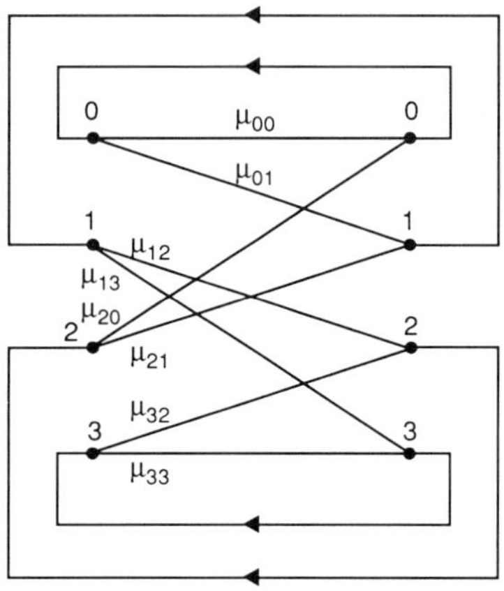


Figure 7.3-1. A section of the trellis of Fig. 4.2-3. (Note that we have interchanged states 1 and 2 of Fig. 4.2-3; this does not really change the encoder.) The recirculation edges are shown to indicate how computations proceed from one stage of the trellis to the next.

next three steps they exchange and route path and metric information among themselves. Now adjacent processors contain best paths to states that enter a common state after a trellis transition. After adding branch metrics to path metrics, pairs of processors, connected by double arrows (the last set of processors in Fig. 7.3-2), compare and store the best metric in their registers. The processors are now ready to begin the next stage of computation.

Figure 7.3-3 illustrates another architecture due to Chang and Yao [7-8]. Before a stage of trellis computation begins, the systolic array is supplied the state metrics $\mu(0), \mu(1), \mu(2)$, and $\mu(3)$ corresponding to the states of the trellis section shown. The metrics enter the left end of the array. Branch metrics are also supplied vertically down into the array as shown in the figure. The infinite metrics, denoted by $\infty$, correspond to branches of the trellis that do not exist. (For example, branch metric between states 0 and 1 is denoted $\mu_{01}$ but that between 0 and 2 is $\infty$). The systolic array is synchronized so that each time the array is clocked, the state metrics and the branch metrics move rhythmically one position right and one position down, respectively. The processors perform simple computations and alter the contents of a metric register associated with them according to

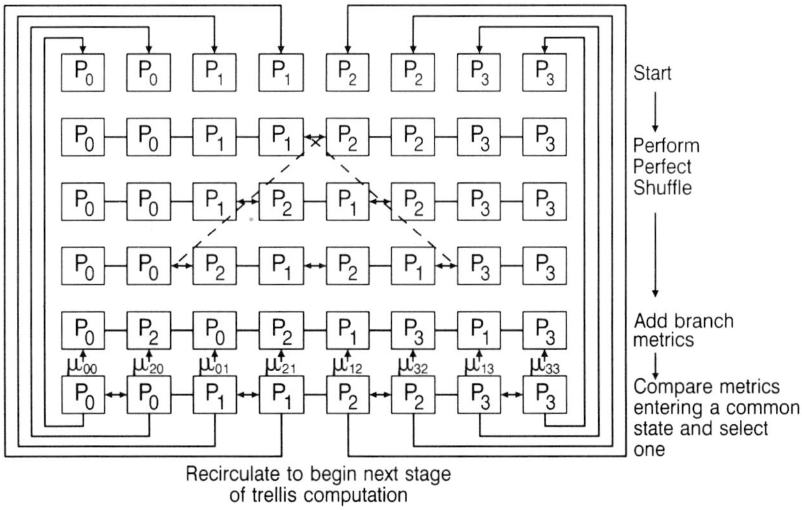


**Figure 7.3-2.** A linear array of processors to implement the Viterbi decoder; $\leftrightarrow$ indicates exchange operations and the dashed lines show the progression of the exchanges among processors in the array; adapted from Gulak, et. al., [7-10].

$$M(P_i) \leftarrow \min\{M(P_i), \mu(\ell) + \mu_{\ell i}\},$$

where $M(P_i)$ represents the contents of the metric register associated with processor $P_i$, $\mu(\ell)$ the metric of the best path to state $\ell$, and $\mu_{\ell i}$ the branch metric entering $P_i$ from above. Just prior to each stage of trellis computations, the $M(P_i)$ are initialized to infinity.

In addition to altering the contents of the metric register, each processor passes along the metric that enters it from the left to its neighbor on the right. When the last state metric to enter the array leaves the array at the right end, a stage of trellis computation is complete. Now each processor $P_i$ has in its metric register $M(P_i)$, the best metric to state $i$. The contents of the metric registers are recirculated into the processor array so that the next stage of trellis computation can begin.

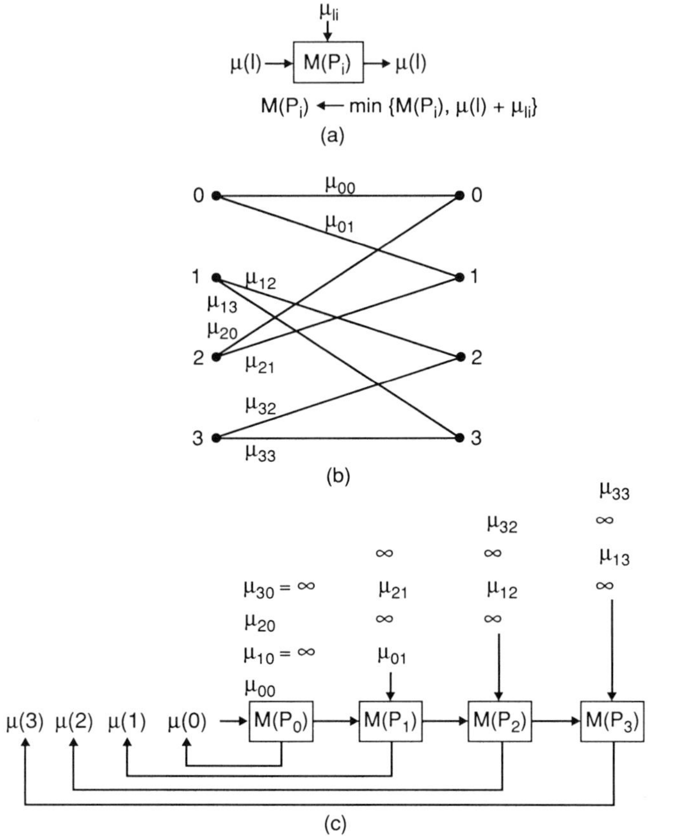


**Figure 7.3-3.** A systolic Viterbi decoder; (a) Computations performed by processor $P_i$; (b) A trellis stage; (c) The systolic array decoder showing state and branch metric flows.

**Example 7.3-1: Systolic Viterbi Decoding.**

Consider the trellis section illustrated in Fig. 7.3-4(a). The figure shows the cumulative metrics to states 0, 1, 2, and 3 computed by the decoder during the preceding trellis stage computations. The branch metrics are also listed on the branches. Before computations begin, the processors in Fig. 7.3-4(b) have a very large number stored in their registers. The figure walks through the computations of the systolic decoder in successive time instants as the array is clocked.

It is clear from Fig. 7.3-4(b) that at most only two processors are

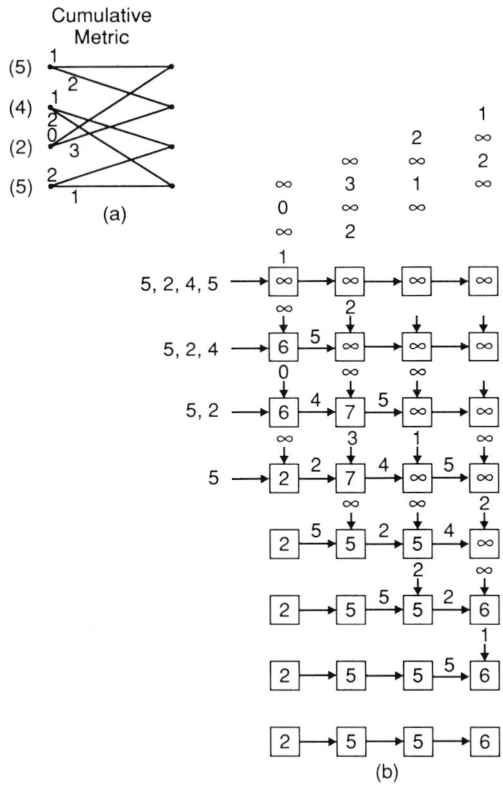


**Figure 7.3-4.** a) A trellis section indicating state and branch metrics; b) Systolic Viterbi decoder computations corresponding to the trellis section in (a).

engaged in meaningful activity. This is due to the sparseness of the trellis. By considering two trellis sections of Fig. 7.3-4(a) jointly, a trellis section is created in which every node at the left end of a trellis section is connected to every node at the right. We can then generate branch metrics for the two-stage trellis and keep the processors fully utilized. In general, for a given rate-$1/a$ trellis with constraint length $\nu$, a strongly connected trellis in which every state can be reached from every other state results from lumping $\nu - 1$ stages of the original trellis [7-8].

## A Hypercube Viterbi Decoder

We describe a scheme due to Pollara [7-9] that enables the Viterbi decoder to be implemented on a hypercube. A hypercube is a system of $2^n$ processors placed on the vertices of a hypercube ($n$-cube) that are interconnected by the edges of the $n$-cube. For simplicity, we restrict the discussion to $n = 3$.

Consider the trellis of a rate-$1/a$ convolutional code with constraint length $\nu = 4$. This code has a trellis with eight states, three sections of which are shown in Fig. 7.3-5(a). Processors are labeled $P_0, P_1, \ldots, P_7$. Initially, processor $P_i$ is assigned state $i$. This means that $P_i$ has in its registers the cumulative metric of the best path to state $i$ as well as the best path sequence. To keep the discussions simple, we will concentrate on metric computations and ignore updating the path map sequence.

From the trellis section in (a), we see that $P_0$ and $P_4$ need to tell each other the metrics to states 0 and 4, add branch metrics, and compare and select one out of the two cumulative metrics. Similarly, pairs of processors $(P_1, P_5), (P_2, P_6)$, and $(P_3, P_7)$ communicate with each other, in order to complete one stage of trellis computations. Processor interconnections on the 3-cube during stage 1 of trellis computations are shown in Fig. 7.3-5(c). The processor-pairs involved in communication are connected by double arrows ($\leftrightarrow$).

At the end of stage 1 computations, the processors represent different trellis states, except for $P_0$ and $P_7$. For example, $P_4$ computes the best metric to state 1, and hence it stores state 1 information at the end of stage 1. Fig. 7.3-5(b) shows the states assigned to various processors at the end of stage 1 computations. Referring to this figure, we observe that processors 0 and 2 have metrics to states 0 and 4, respectively. They therefore communicate with each other to participate in stage 2 trellis computations. Processor connections for stage 2 are shown in (c). Similar observations are made and processor connections established as in (c) to perform stage 3 trellis computations, at the end of which each processor $P_i$ returns to its original state, state $i$.

In addition to showing the state assignment, Fig. 7.3-5(b) also shows interconnects during the three stages, laid out as a flow diagram. What evolves is the FFT flow diagram for decimation in frequency. Forney [7-17] points out the relationship between the FFT flow diagram and the trellis of a convolutional coder; ref. [7-9] makes similar observations.

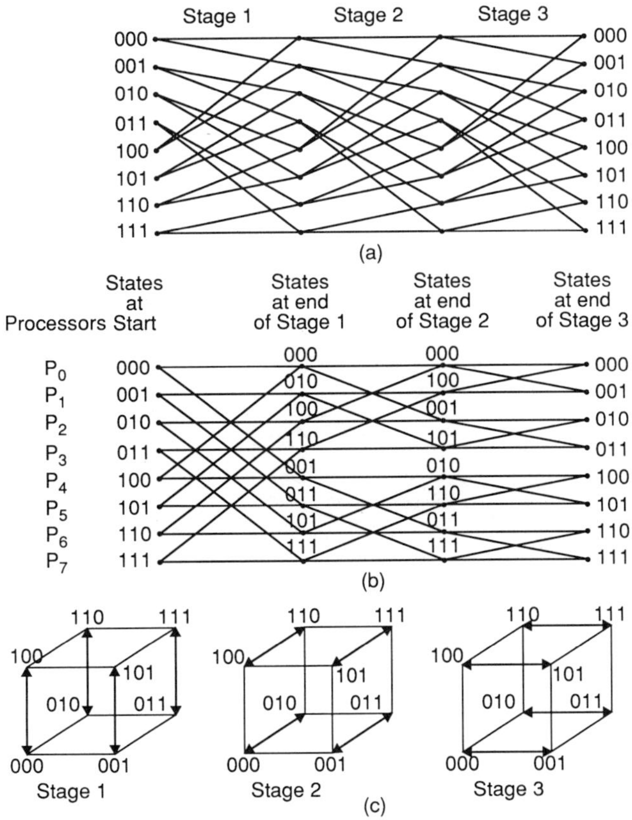


Figure 7.3-5 a) Three sections of an 8-state trellis; b) Assignment of states to processors during the three stages of trellis computations; c) Processor interconnections.

## A Systolic Mesh Decoder

The mesh architecture due to Gulak and Kailath [7-10] consists of a two-dimensional array of size $\sqrt{2^{(\nu-1)}} \times \sqrt{2^{(\nu-1)}}$, consisting of $2^{(\nu-1)}$ processors, where $\nu$ is the constraint length of the convolutional code. Consider, for example, the sixteen-state trellis of Fig. 7.3-6(a). Initially, the sixteen states are mapped onto a set of sixteen processors, arranged in a two-dimensional array of size 4 x 4, as shown in Fig. 7.3-6(b). Observe from the trellis in (a) that during stage 1 of trellis computations, processors communicate in pairs, exchanging state information, adding metrics, and selecting one out of two paths entering a common state. The pairs involved in communication are (0, 8), (1, 9), (2, 10), (3, 11),

(4, 12), (5, 13), (6, 14), and (7, 15). The connections required are indicated in Fig. 7.3-6(c). These communications require two hops. For example, processor 0 communicates with 8 through 4. At the conclusion of these computations, processor 0 contains the best metric to state 0 and processor 8 the best metric to state 8.

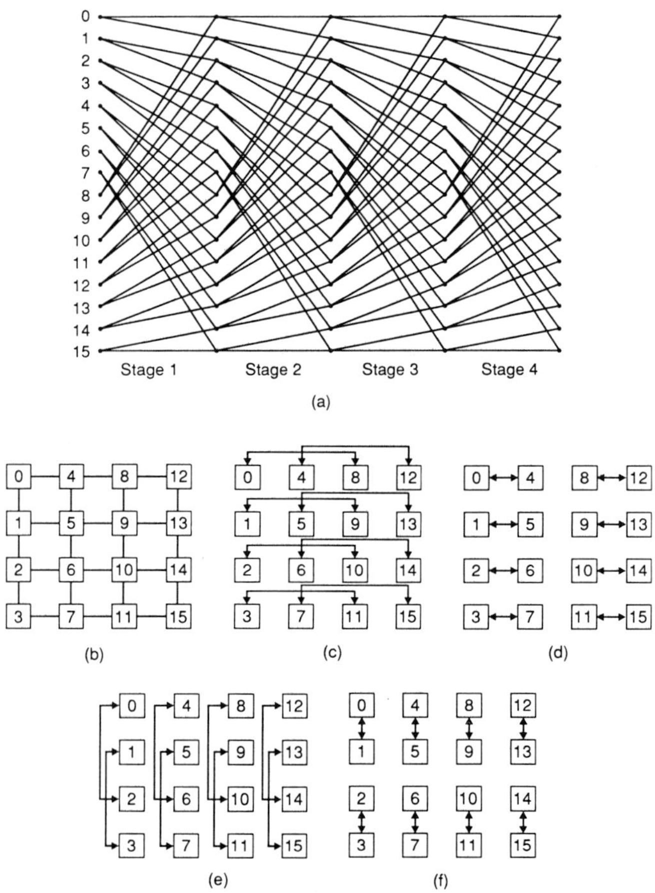


Figure 7.3-6. (a) A sixteen state trellis; (b) The two dimensional square array of 16 processors; (c), (d), (e), and (f): Processors interconnections during stages 1, 2, 3, and 4 of trellis computations.

Communications between other processor pairs result in state assignments, as indicated in Table 7.3-1. Stages 2, 3, 4 require the processor interconnections shown in Fig. 7.3-6(d), (e), (f). For example, at the conclusion of stage 1 computations, processors 0 and 4 contain best metrics to states 0 and 8, respectively. These processors, therefore, communicate during stage 2 computations.

This chapter does not certainly exhaust all the coding architectures that have appeared in the literature. Some more may be found in the references. Besides the breadth-first class of decoders considered here, there are other categories of coding algorithms that lend themselves to systolic array implementation. One example is the vector quantization architecture in reference [7-2]. Another is the metric-first coder architecture in [7-11]. Complexity issues revolve around how difficult it is to realize the various architectures in VLSI circuits.

**Table 7.3-1.** STATES ASSIGNED TO THE SIXTEEN PROCESSORS DURING SUCCESSIVE STAGES OF TRELLIS COMPUTATIONS.

| Processors | State Assignment | | | | |
|---|---|---|---|---|---|
| | Start | End Of Stage 1 | End Of Stage 2 | End Of Stage 3 | End Of Stage 4 |
| 0 | 0 | 0 | 0 | 0 | 0 |
| 1 | 1 | 2 | 4 | 8 | 1 |
| 2 | 2 | 4 | 8 | 1 | 2 |
| 3 | 3 | 6 | 12 | 9 | 3 |
| 4 | 4 | 8 | 1 | 2 | 4 |
| 5 | 5 | 10 | 5 | 10 | 5 |
| 6 | 6 | 12 | 9 | 3 | 6 |
| 7 | 7 | 14 | 13 | 11 | 7 |
| 8 | 8 | 1 | 2 | 4 | 8 |
| 9 | 9 | 3 | 6 | 12 | 9 |
| 10 | 10 | 5 | 10 | 5 | 10 |
| 11 | 11 | 7 | 14 | 13 | 11 |
| 12 | 12 | 9 | 3 | 6 | 12 |
| 13 | 13 | 11 | 7 | 14 | 13 |
| 14 | 14 | 13 | 11 | 7 | 14 |
| 15 | 15 | 15 | 15 | 15 | 15 |

# PROBLEMS

7-1 Consider performing state comparison to eliminate duplicate states using the odd-even transposition scheme. How many odd-even transposition steps are required if the trellis has $b$ branches out of each state and $M$ processors are used? Assume that state sorting has been done prior to state comparison.

7-2 In Problem 7-1, is it possible to combine the state sorting and state comparison blocks into one using the odd-even transposition scheme? If so, discuss the scheme.

7-3 Consider the rate-2/3 code of Fig. 4.2-6. Assume the use of the $M$-algorithm to decode the received sequence, with $M = 4$. We will, therefore, view the trellis section shown in Fig. 4.2-6 as part of a tree. Before trellis computation begins, the four paths in storage have metrics, 10, 7, 6, and 11, ending, respectively, in states 00, 01, 10, and 11. Observe that the four paths in storage need not have distinct states unless state-duplicating paths are eliminated at each stage of trellis computation. Assume the received bits for the trellis section shown are 001. Using the selection algorithm of Section 7.2 (employed by the selection-based architecture of Simmons [7-7]), select four out of the sixteen paths produced during the path extension stage of the $M$-algorithm. Do these four paths have distinct states? Did the algorithm select exactly four paths? Under what circumstances does the algorithm select more than 4 ($> M$ in general)? How can you modify the algorithm so that it always selects $M$ or fewer?

7-4 Perform the perfect-shuffle-based Viterbi decoder computations corresponding to stage 3 of the trellis shown in Fig. 4.2-4. Assume the same received sequence as shown in this figure.

7-5 Repeat Problem 7-4 for the linear array of processors that perform perfect shuffle Viterbi decoding.

7-6 Consider decoding the received sequence given in Problem 4-4 using the systolic Viterbi decoder of Chang and Yao. Show the decoder computations corresponding to the last set of three received bits (001).

# REFERENCES

[7-1] C. Mead and L. Conway, *Introduction to VLSI Systems*, Addison-Wesley, Reading, Mass., 1980.

[7-2] S. Y. Kung, *VLSI Processor Arrays*, Prentice-Hall, Englewood Cliffs, N.J., 1988.

[7-3] D. E. Knuth, *The Art of Computer Programming: Sorting and Searching*, vol. 3, Addison-Wesley, Reading, Mass., 1973.

[7-4] S. G. Akl, *Parallel Sorting Algorithms*, Academic Press, New York, 1985.

[7-5] K. Hwang and F. Briggs, *Computer Architectures and Parallel Processing*, McGraw-Hill, New York, 1984.

[7-6] S. Mohan and A. K. Sood, "A multiprocessor architecture for the $(M, L)$-algorithm suitable for VLSI implementation", *IEEE Trans. Communications*, vol. COM-34, pp. 1218–1224, Dec. 1986.

[7-7] S. J. Simmons, "A nonsorting VLSI structure for implementing the $(M, L)$-algorithm", *IEEE J. Sel. Areas Commun.*, vol. SAC-6, pp. 538–546, April 1988.

[7-8] C. Y. Chang and K. Yao, "Viterbi decoding by systolic array", *Proc.*, Twenty-Third Allerton Conf. on Communs., Control and Computing, Monticello, IL, pp. 430–439, Oct. 1985.

[7-9] F. Pollara, "Concurrent Viterbi algorithm on a hypercube", *ibid.*, pp. 440–449.

[7-10] P. G. Gulak and T. Kailath, "Locally connected VLSI architectures for the Viterbi algorithm", *IEEE J. Sel. Areas Commun.*, vol. SAC-6, pp. 527–537, April 1988.

[7-11] S. Mohan and S. M. Komandur, "Performance of a multiprocessor-based parallel stack algorithm speech encoder", *IEEE Trans. Communications*, vol. COM-35, pp. 463–466, April 1987.

[7-12] K. E. Batcher, "Sorting networks and their applications", in *Proc.*, AFIPS 1968 Spring Joint Comput. Conf., vol. 32, Montvale, N.J., pp. 307–314, 1968.

[7-13] H. S. Stone, *High Performance Computer Architecture*, Addison-Wesley, Reading, Mass., 1990.

[7-14] T. Feng, "A survey of interconnection networks", *Computer*, vol. 14, pp. 12–17, Dec. 1983.

[7-15] J. Hui, "Switching integrated broadband services by sort-Banyan networks", Special Issue on ISDN, *Proc. IEEE*, vol. 79, pp. 145–154, Feb. 1991.

[7-16] P. G. Gulak and E. Schwedyk, "VLSI structures for Viterbi receivers: Part I - General theory and applications", *IEEE J. Sel. Areas Commun.*, vol. SAC-4, pp. 142–154, Jan. 1986.

[7-17] G. D. Forney, Jr., "The Viterbi algorithm", *Proc. IEEE*, vol. 61, pp. 268–278, March 1973.

# APPENDIX A

# Signal Space Theory

This appendix reviews the theory of maximum likelihood detection of signals transmitted through a channel with additive white Gaussian noise. Through the theory we can characterize signals as points in a simple Euclidean vector space and the ML detector as a simple operation in that space. As illustrations we will look at several of the important signals used in the book. A general name for this theory is *signal space* theory. It traces back to the work of Kotelnikov (ref. [1-8] in Chap. 1), work that is still in print, but the classic introduction is the text by Wozencraft and Jacobs [1-10]. More modern, but shorter, treatments appear in the texts [1-4], [1-11], [1-12], [1-14], and [4-22] - [4-24]. We must be even more brief here.

## The Maximum Likelihood Receiver

A *maximum likelihood* (ML) receiver selects the transmitted signal that makes the channel output seem most probable. Suppose that the transmitter sends one of the signal set $s_1(t), ..., s_m(t)$, say signal $s_k(t)$. To this the channel adds the AWGN waveform $\eta(t)$ to form the receiver input, $r(t) = s_k(t) + \eta(t)$. The ML receiver decides in favor of the signal that maximizes the probability $P\{s_i(t)|r(t)\}$ over $i$. In calculating these probabilities, it makes the assumption that all $m$ messages are equally likely.

To compute a probability $P\{s_i(t)|r(t)\}$, we first apply Bayes' Rule to obtain

$$P\{s_i(t)|r(t)\} = \frac{P\{r(t)|s_i(t)\}P\{s_i(t)\}}{P\{r(t)\}}. \tag{A-1}$$

In performing its maximization, the receiver can ignore the signal probabilities $P\{s_i(t)\}$ and the *a priori* probability $P\{r(t)\}$, since both of these are constant over $i$. The receiver may thus find the maximum of just $P\{r(t)|s_i(t)\}$ over $i$. But we can write this probability as $P\{\eta(t) = r(t) - s_i(t)\}$, and so what the receiver needs to perform is

$$\text{Find:}\quad \max_i P\{\eta(t) = r(t) - s_i(t)\}$$

$$\text{Put out:}\quad i \text{ for which } s_i(t) \text{ achieves max.} \tag{A-2}$$

In (A-2), $r(t)$ is observed, $s_i(t)$ is assumed, and the probability measure is that of the noise $\eta(t)$.

## Signal Space

When $\eta(t)$ is an outcome of a Gaussian stochastic process, it is impossible to assign probabilities to individual outcomes in a consistent way. The way out of this difficulty is to construct an orthogonal basis for the functions in (A-2), express the functions as components in a vector space, and then assign probabilities to the components. The vector space here is called the signal space.

Suppose the functions $\phi_1(t), ..., \phi_J(t)$ are such an orthonormal basis function set. Then $s_i(t)$ in signal space is the vector

$$\mathbf{s}_i = (s_{i1}, ..., s_{iJ}), \qquad i = 1, ..., m \tag{A-3}$$

in which the $j$-th component is the inner product

$$s_{ij} = \int s_i(t)\phi_j^*(t)\,dt \triangleq (s_i, \phi_j).$$

The range of integration here is the duration of the signals. We may obtain such a basis set by the usual Gram-Schmidt process applied to the signals, but a good set is often obvious. The size of the set $J$ satisfies $J \leq m$.

In the same way, the noise waveform $\eta(t)$ may be expressed as

$$\eta = (\eta_1, ..., \eta_J, \eta_{J+1}, ...) \tag{A-4}$$

and the received waveform $r(t)$ as

$$\mathbf{r} = (r_1, ..., r_J, r_{J+1}, ...). \tag{A-5}$$

Here we show an infinite number of components beyond the $J$-th, since $\eta(t)$ and $r(t)$ require such a basis for their expression, but it will turn out that the components after the $J$-th do not affect the receiver decision and we thus need not be precise about what they are.

## The ML Receiver in Signal Space

Now we express the ML receiver (A-2) in terms of the vectors (A-3)–(A-5). We get

$$\text{Find:} \quad \max_i P\{\eta = \mathbf{r} - \mathbf{s}_i\}, \tag{A-6}$$

in which all the variables are vectors. For white Gaussian noise, stochastic process theory tells us that the components of $\eta$ are independent, zero-mean variates with variance $N_0/2$, where $N_0/2$ is the power spectral density of the noise in watts/Hz. Eq. (A-6) thus becomes the product form

$$\max_i P\left\{\prod_j (\eta_j = r_j - s_{ij})\right\} = \max_i \prod_j P\{\eta_j = r_j - s_{ij}\}. \tag{A-7}$$

Once again, we may eliminate factors that are constant as $i$ varies; this time they are all factors after the $J$-th in this expression. This fact is sometimes called the *Theorem of Irrelevance*: Noise and received signal components outside of those needed to express the signals are irrelevant to the ML decision.

We can express the probability of the $J$ remaining components in (A-7) directly as the multivariate Gaussian density

$$\max_i (\pi N_0)^{-J/2} \exp\left[-\sum_{j=i}^{J} (r_j - s_{ij})^2\right].$$

Since ln( ) is a monotone function, we obtain the same max by taking

$$\max_i -\frac{J}{2}\ln(\pi N_0) - \sum_{j=1}^{J}(r_j - s_{ij})^2$$

and this is in turn the same as finding

$$\min_i \sum_{j=1}^{J}(r_j - s_{ij})^2. \tag{A-8}$$

Eq. (A-8) is the final expression for the ML receiver; it says *find the closest* $\mathbf{s}_i$ *to* $\mathbf{r}$ *in terms of Euclidean distance in the signal space.*

In signal space the set of signals $\{s_i(t)\}$ forms a collection of points called a *constellation.* Figure A-1 illustrates the constellations of QPSK and 16QAM. Both pictures are divided into *decision regions*, the loci of points $\mathbf{r}$ that lie closest to each point in the constellations. These regions implement (A-8). Simple modulations, parity-check codes and set-partition codes rely typically on two-dimensional spaces like those in the figure, one for each symbol interval, while CPM and intersymbol interference are based on many-dimensional spaces.

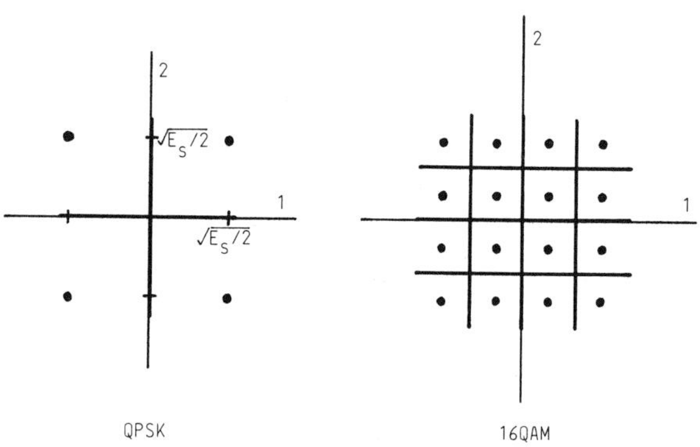


Figure A-1. Signal space constellations of QPSK (left) and 16-QAM (right). Heavy lines show ML decision boundaries. The horizontal axis corresponds to vector component 1, and the vertical to component 2. The scaling in 16-QAM derives from (A-13).

From the properties of the basis functions, we have the Parseval identity which states that

$$\sum_{j=1}^{J}(r_j - s_{ij})^2 = \int [r(t) - s_i(t)]^2 \, dt \tag{A-9}$$

if $r$ and $s_i$ are both signals over a $J$-dimensional space. (We now assume real functions.) By expanding the right hand side of (A-9) we can derive some important alternate receiver structures. The integral becomes

$$\int r(t)^2 \, dt + \int s_i(t)^2 \, dt - 2 \int r(t) s_i(t) \, dt,$$

in which the first term is a constant over the minimization in (A-8). The middle term is the energy of signal $s_i(t)$, a constant $E_i$ that is known *a priori*. For equal-energy signals, all these energies are the same and the minimization in (A-8) is equivalent to the maximization

$$\max_i \int r(t) s_i(t) \, dt. \tag{A-10}$$

This is a correlation integral and a receiver made up of these operations is called a *correlator receiver*. For signals with unequal energies, (A-10) becomes

$$\max_i \int r(t) s_i(t) \, dt - \frac{E_i}{2}. \tag{A-11}$$

With a little more work we can convert the correlation in (A-10) or (A-11) to a linear filtering operation. First truncate and time-shift the functions $s_i(t)$ so that the important part of them lies in the interval $[0, NT]$. Then form $h_i(t) = s_i(NT - t)$. Now the convolution $h_i * r$ evaluated at $NT$ has the same value as the correlation in (A-10); this is a filtering of $r(t)$ by a filter with impulse response $h_i(t)$ and the correlation is the output at the time $NT$. $h_i(t)$ is called a *matched filter* and a matched filter receiver consists of a bank of filters $h_i(t)$, $i = 1, ..., m$, one for each signal. The receiver samples all filters at $NT$ and decides in favor of the index corresponding to the maximum output.

## Quadrature Signal Space Examples

### Example A-1: Rectangular-pulse QPSK.

A quadrature modulation is one based on the two-dimensional signal space set up by the signals $\sin \omega t$ and $\cos \omega t$. A general form for these modulations is

$$s(t) = \sqrt{\frac{2E_s}{T}}[I(t)\cos\omega t - Q(t)\sin\omega t] \qquad \text{(A-12)}$$

in which $T$ is the symbol interval and $I(t)$ and $Q(t)$ are real, independent data-bearing waveforms, called the in-phase and quadrature baseband signals. If we scale the baseband signals so that

$$\mathcal{E}\left[\frac{1}{T}\int_{nT}^{(n+1)T}[I^2(t) + Q^2(t)]\,dt\right] = 1 \qquad \text{(A-13)}$$

then $E_s$ is the average symbol energy for random data. A simple quadrature modulation is quaternary phase-shift keying (QPSK) in which $I(t)$ and $Q(t)$ are made up of length-$T$ square pulses that jointly carry four-level data. Represent the data in the $n$-th interval $nT \le t < (n+1)T$ by the binary pairs $(+1, +1)$, $(-1, +1)$, $(-1, -1)$ and $(+1, -1)$. Then the baseband signals in Table A-1 will satisfy condition (A-13).

An orthonormal function basis for these QPSK signals is $\{\phi_1(t) = \sqrt{2/T}\cos\omega t, \phi_2(t) = \sqrt{2/T}\sin\omega t\}$. The signal space vectors as defined in (A-3) are shown in Fig. A-1. Note that the integrals in the component calculation extend over just the $n$-th interval.

### Example A-2: Raised-cosine QPSK.

Since the bandwidth of rectangular-pulse QPSK is large, schemes with shaped, narrow band pulses have been devised to replace it. One such scheme is raised-cosine (RC) pulse QPSK. Here the square pulse is replaced by $g(t)$, a pulse defined in terms of its square† Fourier transform by

†Because of this square, raised-cosine pulses are sometimes called root raised-cosine.

**Table A-1.** BASEBAND SIGNAL DURING $n$-TH SYMBOL INTERVAL FOR RECTANGULAR-PULSE QPSK, TOGETHER WITH THE IMPLIED RF PHASE.

| Data | I(t) | Q(t) | RF Phase |
|---|---|---|---|
| +1, +1 | $+1/\sqrt{2}$ | $+1/\sqrt{2}$ | 45° |
| -1, +1 | $-1/\sqrt{2}$ | $+1/\sqrt{2}$ | 135 |
| -1, -1 | $-1/\sqrt{2}$ | $-1/\sqrt{2}$ | 225 |
| +1, -1 | $+1/\sqrt{2}$ | $-1/\sqrt{2}$ | 315 |

$$|G(f)|^2 = \begin{cases} T^2, & \text{if } 0 \le |f| \le (1-\alpha)/2T \\ T^2 \cos^2\left[\frac{\pi T}{2\alpha}\left(f - \frac{1-\alpha}{2T}\right)\right], & \text{if } \frac{1-\alpha}{2T} < |f| < \frac{1+\alpha}{2T} \\ 0, & \text{elsewhere.} \end{cases} \tag{A-14}$$

The parameter $\alpha$, called the excess bandwidth factor, satisfies $0 \le \alpha \le 1$; the RF bandwidth of the modulation is obviously $(1+\alpha)/T$ Hz. The transmitted signal is still (A-12) but now

$$\begin{aligned} I(t) &= \sum_n c_n^I g(t - nT) \\ Q(t) &= \sum_n c_n^Q g(t - nT), \end{aligned} \tag{A-15}$$

with the data-bearing weights $c_n^I$ and $c_n^Q$ equal to $\pm 1/\sqrt{2}$ in the same pattern as in Table A-1. In a general QAM scheme the weights take a variety of values specified by the scheme.

It can be shown that successive time-shifts of $g(t)$, namely $g(t), g(t-T), g(t-2T), \ldots$, are orthogonal over $-\infty < t < \infty$, and that consequently an orthonormal basis set is $\{\sqrt{2/T}g(t - nT)\cos\omega t, \sqrt{2/T}g(t-nT)\sin\omega t\}$ for the $n$-th interval signal space. Note that these functions provide the basis for just the $n$-th interval even though they extend over all time. Further calculation shows that the $n$-th interval signal space vectors are again Fig. A-1, *the same constellation as in Ex. A-1.* To verify this, one finds the inner products $(s, \phi_1)$ and $(s, \phi_2)$, in which $s(t)$ is (A-12) with (A-15), $\phi_1$ and $\phi_2$ are as just given, and $c_n^I$ and $c_n^Q$ take the values $\pm 1/\sqrt{2}$. The orthogonality properties of the pulse in (A-14) lead to other important results; for instance, an ML receiver for this QPSK is simply a filter matched to $g(t)$ followed by a sampler (see ref. [1-12], Chap. 10).

## ML Receiver Error Probability

An expression for the overall probability of error is

$$P_e = \sum_i P\{s_i \text{ not received } | s_i \text{ sent }\} P\{s_i \text{ sent }\}; \tag{A-16}$$

recall that $s_i$ is not received if $r$ lies in any other decision region than the $i$-th one. All signals are equally likely, so $P\{s_i\}$ is $1/m$. Furthermore, we can overbound the probability of not receiving $s_i$ by the union bound expression

$$P\{s_i \text{ not received } | s_i \text{ sent }\} \le \sum_{j \ne i} P\{\text{Receive } s_j | s_i \text{ sent}\}.$$

In conclusion, $P_e$ is bounded by

$$P_e \le \frac{1}{m} \sum_{j \ne i} P\{j; i\} \tag{A-17}$$

where $P\{j;i\}$ is shorthand for receiving $s_j$ when $s_i$ was sent.

It remains to evaluate the $P\{j;i\}$; these are called the *2-signal error probabilities.* Suppose that the two signals are separated in signal space by distance $D$. White noise has the property that its projections are distributed uniformly and independently along any set of orthogonal axes in signal space. For our error calculation we will choose a convenient set, namely one with an axis along the line between $s_i$ and $s_j$ and the origin at $s_i$. If $s_i$ is sent, the two-signal ML receiver decides $s_j$ if and only if $r$ lies beyond the horizontal bisector of this line, an event that occurs when the noise projection along the line exceeds $D/2$ toward $s_j$. This is summarized in Fig. A-2.

The noise projection, call it $\eta$, is a zero-mean Gaussian variate with variance $N_0/2$, and the probability that it exceeds $D/2$ is

$$P\{j;i\} = \frac{1}{\sqrt{\pi N_0}} \int_{D/2}^{\infty} \exp\left(-\frac{\eta^2}{N_0}\right) d\eta. \tag{A-18}$$

This is an ordinary Gaussian integral $Q(D/\sqrt{2N_0})$, where $Q(u)$ is the standard integral

$$Q(u) = \frac{1}{\sqrt{2\pi}} \int_u^{\infty} \exp\left(-\frac{u^2}{2}\right) du, \qquad u \ge 0. \tag{1.3-2}$$

$Q(u)$ has the very useful approximation

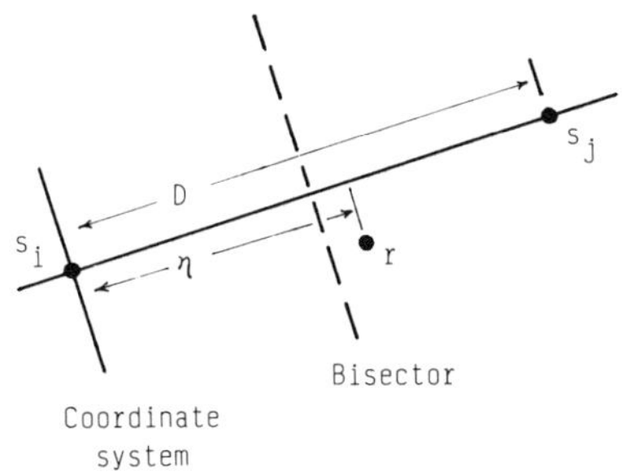


**Figure A-2.** The 2-signal error calculation, showing a received signal $\mathbf{r}$ that will be erroneously detected as $\mathbf{s}_j$.

$$Q(u) \leq \frac{1}{2}\exp\left(-\frac{u^2}{2}\right), \qquad u \geq 0. \tag{A-19}$$

Recall from Chapter 4 that the normalized signal space distance $d$ satisfies $d^2 = D^2/2E_b$; combining this with (A-19), we get

$$P\{j;i\} = Q\left(d\sqrt{\frac{E_b}{N_0}}\right) \approx \frac{1}{2}\exp\left(-\frac{d^2 E_b}{2N_0}\right). \tag{A-20}$$

Returning now to (A-17), the bound on $P_e$, we see that it is a sum of exponentials and that as $E_b/N_0$ grows, (A-17) is strongly dominated by the term with the smallest $d$. This $d$ is precisely the free distance of the signals, the least distance between any signal pair. We can replace (A-17) with just the term or terms with the free distance to obtain the estimate

$$P_e \approx \frac{K_d}{m}\exp\left(-\frac{d_{\text{free}}^2 E_b}{2N_0}\right) \tag{A-21}$$

in which $K_d$ is the number of signal pairs separated by $d_{\text{free}}$. The estimate is a tight underbound when $E_b/N_0$ is of any practical size.

## Capacity and Signal Space

The capacity of a single dimension of signal space is just the capacity of the standard AWGN channel in information theory, the channel with a sequence of independent real-number inputs and real-numbered outputs distributed by additive Gaussian noise variates. This capacity is

$$C = \frac{1}{2}\log_2\left(1 + \frac{2E_d}{N_0}\right) \qquad \text{bits/channel use,} \qquad \text{(A-22)}$$

where $E_d$ is the average energy per channel use, or in our case, per signal space dimension.

To find the capacity of a channel with a certain physical bandwidth $W$ and elapsed time $T$ we need to count the signal space dimensions available in this time and bandwidth. For baseband signaling, it is easy to see that there are at least $2WT$ dimensions. We may send the orthogonal pulse train of sinc $(t/\tau)$ pulses, a new one every $\tau$ seconds, in a bandwidth $(-1/2\tau, 1/2\tau)$ Hz. After time T, $T/\tau$ independent inputs are carried in this way, and substituting $W = 1/2\tau$, we conclude that there are altogether $2WT$ available channel uses. A similar argument applies to carrier-modulated transmission, with $W$ now $B_{RF}$, the double-sideband RF bandwidth. A rigorous proof of the $2WT$ result appears in Wozencraft and Jacobs [1-10].

As a consequence, multiplying (A-22) by $2B_{RF}T$ shows that

$$C = B_{RF}T\log_2\left(1 + \frac{2E_d}{N_0}\right) \qquad \text{bits} \qquad \text{(4.3-1)}$$

is the capacity available in $T$ seconds and $B_{RF}$ Hz. We use this formula in Section 4.3.

APPENDIX B

# Introduction to Algebra

The codes of Chapter 3 rely heavily on the theory of finite fields. This appendix introduces the subject by providing formal definitions of groups, fields, and vector spaces. Examples are provided to illustrate the concepts. Several excellent text books are available that provide in-depth coverage of the subject. Reference [B-1] is a classic text. The interested reader will find additional material in this as well as in the more recent texts [B-2] and [B-3].

## B.1 GROUPS

**Definition:** A *group* is defined as a set $\mathcal{G}$ on whose elements an operation denoted by "$\cdot$" is defined satisfying the following axioms.

1. *Closure.* If $\alpha$ and $\beta$ are in $\mathcal{G}$ then $\alpha\beta$ in $\mathcal{G}$.
2. *Associative law.* The operation $\cdot$ is associative, meaning that for any $\alpha, \beta$, and $\gamma$ in $\mathcal{G}$,

$$\alpha \cdot (\beta \cdot \gamma) = (\alpha \cdot \beta) \cdot \gamma$$

3. *Identity element.* There exists in $\mathcal{G}$ an identity element, denoted by 1, such that

$$\alpha \cdot 1 = 1 \cdot \alpha = \alpha, \qquad \text{for all } \alpha \text{ in } \mathcal{G}.$$

4. *Inverse element.* For every element $\alpha$ in $\mathcal{G}$, there exists an inverse, denoted by $\alpha^{-1}$, such that

$$\alpha \cdot \alpha{-}1 = \alpha{-}1 \cdot \alpha = 1.$$

We call a group *Abelian* group or a *commutative* group, if the operation $\cdot$ is also commutative; that is

$$\alpha \cdot \beta = \beta \cdot \alpha, \qquad \text{for all } \alpha, \beta \text{ in } \mathcal{G}.$$

**Theorem B.1-1.** The identity element in a group is unique and each element in the group has a unique inverse element.

*Proof:* If the identity element is not unique, let there be another identity element denoted by $1'$. Then

$$1' = 1' \cdot 1 = 1,$$

implying that 1 and 1' are identical; that is the identity element in a group is unique.

To show that a group element $\alpha$ has a unique inverse $\alpha^{-1}$, let us assume that $\alpha$ has two inverses, $\alpha^{-1}$ and $\alpha_1^{-1}$. Then

$$\begin{aligned}\alpha_1^{-1} &= \alpha_1^{-1} \cdot 1 = \alpha_1^{-1} \cdot (\alpha \cdot \alpha^{-1}) \\ &= (\alpha_1^{-1} \cdot \alpha) \cdot \alpha^{-1} = 1 \cdot \alpha^{-1} \\ &= \alpha^{-1},\end{aligned}$$

thus establishing the uniqueness of the inverse of any group element.

**QED**

**Example B.1-1.**

(a) Let $\mathcal{G}$ be the set {0,1} with an operation + standing for modulo-2 addition, defined by

$$0+0=0,\ 0+1=1,\ 1+0=1,\ 1+1=0.$$

We see that the closure property is satisfied; 0 is the identity element; 0 and 1 are their own inverses and + is associative. $\mathcal{G}$ with modulo-2 addition on its elements is therefore a group.

(b) The set of integers forms a group under the usual addition operation; 0 is the identity element and for any integer $\ell$, $-\ell$ is its inverse.

(c) The set of all nonzero real numbers forms a group under the usual multiplication operation; 1 is the identity element and for any nonzero real number $\alpha$, $1/\alpha$ is the inverse.

(d) Consider the set of code words of a binary linear $(N, K)$ block code. We know from Section 3.2 that modulo-2 addition of any two code words, defined as the bit-by-bit modulo-2 sum, yields another code word. The all-zeros code word is the identity element and each code word is its own inverse. The associative property also holds. Since all the group properties are satisfied, the set of code words with modulo-2 addition forms a group, the reason why they are also called a group code.

(e) Consider the set of all non-singular matrices of size $N \times N$. The product of any two matrices from the set is another $N \times N$ matrix (Closure). The associative law is easily verified. Since any matrix from the set is non-singular, it has an inverse. The identity matrix serves as the identity element of the group.

The groups in (a) – (d) are commutative, but the one in (e) is not.

We define the *order* of a group as the number of elements in the group. A group is said to *finite* if its order is finite.

Let $\mathcal{S}$ be a subset of the set of elements of a group $\mathcal{G}$. We call $\mathcal{S}$ a *subgroup* if $\mathcal{G}$ if the elements of $\mathcal{S}$ themselves form a group under the same operation as that defined on elements of $\mathcal{G}$. We can verify that $\mathcal{S}$ is a subgroup of $\mathcal{G}$ by just verifying that its elements satisfy the closure property and that an inverse exists for each of its elements.

**Theorem B.1-2 (Lagrange).** The order of a finite group is an integer multiple of any of its subgroups.

*Proof*: See [B-1].

Example B.1-2.

Consider the set $\mathcal{Y}$ of all $N$-tuples, each element of the $N$-tuple being 0 or 1. We can look upon $\mathcal{Y}$ as the of all received words. $\mathcal{Y}$ is group under modulo-2 addition. Consider a subset $\mathcal{C}$ of $\mathcal{Y}$ whose elements are the code words of a binary linear $(N, K)$ block code. Then $\mathcal{C}$ forms a subgroup of $\mathcal{Y}$.

As another example, consider an element $\alpha$ of a finite group $\mathcal{G}$. Let $\mathcal{S}$ be the set of elements $\{\alpha, \alpha^2, \alpha^3, \ldots\}$.‡ Since $\mathcal{G}$ is finite, $\mathcal{S}$ too must be finite. This means that all the powers of $\alpha$ cannot be distinct. There must be some $\ell, m$, with $m > \ell$, such that $\alpha^m = \alpha^\ell$. Hence we have

$$\alpha^m \cdot \alpha^{-\ell} = \alpha^\ell \cdot \alpha^{-\ell},$$

or

$$\alpha^{m-\ell} = 1.$$

Let $k$ be the smallest power of $\alpha$ for which $\alpha^k = 1$. Then $\alpha, \alpha^2, \ldots \alpha^k$ must all be distinct. If not let $\alpha^i = \alpha^j$, $1 \le i < j \le k$. Then $\alpha^{j-i} = 1$. Since $j - i < k$, this contradicts the assumption that $k$ is the least power such that $\alpha^k = 1$. We can now show that $\mathcal{S} = \{\alpha, \alpha^2, \ldots, \alpha^k = 1\}$ is a subgroup of $\mathcal{G}$. $\mathcal{S}$ contains the identity element. Given any $\alpha^j$, $1 \le j \le k$, $\alpha^{k-j}$ is the inverse. Given any two elements, $\alpha^i$ and $\alpha^j$, $\alpha^i \cdot \alpha^j = \alpha^{i+j}$ must be in $\mathcal{S}$ if $i + j \le k$. If $i + j > k$, then

$$\alpha^{i+j} = \alpha^{i+j} \cdot 1 = \alpha^{i+j} \cdot \alpha^{-k} = \alpha^{i+j-k}.$$

Since $i + j - k \le k$, we see that the closure property is satisfied. Clearly, associativity also holds. $\mathcal{S}$ is therefore a subgroup of $\mathcal{G}$.

A *cyclic group* (*subgroup*) is a group (subgroup) whose elements can be generated as powers of some element. Subgroups such as those constructed in the preceding example using powers of an element of the group are therefore cyclic subgroups. The set of code words of a cyclic $(N, K)$ block code is a cyclic subgroup of the group of all $N$-tuples.

---

‡$\alpha^n$ denotes the element $\underbrace{\alpha \cdot \alpha \cdots \alpha}_{n \text{ times.}}$

## B.2 FIELDS

A *field* is a set of elements $\mathcal{F}$ with two operations additions "+" and multiplication "·" such that

1. $\mathcal{F}$ is an Abelian group under addition, with 0 as the identity element;
2. The nonzero elements of $\mathcal{F}$ form an Abelian group under multiplication, with 1 as the identity element; and
3. The operation · distributes over the operation +:

$$\alpha \cdot (\beta + \gamma) = \alpha \cdot \beta + \alpha \cdot \gamma, \qquad \text{for all } \alpha, \beta, \gamma \text{ in } \mathcal{F}.$$

Example B.2-1.

(a) The set of real numbers, with operation + as ordinary multiplication and · as ordinary multiplication, satisfies the above three properties and is therefore a field.

(b) The set consisting of two elements {0,1}, with modulo-2 addition and multiplication given in Table B.2-1 constitutes a field. The field is called a *Galois field* and is denoted GF(2).

**Table B.2-1.** ADDITION AND MULTIPLICATION IN GF(2).

| + | 0 | 1 |
|---|---|---|
| 0 | 0 | 1 |
| 1 | 1 | 0 |

| · | 0 | 1 |
|---|---|---|
| 0 | 0 | 0 |
| 1 | 0 | 1 |

A field with a finite number of elements is called a *finite field*. Finite fields are also called Galois fields.

We will soon show that for any $p$, a prime number, Galois fields exist with $p$ elements. We use GF($p$) to denote a Galois field with $p$ elements. The set of integers modulo-$p$, with modulo-$p$ addition and multiplication, forms a field with p elements. In fact, for any $q$ which is an integer power of a prime number ($q = p^m$, $m$ an integer), there exist fields with $q$ elements, denoted GF($q$). The arithmetic is however not modulo-$q$ arithmetic. For example, the set {0, 1, 2, 3} does not form a field with

Table B.2-2. ADDITION AND MULTIPLICATION IN GF(3).

| + | 0 | 1 | 2 |
|---|---|---|---|
| 0 | 0 | 1 | 2 |
| 1 | 1 | 2 | 0 |
| 2 | 2 | 0 | 1 |

| · | 0 | 1 | 2 |
|---|---|---|---|
| 0 | 0 | 0 | 0 |
| 1 | 0 | 1 | 2 |
| 2 | 0 | 2 | 1 |

Table B.2-3. ADDITION AND MULTIPLICATION IN GF($2^2$).

| + | 0 | 1 | 2 | 3 |
|---|---|---|---|---|
| 0 | 0 | 1 | 2 | 3 |
| 1 | 1 | 0 | 3 | 2 |
| 2 | 2 | 3 | 0 | 1 |
| 3 | 3 | 2 | 1 | 0 |

| · | 0 | 1 | 2 | 3 |
|---|---|---|---|---|
| 0 | 0 | 0 | 0 | 0 |
| 1 | 0 | 1 | 2 | 3 |
| 2 | 0 | 2 | 3 | 1 |
| 3 | 0 | 3 | 1 | 2 |

Table B.2-4. ADDITION AND MULTIPLICATION IN GF(5).

| + | 0 | 1 | 2 | 3 | 4 |
|---|---|---|---|---|---|
| 0 | 0 | 1 | 2 | 3 | 4 |
| 1 | 1 | 2 | 3 | 4 | 0 |
| 2 | 2 | 3 | 4 | 0 | 1 |
| 3 | 3 | 4 | 0 | 1 | 2 |
| 4 | 4 | 0 | 1 | 2 | 3 |

| · | 0 | 1 | 2 | 3 | 4 |
|---|---|---|---|---|---|
| 0 | 0 | 0 | 0 | 0 | 0 |
| 1 | 0 | 1 | 2 | 3 | 4 |
| 2 | 0 | 2 | 4 | 1 | 3 |
| 3 | 0 | 3 | 1 | 4 | 2 |
| 4 | 0 | 4 | 3 | 2 | 1 |

modulo-4 arithmetic since 2 does not have a multiplicative inverse. Tables B.2-2 – B.2-4 give fields with 3, 4, and 5 elements, respectively.† Table B.2-3 clearly shows that GF($2^2$) operations are not modulo-4 operations. Interestingly, GF($2^2$) contains GF(2) as a subfield. In general, GF($p^m$) contains GF($p$) as a subfield.

**Theorem B.2-1.** Let $\mathcal{Z}_p$ denote the set of integers $\{0, 1, \ldots, p-1\}$, with

†Table B.2-3 also appears in Example 3.10-2, which shows a way to construct the table.

addition and multiplication defined as ordinary addition and multiplication modulo-$p$. Then $\mathcal{Z}_p$ is a field if and only if $p$ is prime.

*Proof*: If $p$ is not a prime, then there must be nonzero elements $\alpha$ and $\beta$ in $\mathcal{Z}_p$ such that $p = \alpha\beta$, where we have taken the liberty to omit the operator $\cdot$. We can show that $\beta$ has no multiplicative inverse. For if it does, then there is a $\gamma$ in $\mathcal{Z}_p$ such that $\beta\gamma = 1 \bmod p$ and so $\alpha\beta\gamma = \alpha \bmod p$. But $\alpha\beta\gamma = p\gamma = 0 \bmod p$, implying $0 = \alpha \bmod p$ and contradicting the assumption $p = \alpha\beta$, with $\alpha$ and $\beta$ nonzero. $p$ must therefore be a prime.

Let $p$ be a prime. We need to show the existence of a unique multiplicative inverse for each nonzero element of $\mathcal{Z}_p$. One can verify that $\mathcal{Z}_p$ satisfies all the other properties of a field. Consider any arbitrary nonzero element $\alpha$ in $\mathcal{Z}_p$. Since $p$ is a prime, the greatest common divisor (gcd) of $\alpha$ and $p$ is 1; that is $\gcd(\alpha, p) = 1$. Then there exist integers $m$ and $n$ such that $m\alpha + np = 1$.† This means that $a\alpha \bmod p = 1 \bmod p = 1$, or $a$ is the inverse of $\alpha$.

**QED**

Further developments take place in Sections 3.7 and 3.10. Section 3.7 develops polynomial representation of finite fields, where a field element is represented as a polynomial whose coefficients are drawn from the field GF(2). Section 3.10 extends the development to nonbinary fields.

## B.3 VECTOR SPACES

**Definition:** Let $\mathcal{V}$ be a set of elements with a binary operation "+" and let $\mathcal{F}$ be a field. Further, let an operation "$\cdot$" be defined between the elements of $\mathcal{V}$ and the elements of $\mathcal{F}$. Then $\mathcal{V}$ is said to be a *vector space* over $\mathcal{F}$ if the following conditions are satisfied for all $\alpha, \beta$ in $\mathcal{F}$ and $\mathbf{x}$, $\mathbf{y}$ in $\mathcal{V}$.

1. $\mathcal{V}$ is an Abelian (commutative) group under the addition operation $+$.
2. If $\alpha \in \mathcal{F}$ and $\mathbf{x} \in \mathcal{V}$, then $\alpha \cdot \mathbf{x} \in \mathcal{V}$, for all $\alpha$ and $\mathbf{x}$.
3. *Distributive law*:

$$\alpha \cdot (\mathbf{x} + \mathbf{y}) = \alpha \cdot \mathbf{x} + \alpha \cdot \mathbf{y},$$
$$(\alpha + \beta) \cdot \mathbf{x} = \alpha \cdot \mathbf{x} + \beta \cdot \mathbf{y}.$$

†This result is due to the well known Euclid's algorithm which states that if $\gcd(r, s) = t$, where $r$ and $s$ are nonzero integers, then $t$ can be expressed as a linear combination of $r$ and $s$; that is, $t = nr + ms$ for some integers $n$ and $m$.

4. *Associative law:*

$$(\alpha \cdot \beta) \cdot \mathbf{x} = \alpha \cdot (\beta \cdot \mathbf{x}).$$

5. If 1 denotes the identity element of $\mathcal{F}$, then

$$1 \cdot \mathbf{x} = \mathbf{x}.$$

Let 0 denote the zero element (identity element under +) in $\mathcal{F}$, $\mathbf{0}$ the additive identity in $\mathcal{V}$, and $-1$ the additive inverse of 1, the multiplicative identity in $\mathcal{F}$. The following properties are easily shown:

$$0 \cdot \mathbf{x} = \mathbf{0}$$
$$\mathbf{x} + (-1) \cdot \mathbf{x} = \mathbf{0}$$

We can show that $0 \cdot \mathbf{x} = \mathbf{0}$ as follows.

$$\mathbf{x} = 1 \cdot \mathbf{x} = (1 + 0) \cdot \mathbf{x} = 1 \cdot \mathbf{x} + 0 \cdot \mathbf{x} = \mathbf{x} + 0 \cdot \mathbf{x}.$$

Let $-\mathbf{x}$ denote the additive inverse of $\mathbf{x}$. On adding $-\mathbf{x}$ to both sides of the above equation, we get

$$-\mathbf{x} + \mathbf{x} = -\mathbf{x} + \mathbf{x} + 0 \cdot \mathbf{x}$$

or

$$\mathbf{0} = 0 \cdot \mathbf{x}.$$

To show that the additive inverse of $\mathbf{x}$ is $-1 \cdot \mathbf{x}$, we observe that, since $1 + (-1) = 0$,

$$\mathbf{0} = 0 \cdot \mathbf{x} = [1 + (-1)] \cdot \mathbf{x} = \mathbf{x} + (-1) \cdot \mathbf{x}$$

and so $-1 \cdot \mathbf{x}$ is the additive inverse of $\mathbf{x}$.

**Example B.3-1.**

(a) Consider the set $\mathcal{V}$, whose elements are $n$-tuples of the form

$$\mathbf{v} = (v_0, v_1, \ldots, v_{n-1});$$

each $v_i$, $i = 1, 2, \ldots, n-1$ is from the field of real numbers $\Re$. We can in fact assume a more general case with each $v_i$ drawn from any field $\mathcal{F}$. Define addition of any two $n$-tuples as another $n$-tuple obtained by component-wise addition and multiplication of an $n$-tuple by a number from $\Re$ (or an element from $\mathcal{F}$). Then $\mathcal{V}$ forms a vector space over $\Re$ ($\mathcal{F}$). $\mathcal{V}$ is often denoted $\Re^n$ ($\mathcal{F}^n$). $\Re^n$ is also called the Euclidean $n$-space.

(b) In section 3.2 we showed that the set of $n$-tuples whose elements are from GF(2), again with component-wise addition operation and multiplication by an element from GF(2), formed a vector space over GF(2).

(c) Consider the set $\mathcal{V}$ of polynomials whose coefficients are from GF($q$). Addition of two polynomials is the usual polynomial addition, addition being performed in GF($q$). Let the field $\mathcal{F}$ be GF($q$). Scalar multiplication of a polynomial by a field element from GF($q$) corresponds to multiplying each coefficient of the polynomial by the field element, multiplication being carried out in GF($q$). $\mathcal{V}$ is then a vector space over GF($q$).

(d) Consider the set $\mathcal{V}$ of all $m \times n$ matrices, whose coefficients are from GF($q$) and let the field be GF($q$). Addition of two matrices is the usual matrix addition, but carried out in GF($q$); similarly, scalar multiplication of a matrix from $\mathcal{V}$ by a field element is carried out in the usual manner, but now in GF($q$). $\mathcal{V}$ is then forms a vector space over GF($q$).

Consider a set $\mathcal{V}$ that forms a vector space over a field $\mathcal{F}$. Let $\mathbf{v}_1, \mathbf{v}_2, \ldots, \mathbf{v}_k$ be vectors from $\mathcal{V}$ and $\alpha_1, \alpha_2, \ldots, \alpha_k$ be field elements from $\mathcal{F}$. We call the sum

$$\alpha_1\mathbf{v}_1 + \alpha_2\mathbf{v}_2 + \cdots + \alpha_k\mathbf{v}_k$$

a *linear combination* of the vectors $\mathbf{v}_1, \mathbf{v}_2, \ldots, \mathbf{v}_k$.

The set of vectors $\{\mathbf{v}_1, \mathbf{v}_2, \ldots, \mathbf{v}_k\}$ is said to be *linearly independent* if there does *not* exist a set of field elements $\alpha_1, \alpha_2, \ldots, \alpha_k$, not all $\alpha_i = 0$, such that

$$\alpha_1\mathbf{v}_1 + \alpha_2\mathbf{v}_2 + \cdots + \alpha_k\mathbf{v}_k = \mathbf{0}.$$

Example B.3-2.

The vectors (1 0 0), (0 1 0), and (0 0 1) from $\Re^3$ are linearly independent over $\Re$. In fact, they are linearly independent over any field $\mathcal{F}$. However, the vectors (1 0 2), (2 1 1), and (0 2 0) are linearly dependent over GF(3) since their sum is zero.

Let $\mathcal{V}$ be a vector space over $\mathcal{F}$ and let $\mathcal{S}$ be a subset of the vectors in $\mathcal{V}$. We call $\mathcal{S}$ a subspace if $\mathcal{S}$ is itself a vector space over $\mathcal{F}$ under the same vector addition and scalar multiplication operations applicable to $\mathcal{V}$ and $\mathcal{F}$.

**Example B.3-3.**

Consider the vector space $\mathcal{V}$ over GF(2) given by the set

$$\{(0\ 0\ 0), (0\ 0\ 1), (0\ 1\ 0), (0\ 1\ 1), (1\ 0\ 0), (1\ 0\ 1), (1\ 1\ 0), (1\ 1\ 1)\}$$

Let $\mathcal{S}$ be the subset of $\mathcal{V}$ given by

$$\{(0\ 0\ 0), (1\ 0\ 0), (0\ 1\ 0), (1\ 1\ 0)\}.$$

$\mathcal{S}$ is a subspace over GF(2).

**Theorem B.3-1:** Let $\{\mathbf{v}_1, \mathbf{v}_2, \ldots, \mathbf{v}_k\}$ be a set of vectors from a vector space over a field $\mathcal{F}$. Then the set consisting of all linear combinations of $\{\mathbf{v}_1, \mathbf{v}_2, \ldots, \mathbf{v}_k\}$ forms a vector space over $\mathcal{F}$ and is therefore a subspace of $\mathcal{F}$.

*Proof*: Left to the reader as an exercise.

**Example B.3-3** (*continued*).

Consider the set $\mathcal{V}$ of Ex. B.3-3 and its subset $\mathcal{B}$ given by $\{(0\ 1\ 0), (1\ 0\ 0)\}$. The set of all linear combinations of the vectors in $\mathcal{B}$ is $\mathcal{S}$ given by

$$\{(0\ 0\ 0), (0\ 1\ 0), (1\ 0\ 0), (1\ 1\ 0)\},$$

which is itself a vector space over GF(2). $\mathcal{S}$ is therefore a subspace of $\mathcal{V}$.

The set of vectors in $\mathcal{B}$ is said to span $\mathcal{S}$. The set of vectors $\{(1\ 0\ 0), (0\ 1\ 0), (0\ 0\ 1)\}$ spans the space $\mathcal{V}$ in this example.

A *basis* of a vector space $\mathcal{V}$ is a set of linearly independent vectors that spans the space. The number of vectors in a basis is called the *dimension* of the vector space.

The set $\mathcal{B}$ in Ex. B.3-3 spans $\mathcal{S}$ and is a basis; the set

$$\{(1\ 0\ 0), (0\ 1\ 0), (0\ 0\ 1)\}$$

is a basis of $\mathcal{V}$. The dimension of $\mathcal{S}$ is 2, and that of $\mathcal{V}$ is 3. We can show that any vector space has at least one basis that spans the space and that all bases have the same number of vectors.

# *REFERENCES*

[B-1] G. Birkhoff and S. MacLane, *A Survey of Modern Algebra*, Macmillan, New York, 1953.

[B-2] R. Lidl and H. Niederreiter, *Finite Fields*, Addison-Wesley, Reading, Mass., 1983.

[B-3] R.J. McEliece, *Finite Fields for Computer Scientists and Engineers*, Kluwer Academic, Boston, Mass., 1987.

# APPENDIX C

# Mathematical Details of the Stop and Drop Lines

This appendix contains mathematical details of Chapter 6.

## Proof of Lemma 6.3-1

The lemma states that in an incorrect subset of a random tree of rate $R$ and depth $n$, the minimum distance to the received code word satisfies

$$\frac{1}{n} \log P\{\text{ISS } \leq k \text{ at } n\} \to \left[h_B(y) + R - 1\right], \qquad y = \frac{k}{n} \tag{6.3-15}$$

as $k$ and $n$ tend to infinity.

The proof begins by finding $EN_A(z)$, the expected number of paths that cross metric $A$ in an upward direction before all paths cross $-\zeta$ in a downward direction; see Fig. 6.3-1 for a picture of these two barriers. $EN_A(z)$ is the solution to the same difference equation (6.4-1) as is the quantity $EN_\zeta(z)$ in Section 6.4, except that this time the boundary conditions are

$$EN_A(z) = \begin{cases} \frac{1}{b}, & \text{if } A \le z < A + as \\ 0, & \text{if } -\zeta + as - a < z \le -\zeta. \end{cases} \tag{C-1}$$

As before, $EN_A(z)$ is bounded above by an exponential solution of the form $K\sigma^z$, where $\sigma$ is the smallest real solution to the characteristic equation (6.4-4). From the boundary conditions, we find that $K = (1/b)\sigma^{-A}$, and so

$$EN_A(0) \le \frac{1}{b}\sigma^{-A}. \tag{C-2}$$

In the second part of the proof, we bound the value of the ISS minimum distance at depth $n$ code bits. By checking Fig. 6.3-1, it is clear that $P\{\text{ISS minimum distance } k \text{ satisfies } k \le ns - A \text{ at any } n\}$ is the same as $P\{\text{ISS metric } z \text{ ever exceeds } A\}$; the first event is shaded in the figure and $s$ is given by (6.3-9). At any single $n$, $P\{\text{ISS minimum distance} \le ns - A\}$ must be overbounded by these probabilities. So we have

$$\begin{aligned} &P\{\text{ISS minimum distance} \le ns - A\} \\ &\quad \le P\{z \text{ ever exceeds } A\} \\ &\quad = P\{N_A(0) \ge 1\} \\ &\quad \le EN_A(0) \le \sigma^{-A}. \end{aligned}$$

The rest of the proof optimizes $\sigma$ and $A$ for $k$ and $n$ fixed, in order to obtain the smallest expression for $\sigma^{-A}$. As shown in Fig. 6.3-1, $A$ must be $ns - k$ for a given $k$ and $n$, and so we seek the minimum of

$$\sigma^{-ns+k}, \qquad \sigma \text{ real} \tag{C-3}$$

subject to the constraint that $s$ and $\sigma$ solve the characteristic equation (6.4-4). (Although $s$ is related to the Fano metric with bias $B$, it is better to think of it here as a free parameter). Some calculus shows that the smallest (C-3) within the constraint is

$$\left[\frac{2^{R-1}}{1-k/n}\right]^n \left[\frac{n-k}{k}\right]^k$$

Taking base-2 logs and rearranging, we get

$$-(n-k)\log\left(1-\frac{k}{n}\right) + n\log 2^{R-1} - k\log\left(\frac{k}{n}\right)$$

which is $n[h_B(y) + R - 1]$ , if $y = k/n$.

This shows that the right side of (6.3-15) is an overbound. In reality, it must be a tight bound since otherwise the BSC capacity theorem is violated.

## Proof of Lemma 6.3-2

This lemma states that the correct path distance $k$ at depth $n$ code bits satisfies

$$\frac{1}{n}\log P\{\text{CP near } k \text{ at } n\} \to h_B(y) + y\log p + (1-y)\log(1-p) \qquad (6.3\text{-}16)$$

for $y = k/n$ and $y > p$. Here $p$ is the crossover probability of a BSC. We can evaluate the probability in (6.3-16) with the binomial distribution. The probability of $k$ or more channel errors in $n$ bits is

$$\sum_{i \geq k} \binom{n}{i} p^i (1-p)^{n-i} \qquad \text{(C-4)}$$

A standard estimate of the binomial coefficient states that

$$\binom{n}{k} \approx 2^{n h_B(y)}, \qquad y = \frac{k}{n},\ y > p,$$

where $\approx$ means the logs are equal in ratio. As $k$ and $n$ grow, the log of (C-4) is dominated by the largest term, the one with $i = k$. The base-2 log of (C-4) is thus

$$n\ h_B(y) + k\log p + (n-k)\log(1-p)$$

from which we get (6.3-16).

## Proof of the Pareto character

To show (6.5-4), we assume a random tree code, a first-symbol decoder, and a metric- or depth-first search. The Fano metric has bias $B = R$. (A breadth-first search is impossible because the search region is unbounded). The expected ISS paths satisfy (6.4-10), but this time there is no limit to metric excursions:

$$\mathcal{E}N \leq K' \int_0^{\infty} \sigma^{\zeta} 2^{h\zeta}\, d\zeta. \qquad \text{(C-5)}$$

The constant $K'$ does not affect the argument and we can set it to one. From (6.3-5), the probability that the correct path metric falls below $-\zeta$ is about $2^{h\zeta}$, while from (6.4-6) we know that the ISS paths falling to this metric have expectation close to $\sigma^{\zeta}$. We can demonstrate the Pareto character in two steps. First, suppose that

$$P\left\{N > \sigma^{\zeta}\right\} > \frac{1}{\ln \sigma} 2^{h\zeta}, \qquad \zeta > 0; \tag{C-6}$$

$\zeta$ here is best thought of as simply a parameter. Then

$$\begin{aligned} \mathcal{E}N &= \int_{0}^{\infty} P\{N > x\}\, dx \\ &= \int P\left\{N > \sigma^{\zeta}\right\} \sigma^{\zeta} \ln \sigma \, d\zeta \\ &> \int \sigma^{\zeta} 2^{h\zeta}\, d\zeta. \end{aligned}$$

But the last line contradicts (C-5), and so (C-6) cannot be true. In the second step, we return to (C-5) and reason that if it is, indeed, a good overbound then it must be true that for some $\delta > 0$

$$\mathcal{E}N > \delta \int \sigma^{\zeta} 2^{h\zeta}\, d\zeta.$$

Next, we assume that

$$P\left\{N > \sigma^{\zeta}\right\} < \frac{\delta}{\ln \sigma} 2^{h\zeta},$$

and show by steps similar to those above that this cannot be true. In conclusion, it must be that

$$\delta_1 2^{h\zeta} < P\left\{N > \sigma^{\zeta}\right\} < \delta_2 2^{h\zeta}$$

To see the Pareto character, we make the variable substitutions $\eta = \sigma^{\zeta}$ and $\rho = -h/log_2\sigma$ and obtain

$$\delta_1 \eta^{-\rho} < P\{N > \eta\} < \delta_2 \eta^{-\rho}. \tag{C-7}$$

Making use of (6.4-9) we can write this Pareto exponent as

$$\rho = -\frac{h}{\log_2 \sigma} = -h\left[h + \log\left(\frac{1-p}{p}\right)\right] > 0. \tag{C-8}$$

**QED**

# Index